T0258432

Information Engineering of Emergency Treatment for Marine Oil Spill Accidents

Information Engineering of Emergency Treatment for Marine Oil Spill Accidents

Lin Mu
Lizhe Wang
Jining Yan

CRC Press
Taylor & Francis Group
Boca Raton London New York

CRC Press is an imprint of the
Taylor & Francis Group, an **informa** business

CRC Press
Taylor & Francis Group
6000 Broken Sound Parkway NW, Suite 300
Boca Raton, FL 33487-2742

© 2020 by Taylor & Francis Group, LLC
CRC Press is an imprint of Taylor & Francis Group, an Informa business

No claim to original U.S. Government works

Printed on acid-free paper

International Standard Book Number-13: 978-0-367-25611-1 (Hardback)

Library of Congress Control Number: 2019946243

**Visit the Taylor & Francis Web site at
http://www.taylorandfrancis.com**

**and the CRC Press Web site at
http://www.crcpress.com**

Contents

Preface

Since China transformed from an oil exporter to an oil importer at the beginning of 1990s, both the import volume of crude oil and the foreign-trade dependence have been increasing continuously. In 2009, China became the world's second largest importer of crude oil, with the foreign-trade dependence of crude oil as high as 51.29%, exceeding the warning line of 50%. The huge demand for energy has accelerated the development of China's offshore oil transportation industry and oil and gas exploitation industry, and also increased the risk of oil spill accidents in seagoing ships and oil and gas fields.

On July 16, 2010, the huge explosion of two oil pipelines in Dalian New Port shocked the country and caused a huge crude oil spill, which affected the sea area of nearly 100 square kilometers around the port, of which 10 square kilometers were heavily polluted and the environment was severely damaged, inflicting heavy losses on the local fishing and shipping industries. It can be seen that a major oil spill event often causes great damage to the marine environment and immeasurable losses to the economy.

Marine oil spill accidents mostly are sudden and usually occur in adverse sea conditions. After an accident occurs, the oil drifts and spreads to the surrounding area under the combined action of sea surface wind, sea current and wave, and accompanied by weathering processes such as evaporation, emulsification, dissolution and settlement, the affected area will expand rapidly. Due to its inflammability and toxicity, an oil spill will not only cause damage to the marine environment, but also bring serious harm to the health of coastal residents and public safety. It is imperative to establish a complete technical support system for oil spill emergency response, in order to obtain the accurate trend and weathering situation of oil spills in the first place and ensure relevant departments take timely and effective emergency response actions.

In order to ensure the safety of maritime transport, protect the marine ecological environment and prevent and control pollution from ships, the International Maritime Organization (IMO) has successively formulated and promulgated United Nations Convention on the Law of the Sea, International Convention for the Prevention of Pollution from Ships and International Convention on Oil Pollution Preparedness, Response and Co-operation, which provide comprehensive and detailed provisions for the prevention and control of marine oil spill pollution from ships. As a major maritime country, China has a continental shoreline of 18,000 kilometers and a sea area of about 3,000,000 square kilometers. Meanwhile, as a Category-A member of the IMO and the contracting party of the aforesaid international conventions, China must keep its various systems in line with international practices, operate per international practices and play an exemplary role in implementing international conventions. Therefore, Chinese governments at all levels have a responsibility and an obligation to establish an emergency response system for marine oil spill accidents and increase China's emergency response capacity for marine oil spill accidents.

This book mainly studies emergency response technologies for oil spills. Chapter 1 introduces China's emergency response system for marine oil spill accidents and analyzes the causes of marine oil spill accidents; Chapter 2 discusses remote sensing and monitoring of oil spills on the sea surface; Chapter 3 describes field monitoring of marine oil spills and mainly probes into oil fingerprint identification technologies; Chapter 4 studies the model prediction of marine oil spill behavior and fate; Chapter 5 emphatically introduces the "emergency prediction and warning system for oil spills in the Bohai Sea"; Chapter 6 mainly discusses resources sensitive to oil spill environment and emergency resources and introduces the "oil-spill-sensitive resources and emergency resource management system OSERS2.0"; and Chapter 7 studies emergency treatment technologies for oil spills on the sea surface. This book is greatly supported by the National Key Research and Development Program of China (Grant No. 2017YFC1404700), and it will provide a useful reference for the management, monitoring and emergency response of marine oil spill accidents in China.

As the study of emergency response technology for marine oil spill accidents involves many disciplines such as physical ocean, meteorology, biology, chemistry, geology, environment, transportation and law, the construction of the emergency response system for marine oil spill accidents is a huge and daunting systematic project. This book only covers part of such a project, and the deficiency is inevitable. Readers' criticisms and suggestions are also expected.

Authors

Lin Mu is a professor and doctoral supervisor of the College of Life Sciences and Oceanography, Shenzhen University. He was born in 1977, received a PhD from Ocean University of China, and majored in physical oceanography. Dr. Mu has been devoted to the research fields of informational maritime safety support and applied oceanography, and he has obtained significant achievements in recent years. He has published 3 monographs and over 50 research papers, 20 of which are covered by the Science Citation Index. Prof. Mu is the editorial board member of *Marine Science Bulletin*, committee member of JCOMM Expert Team on Maritime Safety Services (ETMSS), Chinese committee member of IPCC Fifth Assessment (AR5), and member of European Geosciences Union (EGU) and International Society of Offshore and Polar Engineers (ISOPE). As the chief scientist, he has been presiding over important projects such as the National Key Research and Development Program of China, National Natural Science Foundation Project of China and National Science and Technology Support Program of China. In the field of informational maritime safety support, Prof. Mu is specialized in marine oil-spill pollution warning and firstly developed a prediction and warning system of marine oil-spill, search and rescue integration in China, which has been used successfully in a series of accident issues. An expert in marine search and rescue techniques, he studied, predicted and analyzed the drifting trajectory of the debris of Flight MH370, which provided technical support for related emergency responses. In the field of applied oceanography, Prof. Mu proposed a real-time tidal level prediction system based on statistics and dynamic modeling by coupling the real-time monitoring data of meteorology and tidal level, statistical prediction method of tide, atmospheric and marine dynamics model, and real-time visualization technology, which conquered the drawbacks of traditional models and brought clear economic benefits.

Lizhe Wang is a "ChuTian" Chair Professor at the School of Computer Science, China University of Geosciences (CUG), and a professor at the Institute of Remote Sensing and Digital Earth, Chinese Academy of Sciences (CAS). Dr. Wang received BE and ME degrees from Tsinghua University and Doctor of Engineering from University Karlsruhe (magna cum laude), Germany. Prof. Wang is a fellow of IET and the British Computer Society. He serves as an associate editor of the *IEEE TPDS*, *TCC* and *T-SUSC*. His main research interests include high performance computing (HPC), e-Science and remote sensing image processing.

Jining Yan received his PhD in signal and information processing from the University of Chinese Academy of Sciences. Dr. Yan is an associate professor of the School of Computer Science, China University of Geoscience. His research is focused on remote sensing data processing and information service and cloud computing in remote sensing.

Chapter 1

Emergency Response System for Marine Oil Spill Accidents

With the rapid development of the economy, China's energy demand has increased significantly. Since China transformed from an oil exporter to an oil importer in 1993, the import volume of crude oil has been increasing continuously. According to Customs' statistics, in 2006, China's crude oil import reached 1.45×10^8 tons, domestic crude oil output reached 1.85×10^8 tons, and the foreign-trade dependency was 42.7%; in 2007, China's crude oil import reached 1.6×10^8 tons, increasing by 12.4% over the previous year, domestic crude oil output reached 1.87×10^8 tons, and the foreign-trade dependency was 46%; in 2008, China's crude oil import reached 1.79×10^8 tons, increasing by 9.6% over the previous year, domestic crude oil output reached 1.9×10^8 tons, and the foreign-trade dependency was 49.8%; in 2009, China became the world's second largest importer of crude oil, with the foreign-trade dependency as high as 51.29%, exceeding the warning line of 50%; in 2010, China's crude oil import reached 2.4×10^8 tons, increasing by 17.5% over the previous year. The continuous increase of energy demand has accelerated the development of China's offshore oil transportation industry and oil exploitation industry. Accordingly, oil tankers increases continuously, the scale of offshore oil exploration and development and submarine pipeline installation unceasingly expands, and the risk of accident, especially major and extra serious oil spill accidents, also increases greatly.

China's marine oil spill accidents have been recorded since the 1970s. For example, in 1973, an accident caused by a ship collision ("Daqing 36") in Dalian Port caused a spill of crude oil of up to 1,400 tons, which brought serious damage to the marine environment at that time. According to statistics, from 1973 to 2006, there were 2,635 ship oil-spill accidents occurring in the coastal waters of China, among which there were 69 major ship oil spill accidents of more than 50 tons, and the total oil spill volume was 3.7077×10^4 tons, with an average of 2 accidents per year and an average of 537 tons per accident. On July 16, 2010, the huge explosion of two oil pipelines in the New Port in Dalian shocked the country and caused a huge crude oil spill, which affected a large sea area around the port. Some sea area was heavily polluted and the environment was severely damaged, inflicting heavy losses on the local fishing industry. In addition, the accident led to the temporary closure of the New Port in Dalian, which affected the normal operation of ships and caused huge economic losses to the port and shipping industry.

It is obvious that major marine oil spill pollution will certainly damage China's offshore marine environment, affect people's lives and hinder the normal development of the country's economy. Therefore, it is of great urgency to understand the cause of marine oil spill pollution, carry out research on key technologies of marine oil spill emergency response, build and perfect the emergency response system for oil spills, provide decision making support for emergency response and treatment of oil spill accidents, and improve the capacity and technical level of oil spill emergency response. It cannot only provide corresponding technical support for protecting the environmental safety of China's offshore waters, but also provide necessary guarantee for the healthy and stable development of China's oil transportation and exploitation industries.

1.1 Current Situation of Marine Oil Spill Pollution and Cause Analysis

1.1.1 Major Marine Oil Spill Pollution Accidents at Home and Abroad

Since the 1960s, more than 10,000 tons of oil spill accidents have occurred almost every year in the world, and the large amount of oil spill caused by the accidents has seriously affected the marine environment. Before the 1990s, the occurrence rate of oil spill accidents was high, and the oil spill volume was large. Since the 1990s, the occurrence rate of oil spill accidents has decreased, and the oil spill volume has also relatively reduced. However, the deteriorating marine ecological environment and the increasing environmental awareness of governments and people in various countries make marine oil spill accidents receive more and more attention. In recent decades, the catastrophic consequences of major marine oil spill accidents were startling. Some of the accidents occurred due to bad weather and sea conditions, while some were caused by human errors. In any case, the environmental damage and economic losses they caused are huge and immeasurable. Several typical major oil spill accidents at home and abroad will be reviewed below, in order to provide reference for readers.

1.1.1.1 Major marine oil spill accidents abroad

1. Torrey Canyon supertanker accident

The oil spill of Torrey Canyon, a supertanker registered in Liberia, occurring in 1967 was the first large-scale marine oil spill accident in the world. Before the accident, the planned route of the oil tanker was to pass west of the Isles of Scilly, and on March 18, the oil tanker changed course temporarily for some reason. The captain ignored the relevant warnings in the navigation guide, and chose the waterway east of the Isles of Scilly and west of Seven Stones Reef. In the course of navigating to the turning point of the waterway, the supertanker was hampered by several fishing vessels and unable to turn on time, and the operation in the use of automatic pilot was chaotic. As a result, the supertanker hit Pollard's Rock on Seven Stones Reef at a speed of 15.8 knots. The damaged part of the bottom accounted for more than half of the length of the tanker, and 3×10^4 tons of crude oil spilled within 2~3 hours. Within one week of the accident, another 2×10^4 tons of crude oil spilled. Rescue efforts were made after the accident. However, due to the severe sea conditions, the keel broke off on March 26, and another 5×10^4 tons of crude oil spilled out. The total oil spill volume reached about 1×10^5 tons, and rescue efforts also ended in failure. On March 28, the British Air Force was ordered to bomb the tanker. Three days of bombing broke the deck of each oil tank, and the remaining crude oil was set fire. In the meantime, 2.50×10^6 gallons of dispersants were sprayed over marine oil spills, and detergents were adopted on contaminated beaches.

Affected by strong winds, the crude oil changed its direction and drifted across the English Channel to Cape Verde Peninsula in France, contaminating 242 sea miles of shoreline, including south and north banks of Cornwall Peninsula in Britain, Guernsey Island in the English Channel and the north and east of Brittany Peninsula in France. The contamination not only damaged aquaculture and tourism, but also cost a lot of money to clean-up the oil. The economic loss of this accident was estimated at £10,500 British pounds.

2. Exxon Valdez tanker accident

In 1989, Exxon's tanker Exxon Valdez ran aground on Bligh Reef in Prince William Sound, Alaska, and spilled 3.6×10^4 tons of crude oil. Due to the bad weather, emergency measures failed, and 18,600 kilometers of shoreline became a polluted area, with a thick oil slick covering about 1,600 square kilometers of seawater.

This accident caused catastrophic damage to the ecological environment of the Gulf of Alaska. 250,000 seabirds, nearly 4,000 sea otters, 300 spotted seals, 250 white-headed sea eagles and 22 killer whales died. The spawning grounds of salmons and herrings were contaminated by crude oil, and salmon and herring resources almost became extinct, leading to the collapse of the once-thriving herring industry and the closure of fishing grounds and facilities on which tens of thousands of fishermen and local residents depend. Although Exxon paid about $8 billion in fines, clean-up costs, compensation and other costs for the pollution, the accident cost the Alaskan fishing industry nearly $20 billion and the tourism industry nearly $19 billion, and the ecological damage from the oil spill could not be measured in dollars.

3. Prestige tanker accident

The tanker Prestige, made in Japan in the 1970s, was a single-hull oil tanker with a capacity of 8×10^4 tons. On November 13, 2002, the tanker, carrying 7.7×10^4 tons of heavy fuel oil, was hit by a strong storm while passing through Spanish waters. The storm damaged the oil tanks on the tanker, causing fuel oil spill, and an oil spill zone of about 2.5 sea miles wide and 20 sea miles long was formed quickly on the sea surface. Under the influence of wind and waves, the oil spill and the runaway Prestige drifted toward the coast of Galicia, Spain, and the tanker ran aground about eight sea miles off the coast of Galicia. In order to reduce pollution losses, on November 17, the Spanish government decided to tow the Prestige to the waters southwest of the Atlantic. On November 19, the Prestige broke off 150 sea miles off the coast of Spain and sank at a depth of 3,500 meters because of damage to its hull and the impact of wind and waves. According to the information released by the Spanish government, about 1.7×10^4 tons of fuel oil spilled as of the time of the tanker's sinking, and the oil contaminated 400 kilometers of shoreline in Galicia. A thick layer of oil covered the shore, and the rivers, creeks and marshes near the shore were seriously polluted. The sunken Prestige continued to spill oil, and a part of France's shoreline was also polluted.

Ten countries, including Spain, Germany, France and Portugal, participated in the emergency response of the accident, which is an emergency action by far with the largest scope of international cooperation and the largest number of participating countries. In the oil spill emergency response of this accident, relevant departments used a series of high-tech means, including satellite and aerial remote sensing and monitoring and oil spill model prediction, to provide strong technical support and guarantee for decision making in the oil spill emergency response.

According to preliminary estimates, the cost of the preventive measures taken by governments for the oil spill accidents and oil spill clean-up was between 215 and 320 million Euros, the compensation for fishing and breeding industries was 80 and 250 million Euros, and the damage to tourism and the ecological environment was immeasurable.

4. Gulf of Mexico Deepwater Horizon accident

On April 20, 2010, the Deepwater Horizon oil drilling rig leased by BP in the Gulf of Mexico exploded and sank to the seabed, resulting in a large volume of oil drifting into the Gulf of Mexico and quickly spreading.

Due to human misjudgment, the accident caused a much more serious disaster than initially expected. Relevant authorities thought that the amount of oil spilling from the sunken drilling jig was about 138.5 tons per day, but in fact, the amount of oil spill was much higher than the initial estimate. 692.5 tons of oil flowed into the Gulf of Mexico through the damaged pipeline every day [1]. By early May, the volume of crude oil spill exceeded 4.00×10^6 gallons (about 6.86×10^5 tons), the total number of staff involved in the emergency action was more than 17,000 person-times, the total number of ships (including tugs and oil-collecting tankers) participating in the emergency action was 750, 518 kilometers of oil booms (including common oil booms and adsorption oil booms) were laid to contain the oil spill, about 1,476 cubic meters of dispersants were used, and the volume of collected oil-water mixture was about 24,984 cubic meters.

The direct economic loss caused by the accident to BP was as high as 14 billion dollars, but the ecological damage was incalculable. Some experts pointed out that the oil spill pollution caused by the accident was multi-directional. While threatening the coastal ecological environment, the "oil ball" that sank to the seabed would threaten every link of the marine food chain, ultimately affecting large fish and marine mammals. The pollution destroyed 1,609 kilometers of wetlands and beaches along the Gulf of Mexico, damaged the fishing industry, endangered species and threatened nearly 20 national wildlife protection areas in the United States. In terms of the extent of impact, this accident replaced the Exxon Valdez accident in 1989 as the worst oil pollution disaster in American history.

1.1.1.2 Major marine oil spill accidents in China

1. Feoso Ambassador accident

On November 25, 1983, the 207-meter Panamanian tanker Feoso Ambassador, carrying more than 4.3×10^4 tons of crude oil, ran aground on Zhongsha Reef while leaving the Huangdao Oil Area in Qingdao Port, and the cargo hold was damaged and spilled 3,343 tons of crude oil. The maximum thickness of the oil slick in the port exceeded 0.5 meters. The oil spill affected Jiaozhou Bay and 230 kilometers of shoreline near the bay, and seriously contaminated the aquaculture area of 15,000 mu nearby, the scenic area of 9.0×10^5 square meters and the seaside bathing area. The economic loss was up to tens of millions of yuan, and the damage compensation was CNY 17.75 million. Although the government organized a lot of manpower and material resources to clean-up the pollution, its impact continued for a long time.

2. Zhuhai 3.24 accident

On March 24, 1999, Minrangong 2 and Donghai 209 tankers collided near No. 8 buoy on Lingding Waterway between Lingding Island and Qi'ao Island in the water area of Pearl River Estuary. The prow of Donghai 209 was directly inserted into Cabins 2 and 3 on the starboard of Minrangong 2, and Minrangong 2 spilled 589.7 tons of heavy oil. The oil spill spread to Zhuhai, Shenzhen, Zhongshan and other places, and contaminated more than 300 square kilometers of sea areas and 55 kilometers of shoreline, including Jinxing Gate and Qi'ao Island.

The clean-up work lasted six days. The maritime authority dispatched ships 116 times and 376 person-times of cleaning professionals, and used 3 tons of oil absorption felts, 3,460 meters of oil absorption tow cables, 350 meters of oil booms and 21.5 tons of oil dispersants, and more than 160 tons of dirty oil was collected. The government of Zhuhai City mobilized 15,000 thousand person-times of government officers, the masses, students and the armed police and employed 706 person-times of migrant laborers.

Although the municipal government of Zhuhai organized a huge amount of manpower and equipment to clean-up the pollution, nearly 600 tons of heavy oil spilled seriously polluted 55 kilometers of shoreline, 300 square kilometers of sea areas and 190,000 mu of marine farms, and a large number of mariculture products died because of the pollution, leading to a direct economic loss of up to CNY 9,648 thousand. In the accident, the artificial beach built by Zhuhai City with an investment of CNY 3 million was destroyed, and 12.4 square kilometers of farms in Xiangzhou and Qi'ao Island and 113.39 square kilometers of termite breeding area between Qi'aotou and Jiuzhou were also polluted. In addition, more than half of 0.7 square kilometers of mangroves on Qi'ao Island were difficult to survive due to the sticky oil, and the ecological damage was quite severe.

3. Collision accident of Hyundai Advance and MSC Ilona ships

On December 7, 2004, the 182-meter Panamanian container ship Hyundai Advance, sailing from Shenzhen Yantian Port to Singapore, collided with the 300-meter German container ship MSC Ilona, sailing from Chiwan, Shenzhen to Shanghai. The fuel oil tank of the MSC Ilona was damaged and spilled more than 1,200 tons of fuel oil, forming an oil belt of about 16.5 kilometers long on the sea surface. This accident was the largest oil spill accident caused by ship collision in China, and polluted the sea area of Pearl River Estuary, causing a total loss of CNY 68,000 thousand.

After the incident, the Ministry of Transport and the provincial government of Guangdong organized various units concerned to take active actions to effectively control and remove the oil spill at sea without causing shoreline pollution, protecting sensitive resources in the water area of Pearl River Estuary and preventing further losses.

4. 7.16 oil pipeline explosion accident in Dalian New Port

Dalian New Port, located at the northeastern foot of Dagu Mountain at the southern tip of Liaodong Peninsula and southwest of Dayao Bay on the coast of the Yellow Sea, is the largest modern deepwater oil port with the deepest water level in China. Dalian Port is featured by a wide sea area, smooth channel, deep water and small waves, and has beautiful scenery. There is no silting or freezing, and the weather is warm in winter and cool in summer. Therefore, it is a rare natural harbor in China.

On July 16, 2010, the Liberian crude oil tanker Cosmic Jewel, with a capacity of 300,000 tons, unloaded oil in Dalian New Port, and the onshore pipeline of the crude oil storage tank caught fire during adding the catalyst. After the accident, the Cosmic Jewel immediately evacuated and got out of danger. The fire took place on a 900mm-diameter onshore pipeline of the crude oil storage tank, and later another 700mm-diameter pipeline caught fire. The two pipelines on fire ignited a $1.0 \times 10^5 m^3$ crude oil tank next to them. The fire lasted 15 hours and caused a large amount of crude oil spill. The crude oil spill affected the sea area of nearly 100 square kilometers around the port, of which 10 square kilometers were heavily polluted, and the environmental damage was inestimable. The accident inflicted heavy losses on the local fishing and shipping industries.

After the accident, the maritime authority cumulatively coordinated and dispatched more than 20,000 person-times of clean-up workers, over 50 clean-up boats, and vehicles more than 100 times. The oil booms of over 9,000 meters were laid in waters of the oil spill, 32 tons of oil dispersants from Qinhuang-dao, Qingdao and Beijing were transferred to clean-up the pollution, and more than 1,200 fishing ships were mobilized to collect the oil. After more than a month of intensive work, the overall clean-up work was progressing smoothly, achieving the strategic goal of oil spill not entering the Bohai Sea or entering the high seas.

1.1.2 Cause Analysis on Marine Oil Spill Pollution

Marine oil spill pollution can be divided into two categories: oil spill pollution from ships and oil spill pollution from offshore oil and gas fields.

Oil spill pollution from ships refers to the marine environmental pollution caused by crude oil and its products in the process of marine transportation, which accounts for about 47% of the total amount of oil pollution to the sea. Such pollution mainly comes from bilge water of engine rooms of motor ships, ballast water of oil tankers, tank washing sewage, and oil spills from marine accidents and loading and unloading accidents. According to statistics, the oil spill pollution caused by stranding and collision accounts for 18% of the total oil pollution from ships every year, about 5×10^4 tons. Oil spills caused by accidents are sudden and uncertain, and the pollution to the sea is often disastrous. Generally speaking, there is a risk of major oil spills in the sea area through which large-tonnage ships pass, and the probability of this risk is related to a number of factors, such as transport density, water condition, climatic conditions, technical management level, and so on.

A large number of flammable and explosive oil and gas products will be involved in the development of offshore oil and gas fields. At the same time, due to the complexity of development process and equipment operation, there are potential risks of oil and gas leakage, fire and explosion and other major accidents in the development process. The risk of oil spills exists at all stages of oil and gas field development. Possible oil spill accidents include blowout, fire, explosion, pipeline rupture, oil spill from slop tanks, fuel tank rupture and fuel oil spill during transmission. At present, China's offshore oil and gas field development and exploration is mainly carried out in the Bohai Sea and the South China Sea. Among them, there are many offshore oil production platforms and production wells in the Bohai Sea, and the risk of oil spill pollution is relatively high. Meanwhile, because the Bohai sea is a continental sea with poor self-purification capacity, once being destroyed by oil spill accidents, the marine ecological environment is difficult to recover. According to the statistics of the State Oceanic Administration, the oil spill probability of the existing oil pipelines in the Bohai Sea is about 0.1 times a year, and the probability of oil spill accidents caused by fire and blowout of oil platforms in the Bohai Sea is about 0.2 times a year. Although the probability is not high, such accidents will cause a large amount of oil spill, and because the Bohai Sea is an internal sea with poor self-purification capacity, once the oil spill occurs, the threat to the marine ecology of the Bohai Sea is huge. With the increase of offshore oil production platforms, submarine oil pipelines and oil import and export transportation, the South China Sea, especially the Beibu Gulf Zone, has also become a sea area with frequent oil spill accidents. In Beibu Gulf, both China and Vietnam have carried out a lot of oil exploration and exploitation. With the continuous increase of offshore oil production platforms and submarine oil pipelines, most pipelines have aging problems and oil transportation is increasingly frequent. Beibu Gulf has become an area with frequent occurrence of oil spill accidents. In addition, there are more than 1,000 oil drilling rigs in the waters of Nansha Islands, with an annual oil production of more than 50 million tons, and there is a risk of oil spill accidents. Furthermore, the South China Sea is also the oil import artery of Japan, South Korea, Taiwan and mainland China, the oil reserve center of neighboring Vietnam, Philippines, Malaysia and Indonesia, and the transportation artery between Southeast Asian countries. According to relevant data, more than 50% of China's oil imports is transported by this route. All the above-mentioned factors bring a huge oil spill pollution risk to the South China Sea.

1.1.2.1 Cause analysis on oil spill pollution from ships

The cause of oil spill pollution from ships at sea generally has the following two aspects:

1. Operational oil spill

This kind of oil spill is caused by the crew's failure to comply with relevant regulations and arbitrary discharge of waste oil and waste water containing a large amount of oil, such as bilge water, ballast water, and tank washing water.

The so-called bilge water mainly refers to the oily sewage which is the mixture of fuel oil, lubricating oil, fresh water and seawater spilled by machines (main engine and auxiliary engine), equipment (pumps, condensers, heaters, etc.) and pipelines during ship operation. The bilge water discharged by a ship each year accounts for about 10% of its gross tonnage, and the amount of oil in sewage can reach up to 5000 mg/L. Hundreds of thousands of tons of oil are discharged into the sea with bilge water every year around the world. If the sewage is discharged in a port, it will cause long-term pollution to vast water areas of the port. In addition to bilge water, an oil tanker also discharges oily ballast water during normal operation. Under normal weather conditions, when the tanker sails without load, the amount of ballast water is 35%~40% of the carrying capacity, while under severe weather conditions, the amount of ballast water can reach 60%~70% of the carrying capacity. Generally, the oil content of the ballast water in the cargo hold of an oil tanker is 1,000-3,000 mg/L. A 100,000-ton oil tanker will discharge about 100-150 tons of oil per voyage with the ballast water, which is bound to cause pollution to the surrounding environment. In addition to the above-mentioned bilge and ballast water, the tank cleaning water mainly comes from four aspects:

- Residual oil of the previous variety which is cleaned out before the oil tank is filled with oil of another variety;

- Accumulated oily dirt that is cleaned out of oil tanks during periodic maintenance of oil tankers and other ships;

- Oily dirt that is cleaned out of empty oil tanks (for the purpose of safety) when oil tankers and other ships pass through canals and narrow channels;

- Accumulated oily dirt that is cleaned out of cargo holds, oil tanks and bunkers before oil tankers and other ships are repaired and dock.

According to statistics, if all cargo holds and oil tanks of an oil tanker are cleaned once, the oil of about 0.2% of the carrying capacity is discharged with tank washing water, and if the tank washing water flows into the sea due to improper operation, the resulting environmental pollution cannot be underestimated.

The International Maritime Organization (IMO) estimates that at least 3.2×10^6 tons of oil is discharged into the marine environment each year from various sources, of which 3×10^6 tons are discharged each year due to oily sewage in engine rooms and fuel oil residue.

2. Accidental oil spill

This kind of oil spill is mainly caused by accidents, which can be divided into two types, namely, marine accidents such as collision, stranding, fire and explosion and accidents during oil operations.

Marine accidents are generally unavoidable due to their uncertainty, but some factors will affect the probability of such accidents. Lu [3] analyzed these factors, including ship types, ship tonnage, age of ship, ship performance and weather and sea conditions.

(1) Ship type

There are many standards for classification of ships, and they are generally classified according to their use, size and seaworthiness. Ships that may produce oil spill pollution are divided into the following categories: oil tanker, freighter, barge, passenger ship, etc.

According to the statistical data on more than 50 tons of ship and oil spill accidents of coastal ships and wharves in China between 1993 and 2002 [2], oil tanker accidents occur much less frequently than other ships. But the oil spill volume per accident is much greater than that of other ships, and the environmental damage and economic losses arising therefrom are also much greater than that of other ships. Therefore, it is necessary to distinguish ship types for different research objects (oil terminal, anchorage, navigation channel and a certain sea area), because different ship types may have different hazards and destructiveness of oil spills, and oil spill events of oil tankers are often more hazardous and destructive than those of other ships.

(2) Ship tonnage

The tonnage of a ship is usually expressed in gross tonnage, which reflects the size of the ship and also indirectly reflects the carrying capacity of the ship. The deadweight tonnage of an oil tanker is generally used to describe its size, but in the study of ship traffic, the size of a ship can also be measured by the main dimensions of the ship such as length. Different countries have different standards for classification of ship tonnage. Ship tonnage can affect ship maneuverability, which is particularly important in restricted waters or areas where the ship traffic density is high. Good maneuverability of a ship means good direction stability, hunting ability, turning ability and stopping ability, and it is one of the important conditions for safe navigation. In addition, the ship tonnage can also affect the ship steering performance. Generally speaking, ship tonnage has great influence on sea speed and characteristic time.

The effect of ship tonnage on ship oil spill is indirect. Generally, it affects the occurrence of oil spill accidents by affecting the ship maneuverability, but has no direct relation with the frequency of ship oil spills and oil spill volume. It cannot be said that the larger the ship tonnage is, the larger the oil spill volume. Only when the large-tonnage hull is seriously damaged, such as explosion, fire, crushing and other extremely special accidents, the oil spill may be massive.

According to the actual situation of China and the history of oil spill accidents, it can be considered that, small-tonnage ships are more influential on the occurrence of oil spill accidents than large-tonnage ships. However, once a large-tonnage ship spills oil, its oil spill volume may have a greater impact than small-tonnage ships.

(3) Age of ship

The age of the ship reflects the time of ship construction or length of service period and can indirectly reflect the technical level of ship construction, equipment reliability and ship automation degree. According to the regulations of China Ocean Shipping Company, the age of the ship can be divided into several age brackets: (0~5), (5 ~10), (10~15), and (15+).

The age of the ship has a certain influence on the occurrence of oil spill accidents, but the degree of such influence cannot be quantified. It is generally believed that the older the ship, the greater the possibility of oil spill accidents. In the meantime, the older ship is mostly unable to keep up with the more and more strict requirements for prevention and control of oil spills from ships. In addition, with the aging of various equipment and poor technical state, the risk of oil spill accidents is greater. Once an oil spill accident occurs, its control over the oil spill is often powerless and effective containment measures cannot be taken, and then the scale of the accident will expand.

To sum up, the relationship between the age of the ship and ship oil spill accidents can be simplified as follows: the older the ship, the greater the possibility of ship oil spill accidents.

(4) Ship performance

The ship performance can be measured by technical state, automation degree and maneuverability of the ship. The technical state of the ship is the comprehensive reflection of seaworthiness, automation degree and maneuverability. A ship in good technical state has a strong ability to avoid accidents, so the possibility of oil spill accidents is low. The ship automation degree refers to the technical level of automation application of the ship. The higher the automation degree, the lower the dependence on humans. The increase of automation degree can effectively avoid human errors, but on the other hand, the higher the ship automation degree, the more stringent the quality requirements for personnel. Only high-quality personnel can give full play to the advantages brought by the increase of ship automation degree. Ship maneuverability mainly refers to whether a ship has good direction stability performance, hunting performance, turning performance and stopping performance and whether it can be controlled satisfactorily.

Therefore, generally speaking, the higher the technical state, maneuverability and automation degree, the lower the possibility of ship oil spill accidents.

(5) Weather and sea conditions

Weather and sea conditions have great influence on oil spill accidents from ships. Generally, such influence can be summed up in two aspects: inducing effect and worsening effect. Visibility can reflect the effect of weather conditions on visual observation, and it cannot be ignored in the study of ship traffic safety. High visibility of meteorological factors may provide clear recognition of observed objects, which is conducive to the avoidance of danger by ships, and reducing the possibility of ship accidents. Rainfall, snow and storm have a greater impact on visibility.

The wave height reflects the calmness degree of the sea. Different wave heights correspond to different wave levels. The higher the wave level, the greater the corresponding wave height, and the greater the impact on ship safety. With the increase of shipbuilding level, ships have stronger ability to withstand wind and waves, but the impact of waves on the ship safety still cannot be underestimated.

In addition to the above-mentioned main factors, other factors including crew quality, channel conditions, navigation aids, ship traffic density and communication conditions will also affect oil spills from ships. According to statistics, about 70% of oil spill accidents of oil tankers are caused by humans. Therefore, it is particularly important to strengthen the training and improve the quality of the crew [2].

Another kind of accidental oil spill occurs during oil operations, which may occur in three cases: oil loading, in-port oiling and in-ship oil transfer.

(1) Oil spills during oil loading

Oil spills during oil loading usually occur at three critical moments: beginning, changing tanks and pipeline pigging. There are three main reasons for oil spills at the beginning of operations:

- The supply and receiving parties are not coordinated. The supply party has opened the valve first when the receiving party has not opened the valve, causing pipeline burst and an oil spill.

- The oil-conveying pipe cannot bear the normal pressure and bursts due to aging.

- The flanges of the oil-conveying pipe and the oil receiving pipe are not connected firmly and fall off, resulting in oil injection.

The oil spill from changing the tank during oil loading is mainly due to operation errors, such as mistakenly opening the valve and one party arbitrarily changing the flow rate, resulting in pipeline burst, which further causes an oil spill.

There are generally three reasons for oil spills during pipeline pigging at the end of oil loading:

- Because the reserved capacity of the tank is insufficient, the tank is full and spills oil during pipeline pigging.

- The pump is not stopped in time when the oil tank is about to be full, resulting in tank overfilling and oil spilling.

- Because the pressure of pipeline pigging is insufficient, there is still oil left in the pipeline, and the blind plate is not properly sealed when the hose is removed. As a result, the oil left in the pipeline flows into the water and causes pollution.

(2) Oil spills during in-port oiling

Oil spills caused by careless oiling of ships in the port are relatively common, and the oil spill volume is also large. There are mainly five reasons:

- The air pipe of the oil tank is blocked.

- False oil level: Because the oil measuring pipe is easy to be congealed, the oil dipstick is difficult to measure the true oil level, which results in the actual oil level greatly exceeding the oil level measured by the oil dipstick, and an oil spill occurs as oil loading continues.

- The blind plate of the oil-conveying pipe gets loose. There is a tubing joint on each ship board and a section of main pipe in front of the subdivision valve, and there are no stop valves at main pipe orifices of some ships. Oiling is on the larboard side, and the blind plate is not sealed after oiling. Next time, oiling is changed to the starboard side. After the pipeline is connected and the pump is turned on, a part of the oil enters the tank and the other part flows to the larboard side through the main pipe and overflows from the blind plate [3].

- The oil supplier increases the pump output arbitrarily, which results in that the tank get full and oil overflows or the oil gas in the tank cannot be vented and oil overflows.

- Due to valve misoperation, the tank gets full and oil overflows.

(3) Oil spills during in-ship oil transfer

In-ship oil transfer is often carried out in the engine room, such as oiling of daily service tanks and oil transfer between oil tanks. Oil spill accidents often result from improper operation or absence from duty.

According to the statistics from the International Tanker Owners Pollution Federation (ITOPF) about 9234 oil spill accidents of oil tankers, large tankers and barges occurring in 29 years, the data is divided as follows according to oil spill grades and accident causes: there are 7,764 accidents with an oil spill less than 7 tons, including 4,537 operational accidents, accounting for 59%, and 1,025 marine accidents, accounting for 13%; there are a total of 1,135 accidents with an oil spill of 7 to 700 tons, including 407 operational accidents, accounting for 35%, and 585 marine accidents, accounting for 52%; and there are 335 accidents with an oil spill over 700 tons, including 30 operational accidents, accounting for 9%, and 281 marine accidents, accounting for 84%. The data above shows that with the increase of the oil spill grade, the number of marine accidents increases, and the extra major accidents with an oil spill over 700 tons are mainly caused by marine accidents. China has also conducted classified statistics on the causes of 309 ship oil spill accidents, the pattern of ship oil spill accidents in China is basically consistent with the pattern summarized by ITOPF: there are 268 accidents with an oil spill less than 10 tons, including 140 operational accidents, accounting for 52%, and 19 marine accidents, accounting for 7%; there are a total of 22 accidents with an oil spill of 10 to 50 tons, including 2 operational accidents, accounting for 9%, and 17 marine accidents, accounting for 77%; there are 16 accidents with an oil spill of 50 to 700 tons, including 0 operational accident and 13 marine accidents, accounting for 81%; and there are 3 accidents with an oil spill over 700 tons, and all of them are marine accidents. The data above shows that the major accidents with an oil spill over 50 tons are mostly marine accidents, and the extra major accidents with an oil spill over 700 tons are all marine accidents. The above analysis shows that: (1) accidental oil spills from ships are the main reason for oil spill pollution caused by ships to the marine environment; (2) major ship oil spill accidents are mainly caused by marine accidents; and (3) marine accidents are the most serious source of oil spill pollution from ships to the marine environment.

1.1.2.2 Cause analysis on oil spill pollution from offshore oil and gas fields

The risk of oil spills from offshore oil and gas fields exists at all stages of oil and gas field development. Possible oil spill accidents include blowout, fire, explosion, submarine pipeline rupture, oil spill from slop tanks, fuel tank rupture and fuel oil spill during transmission. Facilities that may cause oil spills in the process of offshore oil exploitation and relevant reasons are analyzed below.

1. Drilling platform

The main oil spill risk in the process of drilling is blowout. When the drill encounters the underground abnormal high-pressure oil or gas layer, if the bottom hole pressure is lower than the formation pressure, the formation oil and gas fluid will enter the wellbore and push the drilling fluid to overflow, and then the overflow will occur. At this point, if the control measures for underground oil and gas pressure are not appropriate and the overflow cannot be detected and treated in time, oil, gas and drilling fluid will be rapidly sprayed to the ground, that is, a blowout will occur. Ejective substances of blowout mainly include crude oil, natural gas, formation water and drilling liquid used in drilling. In order to avoid catastrophic consequences caused by blowout, preventive measures such as installing hydraulic blowout preventer in the pipeline for the drilling ship and the drilling platform, adjusting the mud weight in time during the drilling process and balancing downhole formation pressure will be taken [4].

Usually the probability of blowout accidents is low. According to relevant domestic statistics, the probability of blowout accidents in the drilling process during oil exploration and development is about 0.116%. From 1966 to the end of 2001, China drilled more than 1,200 wells in the Bohai Sea, East China Sea, eastern South China Sea and western South China Sea, and only four blowout accidents occurred during drilling of exploration wells, and no blowout accident occurred during drilling of production wells [5].

2. Oil production platform

The oil production platform integrates productive exploitation, treatment and storage, and has numerous devices such as pipelines, valves and storage tanks. It is the most dangerous point source for oil spills at sea. In order to prevent oil spills, downhole safety valves and packers will be installed in production wells (especially in high-production wells). In case of an accident, the well can be automatically shut down. At the same time, various monitoring devices are installed to detect all kinds of accidents in time [4].

The main factor that may cause oil spill at the production stage is explosion caused by leakage of production process facilities. Possible causes of oil and gas leakage on the production platform include valve failure, pipe failure (tee pipes, elbows, flanges, bolts, nuts, gaskets, etc.), weld failure, corrosion, material failure (tubes, pipe fittings and vessel rupture), operating errors, instrument and control failure, etc. [5].

3. Submarine pipeline and riser

External causes of accidents of submarine pipelines and risers include collision of heavy objects on the sea surface, trawling of fishing boats, dredging, pipe laying or anchoring, anchor dragging, pipe abrasion caused by anchor chains, sunken ships and natural disasters, and internal causes include pipe corrosion, material and construction defects, and human errors [5].

1.2   Hazards of Marine Oil Spill Pollution

Marine oil spills are exposed to the air and are flammable. In addition, due to the toxicity of the oil, oil spills will cause serious hazards to human health, public safety and marine ecological environment, which are mainly reflected in the following aspects.

1.2.1   Hazards of Oil Spills to Human Health and Public Safety

Aromatic hydrocarbons in oil spills are quite toxic to humans, and especially polycyclic aromatic hydrocarbons represented by dicyclic and tricyclic aromatic hydrocarbons are more toxic. They can be introduced into human bodies through respiration, skin and mucous membrane exposure, and consumption of food containing oily dirt, and affect normal functions of lungs, stomachs and intestines, kidneys, central nervous systems and hematopoietic systems, causing various toxic symptoms [6]. In addition, benzene and polycyclic aromatic hydrocarbons in oil spills are also carcinogens, and toxic polycyclic aromatic hydrocarbons and toxic heavy metals in oil spills can harm human health through biological concentration and transfer process of food chains.

Crude oil itself is inflammable and explosive, and will threaten personal safety and public safety. Therefore, before oil spill emergency response, it is necessary to fully understand the types of oil spills and inflammability and hazard parameters such as flash point, and carry out fire risk assessment in combination with surrounding conditions, in order to take corresponding preventive measures, especially to strengthen management to avoid fire accidents caused by improper operation or negligence in management.

1.2.2   Hazards of Oil Spills to the Environment

The hazards of oil spills to the environment are mainly shown in the following aspects:

(1) Hazards of oil spills to birds and other animals

Oil spills on the sea surface are the most harmful to birds, especially those diving for food. On one hand, when these birds contact the oil slick, their feathers lose water resistance and insulation ability due to oil absorption, and they also lose their ability to forage, and eventually die of starvation and cold. On the other hand, these birds ingest oil by preening their feathers with their beaks, which can damage their internal organs, and they eventually die of poisoning [7]. In severe cases, birds that are covered by oil will die from suffocation. Therefore, after oil spill accidents occur, it is very important to carry out rescue work for birds from the perspective of natural ecological protection.

(2) Impact of oil spills on plankton

When oil spills enter the sea, they first affect marine life on the surface, especially hyponeuston that live in this thin layer of water. Oil slicks, covering the sea surface, will affect the replenishment of oxygen in seawater and hinder the photosynthesis of phytoplankton in the sea, reducing the original productivity of the sea [8].

(3) Hazards of oil spills to fish

Oil spills may kill fish eggs and larvae, and a large number of bait organisms (zooplankton and phytoplankton) in waters may die from oil spill pollution. Consequently, the bait foundation on which the seed, young, and adult fish survive is damaged, and the energy transferred by the food chain (web) is disconnected, causing the decrease of advanced biomass, regional ecological imbalance, and accordingly interfering with normal physiological and biochemical functions of fish and causing fish disease [9].

(4) Hazards of oil spills to marine mammals

Oil spills are more likely to adhere to the fur of mammals, reducing their insulation ability. In addition, oil spills may damage the epidermis and the parts exposed outside the body. Large amounts of oil can cause death of mammals or organ damage due to physical asphyxiation or chemical toxicity [8].

(5) Hazards of oil spills to coastal beaches

Residual oil floating on the sea surface and oil slicks may gather and form oil aggregates, which will closely adsorb on the coast, mudflats and substrates when meeting the coast and mudflats, suspended particles and sediments in water and form large dark brown solid blocks. This will cause landscape and ecological damage, the impact of which often takes a long time to eliminate [6].

Therefore, the hazards caused by oil spill pollution are very far-reaching, and the research and treatment of oil spill pollution is an important part of marine environmental protection in China.

1.3   Construction of the Emergency Response System for Marine Oil Spills

1.3.1   International Convention System for Marine Oil Spill Emergency Response

During the second world war, ship oil spill accidents occurred frequently, and some coastal countries suffered serious oil spill pollution and heavy losses due to the lack of a complete prevention and control system. This aroused widespread concern around the world, and a series of international conventions were successively issued for limiting oil pollution from ships and dealing with marine oil spills. In 1954, the first international convention on the prevention of marine and coastal environmental pollution, International Convention for the Prevention of Pollution of the Sea by Oil, was adopted, and became the world's first guideline on the control of the discharge of oil and oily sewage from ships into the sea.

Between November 19 and November 30, 1990, the IMO held International Oil Pollution Preparedness, Response and Co-operation Conference in London, and successfully adopted International Convention on Oil Pollution Preparedness, Response and Co-operation (OPRC Convention). The OPRC Convention requires contracting parties to establish national emergency response systems for oil spills and formulate emergency plans for oil spills as their responsibilities and obligations under the Convention. Together with the OPRC-HNS protocol, which came into force in 2007, it constitutes a framework of the preparedness, response and cooperation for marine contaminants, providing a basis for developing national and regional capabilities for the prevention and response of marine pollution accidents and promoting international cooperation and mutual assistance in oil spill accidents.

The OPRC Convention entered into force on May 13, 1995 and, as of January 2010, 100 countries had rectified this Convention. At present, the total tonnage of ships of the contracting parties accounts for 68.20% of the total tonnage of merchant ships in the world.

The OPRC Convention sets out not only the provisions on issues related to appropriate preparedness and response measures taken by the contracting parties to combat oil pollution accidents, either nationally or in co-operation with other countries, but also the specific requirements for the implementation of each provision. The convention is mainly divided into three parts: oil spill preparedness, response and co-operation.

1. Main provisions for oil pollution preparedness

(1) Formulating oil pollution emergency plans

Each state party shall require the ship flying its flag to carry the Shipboard Oil Pollution Emergency Plan stipulated in Article 26 of International Convention for the Prevention of Pollution from Ships, 1973, as modified by the Protocol of 1978 (MARPOL73/78), and offshore units, harbors and oil handling facilities under its jurisdiction shall be also required to have Oil Pollution Emergency Plans. These plans must be coordinated with national emergency systems and shall be verified and approved according to procedures established by national competent authorities.

(2) Building oil pollution preparedness systems

Each contracting party shall establish a national system capable of providing rapid and effective response to oil pollution accidents, including at least a minimal national emergency plan for managing oil pollution preparedness and response actions, reporting and coordination of emergency support, and oil pollution response exercises and training.

(3) Establishing oil pollution emergency equipment stockpiles

Each state party shall, to the extent of its capability, individually or through bilateral or multilateral cooperation, establish a minimal stockpile of oil pollution emergency equipment and the use program, in cooperation with the oil industry or shipping industry and other entities.

(4) Formulating and implementing training programs for oil pollution preparedness and response

The capacity of a country to respond to oil pollution accidents depends on its oil spill combating equipment and trained personnel responsible for oil spill emergency response. The secretary-general of the IMO is obligated to develop a comprehensive training program on oil pollution preparedness and response, in cooperation with relevant governments, related international and regional organizations, and the oil and shipping sectors, and especially, to provide the developing countries with training on necessary technical expertise.

2. Main provisions for oil pollution response

(1) Oil pollution reporting procedures

In case of oil pollution accidents of its ships, offshore units, airplanes, harbors or oil handling facilities, the contracting party shall report the accidents to the nearest coastal competent authorities in the prescribed form and notify the IMO.

(2) Actions after receiving oil pollution reports

Upon receipt of the report, the competent authority of the country concerned shall, as soon as possible, make an assessment of the nature, scope and possible consequences of the oil pollution accident, so as to prepare appropriate measures and promptly inform the countries whose interests are or may be affected. In case of a serious oil spill accident, it shall contact the regional organization to make arrangements and take measures, and inform the IMP of the situation.

3. Main provisions for international cooperation

(1) International cooperation in oil pollution emergency response

Upon receipt of the request of the countries concerned for international cooperation and support in dealing with oil pollution accidents, the contracting parties shall make every effort to provide equipment resources and technical assistance.

(2) Bilateral or multilateral agreement

The contracting parties shall endeavor to conclude bilateral or multilateral agreements on oil pollution preparedness and response, so as to facilitate mutual collaboration and support in the event of oil spill accidents.

(3) Research and development of oil pollution emergency response technologies

The contracting party may, directly or through workshops held by the IMO on oil pollution emergency response technologies and facilities, exchange research findings and development plans, including oil pollution monitoring, control of recovery and elimination, in order to promote the global dissemination of advanced oil spill emergency response technologies.

(4) Technical cooperation

The contracting countries are obliged to provide, upon request, the countries concerned with technical training on oil pollution emergency response and technical equipment, as well as cooperation in basic research, development planning and technology transfer.

In order to effectively implement the OPRC Convention, the IMO has made specific provisions on the implementation of some provisions of the convention in the form of ten resolutions. In the meantime, in order to drive the implementation of the OPRC Convention and provide information and technical advices, education, training, and technical support, the oil pollution coordination center has been established with the participation of representatives from hundreds of countries and international organizations. The OPRC working group has prepared an oil pollution emergency plan guideline for offshore units, harbors and oil handling stations, which has driven each country to formulate oil spill emergency plans for the countries, regions and harbors as well as ships, offshore facilities and handling stations. Furthermore, the working group has also developed a demonstration course for oil spill emergency response training, enabling a number of management and command staff in charge of oil spill emergency response to be trained and providing a model for countries to organize their own training.

Most importantly, OPRC Convention promises worldwide for oil spill emergency response, and the contracting parties are required to provide equipment and technical support to the countries requesting aid. This is a liability under the Convention, enabling the global cooperation in oil spill preparedness and response to come true, and it is also a new development trend of the OPRC Convention with respect to combating oil spills.

At present, coastal countries are implementing or preparing to implement these provisions of the OPRC Convention, and cooperation agreements have been made or are being prepared for international cooperation in combating oil spill pollution. Many countries in the world have established national oil spill contingency plans. Some neighboring countries have established bilateral agreements, and 13 multilateral agreements have come into force or are about to take effect.

After the adoption of the OPRC Convention, relevant international conventions have been amended to maintain consistency with the OPRC Convention. In 1991, Article 26 was added in the amendment to Annex I of the MARPOL73/78 convention, requiring oil tankers with a gross tonnage of 150 tons and more and non-tank vessels with a gross tonnage of 400 tons and more to carry the Shipboard Oil Pollution Emergency Plan approved by the competent authorities before April 4, 1995. This is consistent with Article 3 of the OPRC Convention: Oil Pollution Emergency Plan, and is also a concrete manifestation of taking the lead in implementing relevant provisions consistent with Article 3 of the OPRC Convention before the OPRC Convention came into force.

Moreover, in order to prevent, reduce and control marine environmental pollution, United Nations Convention on the Law of the Sea (UNCLOS) provides contracting parties with standards and requirements for the implementation of international rules and the formulation of domestic laws and regulations. International conventions concerning oil spill pollution damage compensation, including International Convention on Civil Liability for Oil Pollution Damage (CLC1992), International Convention on the Establishment of an International Fund for Compensation for Oil Pollution Damage (IOPC Fund1992&2003) and International Convention on Civil Liability for Bunker Oil Pollution Damage (Bunkers Convention 2001), detail compensation claims for expenses arising from oil spill emergency response. All of these conventions provide strong support for the implementation of the OPRC Convention.

1.3.2 Construction of Emergency Response Systems of Developed Countries for Marine Oil Spills

In the late 1970s and early 1980s, some developed countries successively established and improved their emergency response systems for marine oil spills. The following is an overview of emergency response systems for oil spills of several major developed countries.

1.3.2.1 The United States

America's emergency response system for oil spills mainly consists of the emergency response systems for oil spills established by the national emergency response command center for oil spills and relevant state governments and local organizations. The national emergency response command center is composed of 16 government departments, including the United States Environmental Protection Agency, the Department of the Interior and the Department of Transportation, and is mainly responsible for developing the planning for national marine oil spill prevention and control work, and directing and coordinating the emergency response actions of state governments and local organizations. Each state government is mainly responsible for the planning of oil spill prevention and control in its administrative region and the coordination of emergency cooperation and support of relevant departments, and local emergency response organizations are mainly responsible for directing specific emergency response actions for oil spills as well as for clean-up work.

In the United States, many agencies will respond to oil spill accidents, including United States Environmental Protection Agency, United States Coast Guard, the U.S. Department of Transportation, the national emergency agency Federal Emergency Management Agency and local emergency committees. These institutions will fulfill their respective responsibilities according to relevant laws, and assume command monitoring and management in the process of emergency response according to the specific provisions of national emergency response systems. In addition, they will provide support and assistance in professional clean-up technologies, equipment and staff, so as to ensure that oil spill recovery work will be done rapidly and effectively even when the clean-up company cannot work in a harsh environment. The United States Coast Guard plays an important role in the process of oil spill emergency response. Its national commando, consisting of three national commando forces occupying strategic strongpoints and a coordination center, is mainly responsible for combating oil spills and chemical leakage. National commando forces will participate in emergency response actions in the event of major oil spill accidents, and the coordination center has a national list of emergency equipment combating oil spills, and provides assistance in the formulation and implementation of national emergency response exercises and training plans. There are also other assisting forces, including the National Pollution Fund Center and response groups within the jurisdiction.

In the terms of oil spill preparedness and emergency response, the United States has not only formulated relatively perfect laws and regulations and established a national emergency response system, but also established a scientific oil spill prevention, control and response strategy system, information database system, oil spill identification system and pollution damage compensation system.

1.3.2.2   Japan

Japan's oil spill emergency response force mainly consists of the Japan Coast Guard and Maritime Disaster Prevention Center.

The Japan Coast Guard is mainly responsible for maritime surveillance and supervision. The Coast Guard has its own clean-up and containment facilities and fireboats to combat large-area oil spills, and ensures that they are readily available. In order to guarantee the effective response to marine oil spill accidents, the Coast Guard also establishes basic coastal environmental data and provides relevant information to oil pollution prevention and control agencies via the Internet. In order to deal with marine oil spill accidents, the Coast Guard predicts the drifting direction of oil spills, helps contain and clean-up floating oil spills, and sends patrol ships and aircrafts to monitor pollution at sea, especially in areas where navigation is intensive.

The Maritime Disaster Prevention Center is the core institution of civil maritime disaster prevention in Japan. It receives instructions from the Coast Guard and takes timely measures to clean-up oil spills in case of oil spill accidents. It has four committees, namely oil spill clean-up, ship fire fighting, equipment, and training committees. The center has vessels and equipment for maritime disaster prevention, and it also carries out training for maritime disaster prevention, promotes international cooperation in maritime disaster prevention, and carries out survey and research on marine disaster prevention.

1.3.2.3   Britain

Britain's emergency response to marine oil spills is mainly in charge of the UK Marine Pollution Control Center, which is affiliated with the Coast Guard of the UK Department of Transport and fulfills the IMO's conventions on oil spill response. The center has capabilities of aerial remote sensing surveillance, oil spill volume assessment, prediction of oil spill drifting trajectory, and aerial and shipboard spraying of oil spill dispersants, and has equipment to recycle or transfer oil spills at sea or on shore. The UK Marine Pollution Control Center is mainly responsible for the coordination in marine response and coastal clean-up in major oil spill accidents, and also provides technical guidance to relevant government departments for coordinating coastal oil pollution clean-up.

The Marine Pollution Control Center and local governments carry out emergency response with support from the support system composed of the Coast Guard of the UK Department of Transport (which has 21 rescue coordination centers across the country), the Marine Safety Agency, fishery departments, environment departments, the Ministry of Defence, the Met Office, nature protection organizations, major oil companies, and the British Oil Spill Control Association. In the event of oil spill accidents, the rescue coordination centers under the Coast Guard will be responsible for dealing with minor accidents, and the Marine Pollution Control Center and local governments will be responsible for coordinating relevant departments to deal with major accidents.

1.3.2.4   Germany

Germany's oil pollution clean-up task is undertaken jointly by the federal government and coastal states including Bremen, Hamburg, Lower Saxony and Schleswig-Holstein, and the organization structure is divided into five parts. The first part is the marine/offshore pollution accident committee composed of experts; the second is the oil pollution clean-up team responsible for oil pollution clean-up; the third is the oil pollution accident alarming agency, which is built in the national alarm center, works for 24 hours, and reports oil pollution accidents to international and domestic communication stations; the fourth is the federal task force and the federal offshore task force established in Kugels-Haven, and their missions are to review the plans proposed by the marine/offshore pollution accident committee and to support pollution clean-up teams in case of oil pollution accidents; and the fifth is the consulting institute, which provides knowledge of oil pollution accidents and research plans and recommendations to competent authorities and pollution clean-up teams [10].

1.3.2.5   France

France's emergency response system is divided into two systems: maritime system and land system, and two levels of emergency response organizations are established. The interministerial maritime committee under the supervision of the secretary of state for maritime affairs and the civil safety committee under the supervision of the interior minister are established and respectively responsible for the review and approval of marine and land pollution emergency plans, pollution control, and national emergency exercises. At the local level, the coast defence command (military sector) is responsible for planning and directing marine pollution control operations, and working with local authorities and maritime enterprises to develop contingency plans and organize personnel training exercises [10]. In France, the pollution clean-up work mainly depends on companies and private sectors, usually through temporary hiring or requisition.

1.3.3   Construction of China's Emergency Response System for Marine Oil Spills

China is a major maritime country with a continental shoreline of 1.8×104 kilometers and a sea area of about 3.0×106 square kilometers, and is also the second largest oil importer in the world and the first largest in Asia. Although so far, no extraordinarily serious ship oil spill accident of more than 10,000 tons has ever occurred in China, extra major oil spill accidents are presenting a constant threat. In addition to 69 oil spill accidents of over 50 tons, from 1999 to 2006, seven potential extraordinarily serious oil spill accidents occurred in China's coastal areas. Although no serious pollution was caused due to the timely measures taken by the maritime authorities, the risk of extraordinarily serious ship oil spill accidents is ubiquitous. In addition, as China's sea area is rich in oil and gas resources, the offshore oil exploration industry has been developing for 44 years since 1967, and the number of offshore oil and gas fields has formed a scale, which will increase the risk of oil spill accidents.

To sum up, it is necessary and urgent to strengthen the construction of the emergency response system for marine oil spills in China, and only by improving the whole system can we effectively deal with any possible marine oil spill accident. On the other hand, as a member of the World Trade Organization (WTO), China must keep its various systems in line with international practices, operate in accordance with international practices, and take seriously the issues such as maritime transport safety and marine ecological environmental protection. Ensuring the safety of maritime transport, protecting the marine ecological environment and preventing and controlling pollution from ships have always been the key concern of the international community. The IMO has successively formulated and promulgated United Nations Convention on the Law of the Sea, International Convention for the Prevention of Pollution from Ships and International Convention on Oil Pollution Preparedness,

Response and Co-operation, which provide comprehensive and detailed provisions for the prevention and control of marine oil spill pollution from ships [11]. Therefore, as a Category-A member of the IMO and the contracting party of the aforesaid international conventions, China must play an exemplary role in implementing international conventions, and Chinese governments at all levels have an obligation to establish an effective emergency response system for marine oil spill accidents and increase China's emergency response capacity for marine oil spill accidents.

1.3.3.1 Current situation of China's emergency response system for marine oil spills

The construction of China's emergency response system for marine oil spills started in the mid-1990s. At present, the emergency response system for marine oil spills from ships has been preliminarily established, and the responsibilities of various departments, distribution of sensitive resources under the jurisdiction, composition of emergency forces and emergency action procedures have been clarified.

With great attention from the government and the maritime authorities, China acceded to the OPRC Convention on March 3, 1998. For the good performance of the Convention, the Marine Environment Protection Law of the People's Republic of China came into force on April 1, 2000. It states clear requirements for the establishment of the oil spill emergency response and the implementation of emergency plans, and authorizes national competent maritime authorities to formulate national emergency plans for major marine oil spill pollution accidents of ships and strengthen the organization and management of oil spill emergency response. The Maritime Safety Administration of the Ministry of Transport has carried out the construction of emergency response plan systems at all levels throughout the country, and at present, it has basically completed the construction of national, provincial, port-level and ship-level emergency plan systems. Chinese Marine Ship Oil Spill Emergency Plan has been enacted and executed, covering all water areas in China, and becomes the programmatic document for the construction of marine oil spill emergency plan systems in China.

In order to better implement international conventions and strengthen international ties and cooperation, China has worked with neighboring countries and regions to prepare the NOWPAP (Northwest Pacific Action Plan) Regional Oil Spill Contingency Plan, which has come into force. In order to better integrate domestic resources and strengthen interregional cooperation, the Maritime Safety Administration of the Ministry of Transport has presided over the preparation and implementation of a serious cooperation active plans, including Marine Oil Spill Emergency Plan for the Area of Pearl River Estuary, Shipboard Oil Pollution Emergency Cooperation Plan for the Waters of the Taiwan Strait, and Shipboard Pollution Emergency Linkage and Cooperation Mechanism for the Sea areas of the Bohai Sea.

The Chinese government has also increased investment and construction of oil spill emergency equipment stockpiles and oil spill monitoring and surveillance systems to provide information support for accident detection, oil spill tracking and emergency decision making. At present, the ship oil spill monitoring and surveillance system has covered the sea areas across the country, and advanced technologies, including computer-based oil spill diffusion model, satellite remote sensing, ship-borne radar and air monitoring, has been used to increase the accuracy of oil spill monitoring. Large emergency equipment stockpiles and emergency technical exchange demonstration centers have been established in some key sea areas, and been equipped with advanced devices such as floating oil recovery vessels, oil recovery equipment, oil booms, oil absorption materials and oil storage equipment. Full-time and part-time pollution clean-up teams have been built in coastal provinces and cities in combination with regional characteristics, and by means of special government input, self-input of port and waterway enterprises and support for marketization of professional pollution clean-up companies.

The knowledge, technologies and experiences of organization, command and operation staff in oil spill emergency response are of vital importance to the successful implementation of oil spill emergency response, and this is also a key link in the construction of China's emergency response system for marine oil spills. Chinese Marine Ship Oil Spill Emergency Plan puts forward clear requirements on the contents, requirements and objectives of training of emergency operation and command staff. Every year, competent maritime authorities organize and carry out shipboard pollution prevention and emergency training of different scales in major coastal ports, and have cultivated a number of emergency command staff for oil spills and relatively skilled pollution clean-up workers, and greatly improved China's emergency response capacity for marine oil spills.

1.3.3.2   Deficiency of China's emergency response system for marine oil spills

In recent years, China has accelerated the construction of the emergency response system for marine oil spills, but there are still many problems and the system still cannot fully meet the requirements of national shipping development and environmental protection, which is shown in the following aspects:

1. The management system for marine oil spill emergency response is inadequate

At present, China's maritime emergency management not only involves the functions of multiple departments including marine sectors, fishery sectors, environmental protection sections, transport and maritime sectors, customs and border defence sectors, but also the coordination mechanism between different departments is ambiguous. Besides, coastal provinces, cities and departments only manage their adjacent sea areas by themselves, and it is difficult to form the integrated management pattern among departments, provinces and cities.

If things continue this way, the comprehensiveness and uniformity of China's ocean management functions will be weakened, and it will be difficult to form an efficient and scientific management system for oil spill emergency response.

2. The capacity of emergency response to marine oil spill accidents is limited

China's existing emergency response force for marine oil spills is still fragile. The scale of emergency response centers invested by the state and local governments is small, and the social pollution clean-up force is weak. There is a general lack of professional emergency ships, equipment and personnel combating oil spills from the state to sea areas and ports, and there is even no emergency equipment to combating marine oil spills. China's emergency response force can only combat oil spill pollution accidents occurring in ports and offshore areas, and there is no capacity of dealing with oil spill pollution accidents, especially major oil spill pollution accidents, occurring in sea areas under the jurisdiction of China [12].

3. The allocation level of emergency facilities is not high and the construction of emergency team is insufficient

At present, the allocation of oil spill emergency equipment in China has obvious defects. On one hand, the allocation of emergency facilities and equipment is unreasonable. Facilities for containing oil spills are much more than those for oil spill recovery and clean-up, causing the imbalance of emergency response force and directly affecting the rapidity and effectiveness of emergency response actions. On the other hand, the scientific and technological strength of emergency equipment is weak, and the research and development of oil spill clean-up technology and emergency equipment and facilities progresses slowly, resulting in the low technological content of emergency equipment. Due to the complexity of emergency treatment of oil spills, the support from a variety of disciplines, such as environmental science, information science and medicine, is quite necessary, and emergency response personnel are required to understand relevant knowledge in advance. In addition, experts in relevant fields are needed to provide guidance and technical consulting services at crucial times during emergency response actions. At present, China is not only short of professional training for emergency response personnel, but also lacks talent reserve in relevant professional fields.

4. The decision making and command system for marine oil spill emergency response is relatively outdated

The decision making and command system for marine oil spill emergency response is the neural network and command center of the marine oil spill emergency response work. However, the decision making and command systems of maritime search and rescue centers at all levels in China are still backward and seriously short of command equipment, and depend on manual operation and experience-based judgment to a great extent. The lack of maritime communications, poor capacity of dispatching and command of emergency force, low intellectualization of maritime emergency response command and low automation cannot guarantee the smoothness of information transmission and the scientificity of the measures to be taken. During the actual combat against ship oil spill pollution accidents, on-site information, extent of pollution damage and effect of the measures taken cannot be transmitted to the command center in real time. All the above factors seriously restrict scientific decision making for emergency response [12].

1.3.3.3 Comprehensively strengthen the construction of China's emergency response system for marine oil spills

In view of the deficiency of China's emergency response system for marine oil spills, the construction of China's emergency response system for marine oil spills should be comprehensively strengthened from the following aspects [11]:

1. Improving the decision making and command system for oil spill emergency response

The decision making and command system for oil spill emergency response mainly relies on maritime search and rescue centers as the command hub of oil spill emergency response. It utilizes their convenient communication and traffic command systems, and integrates various information resources to make rapid and accurate judgments about the situation of oil spill emergency response, so as to provide decision making support for oil spill emergency response. Emphasis should be laid on the establishment of the decision making information system for oil spill emergency response, including oil spill resource pool, oil spill emergency response action optimization model, sensitive resources protection priority program, oil spill emergency response expert pool, and oil spill pollution damage evaluation program.

2. Developing an oil spill model system

The oil spill model system can not only predict the diffusion trajectory of oil spills, enabling emergency decision makers to plan ahead for the drift and fate of oil spills, but also assess the risks of oil spills in major waterways and ports in sea areas within the jurisdiction, so that professional forces such as oil spill emergency equipment stockpiles and oil spill emergency ships can be distributed reasonably according to the distribution of risks, accordingly increasing the overall response efficiency for oil spill accidents.

3. Improving three-dimensional offshore and aerial monitoring systems for oil spills

The offshore monitoring system includes patrol ships, vessel traffic service (VTS) radars, ship-borne radars, and oil spill position-indicating radio beacons. Patrol ships are the traditional means of monitoring oil spills at sea. Currently, medium and large patrol ships are still insufficient in China. Under poor sea conditions, patrol ships are often unable to reach the sites of oil spill to complete monitoring tasks. Monitoring oil spills by VTS radars is still in the stage of technical application development and promotion. It is an ideal technology for monitoring oil spills near the coast and should be promoted more rapidly. The technology of monitoring oil spills by ship-borne radars has been applied in foreign countries and should be applied in China as soon as possible. Oil spill position-indicating radio beacons now are in the process of product development, and investment should be increased to put them into use as soon as possible.

The aerial monitoring system includes satellite remote sensing, airborne remote sensing and manual observation from airplanes. Satellite remote sensing images are generally purchased from professional institutions, and it often takes a long time to purchase images after accidents occur. At present, China's maritime departments have tried to carry out routine monitoring through satellite remote sensing. After practice, the monitoring density should be strengthened in key sea areas, so as to give full play to the role of satellite remote sensing monitoring. The airborne remote sensing technology has been applied experimentally in China and has achieved good results. Manual observation from airplanes often requires fixed-wing aircrafts or helicopters, but it is only used to understand the situation of oil spills at sea in the most intuitive and fastest way in appropriate weather and in tight situations. Recent accidents indicate that oil spills may be sub-submerged in seawater and cannot be observed from the sea surface, which increases the difficulty in oil spill monitoring. Current monitoring methods are not effective in dealing with this situation, which requires more in-depth research on the status and behavior of oil spills in seawater, and also requires development of new monitoring methods and equipment.

4. Strengthening the construction and technical research of oil spill recovery and disposal equipment

The state should strongly encourage major ports to build efficient oil recovery vessels with functions such as storage of oil booms, oil dispersant spraying and oil spill recovery, transform existing pollution clean-up ships, improve the ability of a single ship to combat oil spills, and develop efficient multifunctional recovery ships for offshore oil spills. With respect to the laying of oil booms, mathematical models applicable to various oil booms and computer simulation systems used for laying oil booms should be optimized, including buoyancy-weight ratio, counterweight, flowability, tensile strength, and operation parameters such as reasonable length of oil booms under specific sea conditions and towing speed, in order to develop the new-type connecting devices between oil booms and ships. The technology of onshore oil spill recovery should also be studied, including technologies for cleaning up oil spills along sandy or rocky shorelines. Furthermore, new-type environment-friendly products including oil spill dispersants, ship-borne sprinklers and airborne sprinklers should be developed, and sufficient oil absorption equipment should be kept in reserve.

5. Build and improve the marine oil spill coordination center

It is necessary to build and improve the marine oil spill coordination center consisting of maritime sectors, marine sectors, transport sectors, public security sectors, agriculture sectors, customs, civil aviation sectors, diplomacy sectors, journalism sectors, finance sectors, safety production sectors, armed forces and oil production corporations, which will rapidly coordinate the treatment of major marine oil spill accidents and reasonably allocate existing emergency response resources, so as to minimize accident losses. In the meantime, it will formulate work guidelines and policies for the emergency response of major marine oil spills during offshore oil exploration and development, review emergency response plans and annual work plans for major marine oil spills during offshore oil exploration and development, and organize the formulation and implementation of national emergency plans for major marine oil spills during offshore oil exploration and development.

Chapter 2

Remote Sensing Monitoring of Marine Oil Spills

Due to the contingency and complexity of marine oil spill accidents, it is particularly important for the oil spill emergency response departments to monitor the sources of oil spill accidents such as ships, oil pipelines, and drilling platforms. It is the goal of many coastal countries in the world to obtain the real-time situation of oil spill occurrence and development so as to adopt timely and effective treatment measures. A number of countries have realized the effective monitoring and management of marine oil spills through constructing the three-dimensional monitoring and management system which combines artificial satellite, remote sensing monitoring aircraft, and offshore patrol ship.

Remote sensing technology is a kind of detection technology emerging in the 1960s. It mainly uses various remote sensors to collect, process and finally image the electromagnetic wave information radiated and reflected by remote targets, so as to achieve the purpose of detecting and recognizing various objects on the ground. Carriers carrying remote sensors are also known as remote sensing platforms, such as aircraft and artificial satellites. In general, remote sensing can be divided into aerial remote sensing and satellite remote sensing according to the types of remote sensing platforms.

At present, aerial remote sensing is one of the most important and effective surveillance measures. Under the comprehensive action of various marine environmental dynamic factors, oil spills have a random dynamic nature. The flexible mobility of aerial remote sensing and the selectivity of remote sensors during deployment make it perform well in capturing the information about oil spills on the sea surface, and also decide that aerial remote sensing often plays a leading role in emergency response to oil spill accidents. However, in terms of observation range, aerial remote sensing has some limitations. Therefore, the advantages of satellite remote sensing are highlighted. Satellite remote sensing can provide macroscopic images of entire water areas contaminated by oil spills in terms of determining the location and area of oil spills. In the event of large-scale and catastrophic oil spill accidents, both satellite remote sensing and aerial remote sensing are used simultaneously to track and monitor the drift and diffusion of oil spills. In the event of chronic oil spills, more countries use satellite remote sensing to monitor the sources of pollution.

There are many international examples of developed countries that use satellite remote sensing for oil spill monitoring. For example, from 1972 to 1975, the United States cooperated with European countries to monitor oil pollution in the Mediterranean by using the data of Landsat multi-spectrum scanner (MSS), and determined the pollution area, pollution speed and diffusion direction. In 1985, the United States used satellite images of ERTS-1 to monitor the oil film 100 kilometers long in the southeast sea of

Assateague Island, Virginia, and estimated the pollution area and inferred the source of oil pollution [13]. Chinese scientists have also carried out some research to identify marine oil spills from the image data of Landsat-TM, NOAA-AVHRR, and Terra/AQUA-MODIS, and achieved some results [14]. For example, [15] tested and analyzed the spectral characteristics of marine oil spills, and proposed to use Landsat-TM and NOAA-AVHRR data to monitor the optimal band combination of kerosene, light diesel oil, lubricating oil, heavy diesel oil, and crude oil. Zhao et al. [16] made a comparative analysis of the spectral characteristic curve of ground objects in the near infrared band of visible light of crude oil, diesel oil, and lubricating oil in the near infrared band, revealing the spectral characteristics, oil-water contrast law and absorption characteristic parameters of oil films varying with thickness.

Among all remote sensing technologies, Synthetic Aperture Radar (SAR) is unique. SAR, as a high-resolution active remote sensing system, has been widely used in oil spill monitoring. It can realize semi-automatic extraction of marine oil spill information, and has the advantages of all-weather and all-day monitoring and strong penetrating power. Nowadays, it has gradually become a major means of monitoring marine oil spills.

2.1 Remote Sensing Monitoring

2.1.1 Electromagnetic Wave and Remote Sensing Technology

Electromagnetic wave is formed when electric and magnetic fields that oscillate in phase and are perpendicular to each other travel through space in the form of waves. Its propagation direction is perpendicular to the plane formed by electric and magnetic fields, which can effectively transfer energy and momentum. In general, electromagnetic waves can be emitted by objects whose temperature is not absolute zero. People arrange electromagnetic waves in the order of wavelength or frequency to form the electromagnetic spectrum. The electromagnetic spectrum is roughly divided into radio wave, microwave, infrared, visible light, ultraviolet, χ-ray and γ-ray. Radio waves, which usually transmit signals from televisions, mobile phones and radio broadcasts, range in wavelength from 0.3 meters to several thousand kilometers.

Microwaves, which are mainly used in radars or other communication systems, range in wavelength from 10^{-3} meters to 0.3 meters. Infrared has significant thermal effects, and has a wavelength range of $7.8 \times 10^{-7} \sim 10^{-3}$ meters. Visible light is the electromagnetic wave emitted by electrons in an atom or molecule when their state of motion changes. It has a narrow wavelength range of about $4 \times 10^{-7} \sim 7.6 \times 10^{-7}$ meters. The wavelength of ultraviolet is shorter than that of visible light, and its wavelength range is $10^{-8} \sim 3.8 \times 10^{-7}$ meters. It has significant chemical effect and fluorescence effect. The generation of ultraviolet light wave is similar to that of visible light wave, which is often emitted when discharging. Because it has the same amount of energy as that needed by a general chemical reaction, the chemical effect of ultraviolet is the strongest. Neither infrared nor ultraviolet is visible and can only be detected by special instruments. Visible light, infrared and ultraviolet are all produced by the excitation of microscopic particles such as atoms or molecules. In recent years, on the one hand, due to the development of the ultrashort wave radio technology, the scope of radio waves is continuously developing toward shorter wavelength. On the other hand, due to the development of infrared technology, the scope of infrared is constantly expanding toward longer wavelength. As a result, the boundary between ultrashort waves and infrared no longer exists, and their wavelength ranges overlap to some extent. χ-ray is emitted by the electrons of an atom as they jump from one energy level to another or as they slow down in the atomic nuclear field, and its wavelength range is $10^{-11} \sim 10^{-8}$ meters. With the development of χ ray technology, its wavelength range continuously expands in two directions. At present, χ-ray in the long waveband has overlapped with ultraviolet, and χ-ray in the short waveband has entered the scope of γ-rays. γ-rays are electromagnetic waves with a wavelength range of $10^{-14} \sim 10^{-10}$ meters. This invisible electromagnetic wave is emitted from the inside of the nucleus, and such radiation often occurs in radioactive materials or nuclear reactions. γ-rays have strong penetrating power, and can greatly destroy living things.

The reflectivity corresponding to the wavelength of each spectrum in electromagnetic wave is called spectral reflectivity or reflectivity spectrum. The spectral reflectivity of an object varies according to its type. As the spectral radiance emitted by objects is affected by spectral reflectivity, it is possible to identify distant objects by observing the spectral radiance. In remote sensing technology, the type, nature, state and surrounding environmental conditions of objects can be distinguished through observing the spectral radiance of the electromagnetic waves radiated or reflected by remote sensors. At present, electromagnetic waves that can be used for remote sensing mainly include visible light, microwave, infrared and ultraviolet, and electromagnetic waves reflected or radiated by various substances are mostly within the band range of these waves. Nowadays, remote sensing technology has been widely used in the fields of military reconnaissance and surveying, missile early warning, environmental pollution monitoring, marine monitoring and meteorological observation. In addition, it can also be applied in general survey of earth resources, vegetation classification, land use planning, survey of crop diseases and pests and crop yields, and earthquake monitoring. In the future, the general development trend of remote sensing technology is to improve the resolution and comprehensive utilization of information. More advanced remote sensing devices and information transmission and processing equipment will be used to realize all-weather work and real-time information acquisition of remote sensing systems, and the anti-interference ability of remote sensing systems will be further enhanced.

2.1.2 Remote Sensing Monitoring Technology for Oil Spills

When there is an oil spill on or near the sea surface, the physical properties (emission rate, radiation, color, etc.) of the sea surface covered by the oil film are different from that of the pure sea surface due to the differences between the spectral characteristics of oil pollution and seawater. Therefore, the reflection and radiation spectrum of the sea surface to electromagnetic waves also changes. Through detecting the changes of spectral information in the oil spill area, the remote sensor can distinguish the oil spill area as well as the type, thickness and area of oil pollution, and analyze the moving direction and speed of oil pollution based on several consecutive days of image data to infer the source of the oil spill and estimate oil spill volume. The ability of remote sensors to monitor and track oil spills also depends on several factors, such as the type of remote sensor (telemetry angle, observation frequency, operation cycle, bandwidth, telemetry distance, etc.), weather and sea conditions during and after the oil spill accident, type and composition of oil pollution, weathering degree and duration of oil pollution, etc. In the process of monitoring oil spills with remote sensing technology, remote sensing with ultraviolet, visible light, infrared, microwaves and laser fluorescence is used the most. Spectral characteristics of the sea in these spectral bands are the key to image processing and information extraction of oil spills.

2.1.2.1 Ultraviolet remote sensing monitoring

Ultraviolet is divided into ultraviolet A (UVA), ultraviolet B (UVB) and ultraviolet C (UVC) according to the wavelength. When passing through the surface stratosphere of the earth, UVC is absorbed by the atmospheric ozonosphere and can hardly reach the sea surface. At the same time, due to the strong reflection and radiation of oil films on the sea surface, it will produce a certain fluorescence. However, when the spectral band is larger than that of UVC, it is affected by atmospheric aerosol. Therefore, ultraviolet remote sensing is generally carried out in the effective band of 0.3 ∼ 0.4 microns. UV remote sensors work relying on the difference between reflection characteristics of the sea surface and oil films to effective ultraviolet spectra and the fluorescence produced by oil films under the action of ultraviolet radiation.

The reflection intensity of the oil film in the ultraviolet spectral region is related to the thickness of the oil film. The oil film with a thickness of less than 5 microns is very sensitive to reflection in the ultraviolet spectral region. Its reflectivity is 1.2 ∼ 1.8 times higher than that of the seawater, making the thin oil film look bright white. According to a number of experiments, it is concluded that: in the ultraviolet spectral region of 0.28 ∼ 0.4 microns (especially 0.32 ∼ 0.38 microns), the reflection of the oil film is much stronger than that of the seawater, and the oil film and the seawater can be clearly distinguished.

Ultraviolet remote sensing can be used to detect thin oil films on the sea surface, even very thin films (less than 0.05 microns). Meanwhile, due to the obvious response of ultraviolet remote sensing to the heat mutation in the monitored sea surface, it is possible to distinguish the areas of thermal pollution and oil pollution, so the ship's wake can be recognized and the ship in violation of regulations can also be identified. The disadvantage of ultraviolet remote sensing is that it is vulnerable to the interference of external environmental factors and may generate false information (such as solar flares, sea-surface bright spots caused by wind and interference of biological substances, etc.), and it cannot work at night. Poor weather conditions will also make it difficult to use.

Because solar flares, sea-surface bright spots caused by wind and interference of biological substances produce different effects in the ultraviolet band from those produced in the infrared band, the composite analysis of infrared and ultraviolet data generally can obtain better results than the detection in a single band.

2.1.2.2 Visible-light remote sensing monitoring

As is shown in Figure 2.1, When visible-light remote sensing is used to monitor a marine oil spill, although the emission rate of the oil film is much higher than that of the clean sea surface, the reflection intensity of the oil film will also change with the wavelength of lighting light, observation angle, oil type and the transparency of background water. Therefore, detecting the color and brightness of the designated oil spilled in the turbid water bodies at middle latitudes is different from that in clean water bodies in the tropics. In other words, in visible light, the oil spill lacks an effective characteristic spectrum that differs from background information. Taylor [17] carried out laboratory and field observation and research on the spectrum of crude oil. The study found that the visible spectrum curve is flat, there is no characteristic spectrum that can be used for crude oil information extraction, and there is a lot of interference or false information. For example, solar flares and sea-surface bright spots caused by wind and biological substances such as sea-surface seaweed and underwater seaweed beds may be confused with oil spills, and it is often difficult to distinguish coastal oil spills from other substances such as seaweed. At present, hyperspectral data on visible light are also used for oil spill detection, and the ability of detecting and recognizing oil spills can be improved through subdivision of spectral bands. In summary, the ability of visible-light monitoring on oil spills is limited, but it is an economical and practical means to provide qualitative description and relative position of oil spills.

Common visible-light monitors include cameras, video cameras, visible scanners, etc. In recent years, with the development of global positioning system technology, photogrammetry using cameras has also made great progress. The geographical location of photography can be obtained directly through GPS measurement, which makes the obtained information more useful. Optical filters are often used in photography to improve the contrast of images. Due to the low contrast between oil spills and the background, this technology is not very effective for oil spill detection; it also lacks the active recognition capability. Visible scanners are visible remote sensors which are commonly used, and a kind of instrument with high sensitivity and advantages in signal digitization prior to the emergence of charge coupled device (CCD) technology. The CCD imager emerging in recent years is more advantageous than the old-type scanner. It scans and images the object by means of pushing the broom, and pixels on the scan line perpendicular to the navigation direction can be imaged simultaneously, which can reduce various errors. It is more reliable than mechanical scanners.

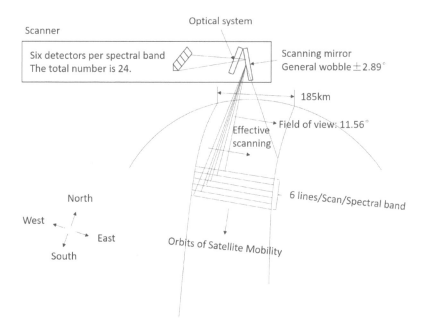

FIGURE 2.1: Visible-light remote sensing monitoring.

2.1.2.3 Infrared remote sensing monitoring

Remote sensors working in the thermal infrared spectral band, such as infrared radiometer and scanner, mainly receive and record the infrared thermal radiation energy of the targets, including thermal radiation of surroundings and atmospheric path radiation energy reflected by a few targets. The latter two can be ignored or corrected with a mathematical-physical model under certain conditions.

The relative magnitude of the object's thermal radiation capacity at a given temperature is described as spectral radiance ε. In very narrow spectral bands (e.g., $8 \sim 14$ microns), when the measured object is close to the black body, the Planck formula and the integral mean value theorem are used to calculate the spectral radiance ε, and the expression is as follows:

$$\varepsilon = \frac{c\delta T}{\lambda T^2} \tag{2.1}$$

where, c is a constant (1.44×10^{-2} meters $\cdot$ Kelvin); λ refers to the central wavelength of the spectral band; δT refers to the difference between the radiation temperature of the object and the thermometrical temperature (measured temperature); and T refers to the measured temperature of the object. In the case of marine remote sensing, since the radiance of the seawater is very low, close to 1, the seawater can be regarded as a black body. If the radiation temperature difference between the target (such as oil film) and the background seawater (δT) and the actual temperature of the seawater (T) are measured, the radiance of the target material can be calculated using the formula.

In the infrared remote sensing image, the oil film has a higher gray level than the surrounding seawater, and looks blackish gray. In the thermal infrared band of $8 - 14$ microns, the oil pollution on the sea surface can be clearly detected. Satellite-borne infrared remote sensors can distinguish the scope of large-area marine oil spills, and aerial infrared remote sensors can reliably distinguish the coverage area of oil films and their drift and diffusion, and the images are clearer than those obtained in the visible light band. The radiance of oil films with thickness of less than 1 millimeter increases with the increase of the thickness. Therefore, the infrared image can also reflect the change of gray level with thickness, and the thickness and distribution of oil films can be calculated and the total oil spill volume can also be calculated [18].

In most cases, thermal infrared remote sensors cannot detect emulsified oil pollution because the oil contains 70% seawater, and the temperature change cannot be detected in the infrared image. Besides, thermal infrared remote sensors are a kind of passive receiving devices, there will be some interference from wrong targets in the image, such as seaweed, coastline, and so on. However, the thermal infrared remote sensor is still the most important device for oil spill monitoring at present.

2.1.2.4   Microwave remote sensing monitoring

Since the wavelength of microwave is much larger than that of visible light, near infrared, and thermal infrared, the nature of microwave is different from that of visible light, near infrared, and thermal infrared. Microwave has excellent atmospheric transmittance and the ability to penetrate clouds, fog, and light rain, and solar radiation has little effect on measurements. Therefore, microwave remote sensing can not only work under severe weather conditions, but also play its role during the day and night, and it has a strong ability of all-weather and all-day working. Consequently, compared with other remote sensing technologies, microwave remote sensing has an incomparable monitoring capability.

There are two types of microwave remote sensors: (1) active remote sensors, including Synthetic Aperture Radar (SAR), Side-Looking Airborne Radar

(SLAR), Microwave Scatterometer, etc.; (2) passive remote sensors, including Microwave Radiometer, etc.

The active remote sensor mainly uses the characteristics of sea surface reflecting radar waves to monitor marine oil spills. Since capillary waves and short gravity waves of the sea surface may reflect radar waves, a "bright" image called sea clutter can be created. If there is an oil spill on the sea surface, the oil film can exert a damping effect on capillary waves and short gravity waves. Therefore, the area where the oil film exists will appear in the "dark zone" due to the lack of clutter [19]. However, this phenomenon is not only caused by oil spills, and it also occurs in the fresh water layer, in wave shadow, and on the calm sea surface, which brings uncertainty to oil spill detection by such remote sensors. If the sea surface is too calm, it is difficult to contrast with the oily sea surface. If the sea surface is too rough, the radar waves will be scattered, thus affecting the detection effect. The wind speed suitable for the radar detection of oil spills is generally $1.5 - 6m/s$, which to some extent limits the application of radar in marine oil spill detection. Nevertheless, radar is still a very important remote sensor for oil spill detection, which can conduct a wide range of target search and work at night or under extreme cloudy and foggy weather conditions.

The development of radar technology is limited by the high cost of development. At present, there is no commercial oil spill SAR, and only Ericson and Terma companies are producing commercial SLARs. While Motorola and Texas Instruments can provide SLAR customization services, GER/Raytheon and Motorola can customize commercial SARs to meet the requirements for a variety of applications.

The passive remote sensor mainly uses the characteristics of sea surface emitting microwave radiation to monitor marine oil spills. The oil layer on the sea surface emits a stronger microwave signal than the water itself (the emissivity of water is 0.4, while that of oil is 0.8), so it presents a bright signal under the dark background (the normal sea surface without oil spills). The passive remote sensor can detect this difference, so it can be used for marine oil spill monitoring. Since the signal itself varies with the thickness of the oil layer, it can be used in theory to measure the thickness of the oil layer. However, in order to detect the thickness, it is necessary to know multiple environmental parameters and specific oil parameters, and the spatial resolution of the instrument is low, so this method is limited to some extent in practical application.

Switzerland Space Agency has developed similar systems, including double-band (22.4 MHz and 31 MHz) and single-band (37 MHz) instruments. However, the research made by [20] shows that the signal is weakly related to the oil thickness, and in addition to the oil thickness, other factors can also affect the signal strength. Since changes in the thickness of the oil film will affect the reception of the signal, many experiments in this field have not been successful. At present, people begin to study the polarization contrast strength measured in two orthogonal polarization directions to measure the thickness of the oil layer.

The advantage of microwave remote sensor is that it is free from the interference of cloud, fog and smoke, can perform the monitoring work under any weather conditions, and can work at day and night without the influence of light conditions. The disadvantage is that it is vulnerable to the interference of marine organisms, has a low signal-to-noise ratio and is difficult to achieve high spatial resolution. However, in the long run, the microwave radiometer is a kind of all-weather oil spill detector with an application potential.

2.1.2.5   Laser-fluorescence remote sensing monitoring

The laser emits a laser beam to the sea surface to induce the seawater or substances on the surface to produce fluorescence. As some components of oil absorb ultraviolet light and excite internal electrons, the excitation energy can be rapidly released through fluorescence emission (mainly in the visible light region). Few other components have this characteristic, and the fluorescence wavelength of natural fluorescent substances (such as chlorophyll) is greatly different from that of oil and easy to distinguish. The fluorescence intensity and spectral signal intensity of different oils are different. Therefore, it is possible to use this characteristic for remote sensing identification of oils.

The laser remote sensor used for oil spill monitoring is mainly composed of a laser, a spectrometric receiver, a GPS and a computer. Most laser remote sensors used for oil spill detection have a working wavelength range of 300 to 355 nanometers to distinguish them from the fluorescence areas of other substances. For example, the fluorescence area of yellow substances is 420-nanometer center wavelength, and the fluorescence peak of chlorophyll is 685 nanometers. The fluorescence area of crude oil is 400 ～ 600 nanometers, and the peak center wavelength is 480 nanometers.

The thickness of the oil layer may be estimated with Raman scattering, which involves energy transfer between incident light and water molecules. The XeCI laser has an emission wavelength of 308 nanometers, while water produces Raman signal at 344 nanometers. The Raman signal emitted by water is very useful for wavelength correction of the laser and can be used to estimate the thickness of the oil layer. However, due to the strong absorption of oil, the signal is suppressed and its application is limited to some extent.

The disadvantage of laser fluorescence devices is that they are large, bulky and expensive, so they have been produced in small quantities in recent years. Recently, Canada Environment and Minerals Agency has produced two kinds of laser remote sensors: LEAF and SLEAF, which have a great application potential, because they may be the only remote sensors capable of distinguishing seagrass from oil pollution and detecting oil spills on the beach, and also the only reliable means to detect ice and snow oil pollution. Therefore, they are powerful tools in various oil spill monitoring applications.

2.1.3 Remote Sensing Monitoring Means for Oil Spills

In the 1960s, aerial remote sensing with aircraft as the platform was first applied to oil spill pollution monitoring. Since the 1980s, aerial remote sensing has formed a practical remote sensing monitoring method for marine oil spills in northern Europe. Satellite remote sensing monitoring of the marine environment began in the early 1970s. With its advantages of large coverage area, multiphase, continuity and low price, it has rapidly become the means of remote sensing monitoring vigorously developed by all countries in the world. The main remote sensing monitoring means for oil spills are separately described below, including aerial remote sensing monitoring and satellite remote sensing monitoring.

2.1.3.1 Aerial remote sensing monitoring

Aerial remote sensing has the advantages of flexibility and maneuverability, and is the most used and effective technical platform in oil spill accident monitoring. Commonly used aerial remote sensors include side-looking airborne radar (SLAR), infrared/ultraviolet (IR/UV) scanners, microwave radiometer (MWR), aerial cameras, television cameras, and remote sensor control systems with real-time image processing capability that match these instruments. With respect to ocean and aerial remote sensing technologies, some developed countries have paid attention to the research on high resolution, high spectral resolution, high temporal resolution and all-weather, all-day and full spectrum monitoring, and have developed various remote sensors specially used for ocean monitoring with the comprehensive utilization of ultraviolet, visible-light, near-infrared, shortwave-infrared, thermal-infrared, submillimeter-wave and microwave remote sensors and laser remote sensors. Meanwhile, monitoring systems and information processing systems applicable to different monitoring objects have been established [21].

The situation of aerial remote sensing monitoring of marine oil spills in some developed countries is described as below:

- The United States' USCG is a series of Dassault Falcon-20 jets for daily maritime patrol flights. Among them, one is equipped with Aireye maritime detection system of Swedish Space Corporation, which consists of a SLAR and an RS-18C IR/UV line scanner. Another jet is equipped with a turreted WF-360TL forward-looking infrared (FLIR)/CCD camera.

- Germany's federal maritime pollution control agency has two Dornier Do-228 aircrafts for surveillance missions. The two aircrafts are equipped with SLARs to determine the location of oil pollution, IR/UV line scanners to determine the range of oil pollution, MWRs to detect the thickness of oil films, and LFSs to determine the type of oil pollution. In addition, the aircraft is also equipped with a downview color camera and a camcorder for remote sensor operators.

- Norway's Statens Forurensningstilsyn (SFT) has deployed a Fairchild Merlin IIIB twin-turboprop aircraft for maritime monitoring. The aircraft is equipped with the latest maritime monitoring system MSS 5000, which consists of SLAR, IR scanner, and UV scanner, increasing the possibility of correctly detecting oil spills and avoiding possible errors that a single instrument may emit. Meanwhile, The MSS5000 also utilizes the GIS and GPS to enable the acquired data to be transmitted to the SFT headquarter in real time.

- In Canada, the Canadian Coast Guard (CCG) has deployed a series of twin-turboprop aircrafts in coastal waters to detect marine pollution. In the Pacific Ocean, the CCG's National Aerial Surveillance Program (NASP) deployed an aircraft (DeHavilland, DHC 6 series 300 Twin Otter, and C-FCSU), which is equipped with a Nikon F4 35cm camera, a SONY Hi-8 video camera and a cockpit voice recorder used to record the location of oil spills.

  As the monitoring and treatment department for large-volume oil spills, the Emergency Science Department (ESD) of Environment Canada has two remote-sensing detection aircrafts. One is DC-3, C-GRSB, and the other is a large twin-turboprop aircraft, Convair 580, C-GRSC. The DC-3 aircraft is equipped with four large remote sensor supports and one small remote sensor support. At present, the main airborne equipment is the Scanning Laser Environmental Airborne Fluorosensor (SLEAF), which can process and analyze the data of the fluorosensor in real time and determine whether there is oil spill pollution in the detected sea or coastal zone in a timely manner. ESD's Convair 580 aircraft is equipped with SARs with C-band and X-band [22].

In China, as early as the 1970s, the fluorescence spectrophotometer is used for aerial remote sensing to determine the content of oil in water, and field surveys are conducted in the waters of Dalian Bay and Qinhuangdao Port. In the early 1980s, the aerial remote sensing technology is adopted to study the oil pollution in Dalian Bay. At present, in China, the Maritime Safety Administration of the Ministry of Transport has also deployed marine monitoring aircrafts for remote sensing monitoring of marine oil spill accidents. However, due to objective economic conditions, China is still a late starter in the field of aerial remote sensing monitoring of oil spills, and there is a big gap with developed countries. The airborne remote sensors in China are only equivalent to the level of the second-generation sensors abroad. The equipment integration degree is low, the monitoring level is still in the qualitative stage, the data real-time transmission is almost blank, the information processing capacity is weak, and there is a lack of special remote sensors applicable to marine aerial remote sensing.

The aerial remote sensing monitoring of oil spills is of high timeliness and can be used to track and monitor the dynamic development of the accident.

However, aerial remote sensing is expensive, and it is difficult to meet operational monitoring requirements because of its incapability in remote coastal areas. In addition, aerial remote sensing is also restricted by weather conditions, because most oil spill accidents occur in severe weather conditions, which poses a threat to aircraft safety [19].

2.1.3.2 Satellite remote sensing monitoring

The remote sensing satellite refers to the artificial satellite which carries a remote sensor to observe the earth from space. A remote sensing satellite usually consists of observation remote sensor, data recorder, attitude control system, communication system, power system and thermal control system. Among them, the observation remote sensor and data recorder are called functional equipment. The remote sensing satellite completes the acquisition and transmission of remote sensing information through various tracking control systems, operation control systems and data acquisition and processing systems composed of the above six parts.

At present, common satellites in the world include radar satellites of Radarsat-1, Envisat-1 and ERS-2, Landsat satellites of Landsat series and SPOT series, weather satellites of NOAA series, FY-1C, and ocean satellites of SeaWIFS.

- Radarsat-1 radar satellite. Radarsat-1 is a commercial radar satellite launched by Canada in 1995, and officially began services in April 1996. Radarsat-1 runs on a sun-synchronous orbit, its remote sensor is SAR, and it operates in the c-band (5.3 MHz). The most outstanding feature of Radarsat-1 is its ability of multi-mode and multi-beam imaging, and its work is very flexible. Users can choose the incidence angle, resolution and width. The range of incidence angle is 20–50 degrees, the range of resolution is 10–100 meters, and the range of width is 45–500 kilometers. Users can change the scanning width and resolution via the ground control instruction according to different needs to meet the requirements. Because this satellite is designed based on the principle of microwave remote sensing, it can work all day in all weather conditions, regardless of time and climate conditions. Although the orbit revisit cycle of the satellite is 24 days, by selecting a working mode, the image width can be controlled to cover the arctic region (north of 73 degrees north latitude) every day, and it can also provide some users with repeated observation for seven days. Main characteristic parameters of the synthetic aperture radar (SAR) are listed in Table 2.1. In order to maintain the continuity of data, Canada has planned to produce Radarsat-2, which is basically the same as Radarsat-1.

- ERS-2 radar satellite. Launched in 1995, the ERS-2 is a subsequent satellite to the ERS-1 (which was launched in 1991 and expired in 1996). The satellite adopts an elliptic sun-synchronous orbit, with an orbit altitude of 780 kilometers and a flight cycle of 100.465 minutes. The local solar time of descending node is 10 : 30, the azimuth spatial resolution is 30 meters, the range resolution is 26.3 meters, the width is 100 kilometers, and the revisit cycle is 35 days. The ERS-2 carries a SAR. At present, ERS-1 and ERS-2 have successfully conducted pollution monitoring for several oil spill accidents.

- Envisat-1 radar satellite. Envisat-1, a commercial radar satellite launched by the European Space Agency in March 2002, uses a near-polar sun-synchronous orbit with a polar orbit altitude of 768 km and a revisit cycle of 35 days. The satellite carries an advanced SAR, which can produce high-quality images of oceans, coasts, polar caps, and land, providing scientists with higher-resolution images for studying the oceans. The satellite is the most advanced radar satellite to date and has a wide application prospect.

- Landsat series. Since the United States launched Landsat 1 on July 23, 1972, seven Landsat satellites have been launched, among which the launch of Landsat-6 failed. Currently, Landsat satellites in operation are Landsat-5 launched in 1985 and Landsat-7 launched in 1999. Landsat satellites have a large coverage area during operation, and can cover the vast majority of the world, and the ground images obtained by them are of high resolution.

  Landsat-5 uses a sun-synchronous quasi-recursive orbit with a flight altitude of 705 km and an inclination of 98 degrees, that is, the local time at each point in the orbit will remain constant, and the time when the satellite passes over the equator will be 9 : 39 a.m. The landsat moves in a predetermined orbit, either opposite or in the direction of the earth rotation, depending on the inclination of its orbit. The satellite orbits the earth for about 99 minutes. The scanning width is about 185 km, and the revisit cycle is 16 days. Orbit parameters in connection to Landsat5/7 are shown in Table 2.2.

- SPOT series. Five SPOT satellites have been launched since France launched the first SPOT satellite in February 1986, and SPOT 1/2/4/5 are currently in operation. SPOT satellites run on a sun-synchronous quasi-recursive orbit with an altitude of 832 km and an inclination of 98.7 degrees. The local time when the satellites pass over the equator is 10:30 a.m. and the visit cycle is 26 days. In front of each detector of the SPOT, there is a swingable reflector, which can be used to conduct tilt observation of the earth within a certain angle through reflector swinging, shortening the revisit cycle. The nominal revisit cycle of the SPOT is 26 days. Now, four satellites with the same orbital parameters are running simultaneously and each satellite can conduct tile observation. SPOT can actually repeat observations of the same spot within one to two days, which is of great significance in dealing with emergencies such as natural or man-made disasters.

SPOT5 was launched in May 2002. This satellite raised its panchromatic resolution to 2.5 meters and multispectral resolution to 10 meters while its field of view remained unchanged at 60 kilometers. Main parameters of SPOT satellites are shown in Table 2.3.

- NOAA series. The U.S. NOAA satellites are weather satellites. The orbit is an approximately-circular, sun-synchronous orbit with an altitude of $833 \sim 870$ km and an inclination of 99.092 degrees. The operation cycle is 102 minutes and the revisit cycle 9 days. The satellite passes over the equator at 7:00 a.m. and 2:00 p.m. The sub-satellite point resolution is 1.1 km, and the width of scanning image is 3020 km. Each satellite can cover the earth once every day.

- The U.S. ocean color satellite SeaWIFS. The U.S. ocean color satellite aims at quantitative remote sensing exploration of global ocean color environment and primary productivity, and is designed for the purpose of application in rapid marine fishery reporting, environmental assessment of estuary and harbor engineering, marine pollution monitoring, environmental protection for offshore oil and gas exploration and development, opening of new sea routes and agricultural production estimate. This satellite was successfully launched on August 1, 1997, and only equipped with a wide-view-field ocean color observer, and it covers the earth once every $2 - 3$ day. The range of wave band is $412 \sim 865$ nanometers, and is divided into eight wave bands, and the scanning angle is ± 58.3 degrees. With high signal-to-noise ratio of data and good detection sensitivity, the satellite is the most advanced ocean visible light remote sensor in the world to date.

- MODIS satellite. On December 18, 1999, the United States successfully launched the first advanced polar-orbiting environmental remote sensing satellite Terra (EOS-AM1) of the Earth Observing System (EOS). The Moderate Resolution Imaging Spectroradiometer (MODIS) installed in the satellite has 36 spectral bands ranging from 0.4 to 14.4 microns. The resolution of the first and second bands is 250 meters, the resolution of the third to seventh bands is 500 meters, the resolution of the remaining bands is 1000 meters, the scanning width is 2330 kilometers, and 6.1 megabits of information from the atmosphere, ocean and land surface can be obtained at the same time every second. MODIS has made many improvements in channel refinement, resolution, satellite parameters and life span, but there are still band and continuity constraints, especially for satellite remote sensing oil spill channels. AVHRR data of NOAA satellite and TM data of Landsat-5 satellite are still the main information sources of remote sensing monitoring of oil spills.

- Other satellites. At present, high-resolution commercial satellites mainly include Quickbird, IKONOS, Orbview-3, etc., and the resolution of the three satellites is 61, 82 and 82 cm, respectively. Main technical parameters are shown in Table 2.4.

Because the resolution of the three satellites is high, equivalent to the quality of 1:5000 aerial photo, and the image is in natural color after band combination, no interpretation is required for this kind of satellite image, and oil spills and pollution can be clearly identified through visual inspection. Monitoring oil spills using the three satellites also has its disadvantages: first, the width is too small, up to 20 kilometers, and cannot meet the requirements of monitoring large-scale oil spills; second, the cost is too high; third, oil spills cannot be monitored at night or in cloudy or foggy weather.

TABLE 2.1: Radar beam parameters of Radarsat-1.

Radar Beam	Standard Beam	Wide Beam	High-resolution Beam	Wide Scanning Beam	Experimental Beam
Incidence angle/(◦)	20 ~ 49	20 ~ 40	37 ~ 49	20 ~ 49	49 ~ 59
Ground width/km	100	150	45	300/50	75
Resolution/m	28 × 25	28 × 35	10 × 10	100 × 50, 100 × 100	28 × 30

TABLE 2.2: Main technical parameters of Landsat5/7.

Satellite	Launch Time	Orbit Altitude/km	Remote Sensor	Waveband/micron	Spatial Resolution/m
Landsat-5	1984	705	TM	0.45 ~ 0.52	30
				0.52 ~ 0.60	30
				0.63 ~ 0.69	30
				0.76 ~ 0.90	30
				1.55 ~ 1.75	30
				2.08 ~ 2.35	30
				10.4 ~ 12.5	30
Landsat-7	1999	705	ETM+	0.45 ~ 0.52	30
				0.52 ~ 0.60	30
				0.63 ~ 0.69	30
				0.76 ~ 0.90	30
				1.55 ~ 1.75	30
				2.08 ~ 2.35	30
				10.4 ~ 12.5	30

TABLE 2.3: Main parameters of SPOT satellites.

Satellite	Launch Time	Orbit Altitude/km	Remote Sensor	Spectral Coverage (micron)	Spatial Resolution/m
SPOT-1 SPOT-2 SPOT-3	1986 1990 1993	832	HRV	$0.50 \sim 0.59$	20
				$0.61 \sim 0.68$	20
				$0.79 \sim 0.89$	20
				$0.51 \sim 0.73$	10
SPOT-4	1998	832	HRVIR VEGETATION	$0.50 \sim 0.59$	20
				$0.61 \sim 0.68$	20
				$0.79 \sim 0.89$	20
				$1.57 \sim 1.70$	20
				$0.43 \sim 0.47$	1150
				$0.61 \sim 0.68$	1150
				$0.78 \sim 0.89$	1150
				$1.58 \sim 1.75$	1150
SPOT-5	2002	832	HRG VEGETATION	$0.50 \sim 0.58$	20
				$0.61 \sim 0.67$	20
				$0.78 \sim 0.89$	20
				$1.57 \sim 1.70$	20
				$0.49 \sim 0.715$	5
				The same as SPOT4	1150

TABLE 2.4: Main technical parameters of QuickBird, IKONOS and OrbView-3 satellites.

Satellite	IKONOS-2	QuickBird-2	OrbView-3
Company	Spatial Imaging	Earth Observation	Orbital Imaging
Orbit Altitude (km)	680	600	470
Orbit Inclination (degree)	98.1	66	97.25
Revisit Cycle (day)	$1 \sim 3$	$1 \sim 5$	Greater than 3
Launch Time	1999	1999	2000
Resolution (m)	Panchromatic 0.82 Multi-spectral 3.28	Panchromatic 0.61 Multi-spectral 3.28	Panchromatic 0.82 Multi-spectral 4
Width (km)	11	22	8

In China, a lot of research has been made on satellite remote sensing monitoring of oil spills, and most of them have also been applied to the pollution monitoring of oil spill accidents. However, the theoretical research on oil spill information extracted from satellite remote sensing data is not mature, and weather satellite remote sensing data is basically used in the monitoring of oil spill accidents. Weather satellites receive signals depending on solar radiation, their working band is not sensitive to the characteristic spectrum of oil spills, and the resolution is low, generally 1.1 km, which makes it difficult to effectively monitor oil spills. Two decades of foreign research shows that SAR satellite is the most effective satellite remote sensing tool for monitoring oil spills [23]. At present, the research on extracting oil spill information through interpreting radar satellite data is still in its infancy in China. In order to make full use of radar satellite remote sensing to monitor oil spills, the following two measures can be taken:

- Strengthen the research on extracting oil spill information through interpreting radar satellite data. Users can directly analyze the data obtained from the satellite data receiving department to extract oil spill information.

- Establish a fast access channel for satellite remote sensing data, and conduct timely and continuous monitoring of oil spills. China is not a member of the International Charter "Space and Major Disasters," and cannot require to activate the charter to serve us. However, we can enter into paid service agreements directly with foreign space departments and domestic satellite receiving departments or agents, and the working style of referring to the "charter" may enable us to quickly obtain oil spill monitoring data when oil spill accidents occur.

2.1.4 Satellite Remote Sensing Image Processing for Oil Spills

2.1.4.1 Acquisition of satellite images

Satellite remote sensing images are usually purchased from domestic satellite data receiving and processing departments and their image data agents. The Remote Sensing Satellite Ground Station (RSSGS) of the Chinese Academy of Sciences (CAS) is responsible for receiving, processing and distributing satellite remote sensing image data. The satellite data, which is allowed to enter the Chinese market but the RSSGS of the CAS has no right to receive, can be purchased via authorized domestic agents or foreign agents.

The RSSGS now has the right to receive and sell data from U.S. Landsat-5/7, French SPOT-1/2/4, ESA ERS-1/2, CBERS-1, Canadian Radarsat-1, Japanese JERS-1. In addition, the ground station can also sell the data from QuickBird.

The data of some satellites can only be obtained through specialized channels. For example, only Beijing Spot Image Co. Ltd., jointly established by France SPOT Company and the RSSGS, has the right to distribute and sell SPOT data. The Second Institute of Oceanography, which is directly associated with the State Oceanic Administration (SOA), has established the data receiving station for the U.S. ocean color satellite SeaWIFS in Hangzhou, and is responsible for data distribution in China. SeaWIFS data is free of charge and can be used only for non-commercial purposes after authorization.

The data of the U.S. NOAA satellites can be received free of charge. In China, NOAA satellite images of our country and surrounding areas can be obtained from all NOAA satellite data receiving stations. At present, a number of NOAA satellite data receiving stations have been established in China, especially in some universities and scientific research institutions.

Due to the difference of working principle, resolution, width, revisit cycle and price of each satellite, the available satellite data should be purchased and determined according to the location, time and purpose of the required data and in combination with the above factors.

2.1.4.2   Satellite remote sensing digital image processing method

The original remote sensing image data provided by the satellite contains a large amount of information, including remote sensing information of various ground objects as well as interference and noise. Such original images cannot be directly applied in the actual work without any processing. Remote sensing users must eliminate interference, correct distorted signals, suppress unnecessary ground object information and highlight useful information according to their own needs, so as to achieve the purpose of practical application. All such work is called "remote sensing image processing." Basic contents of remote sensing image processing include:

- Preprocessing. Preprocessing is the process of correcting or compensating the radiation distortion, system noise, random noise, geometric distortion and high-frequency information loss of the original image caused in the imaging process, and calibrating the image and restoring its original appearance. Due to the influence of atmospheric absorption and scattering, frequency characteristics of the remote sensing monitoring system and other random factors, the image is fuzzy and distorted, and the resolution and contrast are relatively reduced. In addition, the changes of attitude, speed and height of the remote sensing aircrafts and the platforms will also cause distortion in the geometric position of the image. Preprocessing mainly includes three parts: radiation correction, geometric correction, and registration and adjustment of multi-spectral image. Preprocessing is usually carried out at the satellite ground receiving station, and the remote sensing image data obtained by the user has been preprocessed.

- Image enhancement. Image enhancement is an image enhancement process based on the observation characteristics of human eyes to improve the visual effect of ground objects in the image. The number of brightness levels of an unenhanced image is very small, the average brightness value is low, and the contour and boundary are not clear. The purpose of image enhancement processing is to highlight the useful information in the image, expand the difference between different image features, and change the quality of the image, so that the information can be adapted to human visual features, and be accepted and felt by human eyes, so as to achieve the best effect.

- Classification and recognition. Classification and recognition is a process of carrying out statistical discrimination of the spectral characteristics of ground objects in the image, and extract distribution information of categories of ground objects (such as water, plants, geological features, large artificial buildings, oil pollution in the sea, etc.). According to whether there is prior knowledge about a known category or not, classification and recognition can be divided into two categories: supervised classification and unsupervised classification. Supervised classification refers to image data features of all kinds of ground objects that can be extracted from the data of the training area during classification. A training area is a small area in an image the category of which is known. The statistical features of various ground objects can be obtained based on statistical analysis of the image data of these small areas. Then, the data of other parts of the image is compared with the statistical features of each category to determine which category the pixel belongs to. Unsupervised classification refers to the classification carried out without knowing which category an area in the image belongs to.

2.1.4.3   Methods for extraction of oil pollution information and interpretation of satellite images

- Extraction of oil spill information. The spectral reflectivity of seawater will definitely change after it is polluted by a certain oil product. Therefore, there will be tonal changes reflected in the satellite image, which is the basis for judging and recognizing oil spill information, as well as the basis for further extracting oil spill information. The key to satellite interpretation is to design different information extraction patterns according to the spectral characteristics of different oil spill contaminants.

- Methods for interpretation of satellite images of oil spills. There are three methods for image interpretation: direct determination method, comparative analysis method, and logical reasoning method.

  Direct determination method: recognize the ground objects directly according to comprehensive analysis of various signs. For example, tone, shape, texture, structure, size and position are all important basis for direct interpretation.

Comparative analysis method: comparatively analyze satellite images of different wave bands and different phases (comparison of images with and without oil spill information) and known data (nautical charts and electronic images) to obtain more accurate information.

Logical reasoning method: use the logical reasoning method to interpret more abundant information on the basis of respecting the objective facts of the image and according to the internal relations between things and every subtle feature of the image, and meanwhile, verify the information through field observation to ensure the quality and accuracy of interpretation.

The interpretation of oil spill information basically adopts the above methods. However, according to the particularity of oil spill pollution, some specific practices are summarized below for reference. First, oil spills may change physical properties of the seawater. According to the differences in solar radiation characteristics between seawater and oil films, the gray values are different in the image, so there is great difference in color and texture. The seawater and oil films can be distinguished based on this point. Several consecutive days of satellite image data is used to analyze whether the abnormal gray value area in the polluted sea area is changing, drifting or diffusing, so as to determine whether it is a contaminant. It is more objective, scientific and accurate than visual interpretation to distinguish contaminants and seawater by using the size of the trichromatic reflectivity. Finally, the existence of contaminants is confirmed on the basis of on-site evidence.

2.2 Application of SAR in Marine Oil Spill Monitoring

The Synthetic Aperture Radar (SAR), based on the principle of active microwave remote sensing, has the advantages of all-weather and all-day monitoring and strong penetrating power. Nowadays, it has become a major means of monitoring marine oil spills and has been adopted by a lot of countries [24]. On November 13, 2002, three days after the Prestige accident, the ESA extracted oil spill information through processing Envisat-1 ASAR image of the waters around Galicia, Spain, and clearly recognized the oil spill pollution area [23]. In the research on SAR oil spill monitoring, so far, a number of experts and scholars at home and abroad have proposed some algorithms for automatic oil spill monitoring, mainly including oil spill segmentation, feature extraction and classification of marine oil spills and suspected oil spills [25, 26] used the valley value of bimodal histogram of SAR image as the segmentation threshold to process ERS-1/SAR image. This method can well segment the oil spill phenomenon with thicker oil films. Manore et al. [27] applied this method to RADARSAT images. Change et al. [28] combined Laplace of Gaussian (LoG) operators and multi-scale method to detect the edge of the oil spill area. Liu et al. [29] used wavelet transform to extract numerous marine characteristics including oil spills. Solberg et al. [30] used the value kdB lower than the average value of backscattering coefficient in the moving window as the segmentation threshold, and combined the multi-scale method to segment the oil spill areas in SAR images, and the value of k was related to the size of local wind speed. Kanaa et al. [31] used the double-threshold method to detect the oil spill areas in ERS images. Mercier et al. [32] combined wavelet packet transform with the hidden Markov chain to detect local changes in the spectrum, thus detecting oil spills in SAR images. Moreover, other image processing methods including mathematical morphology and level set method have been successively applied in this field [25].

In China, there are many achievements in the research on SAR oil spill monitoring. For example, [33] studied the segmentation method of oil spill areas by using wavelet transform; [24], taking the ASAR image of Envisat-1 as an example, discussed the method of semi-automatic detection of marine oil spills using the backscattering coefficient of the radar image, presented the process system of marine oil spill detection, and estimated the area of marine oil spills in combination with the calculation formula of oil pollution diffusion area; [25] systematically introduced the whole process of identifying oil spill phenomena and suspected phenomena in SAR images, including image preprocessing (radiation correction and speckle noise filtering), image segmentation, feature extraction, feature screening and neural network identification, and discussed the method of using SAR to monitor marine oil spills; [34], taking Envisat-ASAR data as an example, analyzed the process of SAR data processing and information extraction, and monitored the oil spill occurring in South Korea in December 2007.

Relevant knowledge of the SAR and its application in marine oil spill monitoring will be highlighted below.

2.2.1 Introduction to SAR

A SAR is an airborne or spaceborne active microwave detector. The former is called the airborne synthetic aperture radar (airborne SAR), and the latter is called the spaceborne synthetic aperture radar (spaceborne SAR). The so-called active type means that the radar is equipped with a device transmitting and receiving signals, so it not only receives the signal, but also actively sends the signal, and connects the returned signal with the sea surface roughness to obtain the sea surface information. The electromagnetic wave emitted by the SAR is in the microwave band, so it can penetrate clouds directly, without being affected by light, and the device can work even at night or under bad weather conditions.

The "synthetic" aperture principle of the SAR is that the radar can move along the path of the radar to form a considerable virtual aperture, and this synthesis is digital synthesis rather than physical synthesis. The large synthetic aperture provides it with very high resolution, reaching an order of magnitude from several meters to several tens of meters. Therefore, the SAR image can clearly show the complex spatial structure of marine phenomena at a magnitude of more than 10 meters on the sea surface.

Satellites carrying SARs have their specific orbital parameters, so they scan the area again a period of time after observing a target area. This period is called repeated observation cycle. The shorter the repeated observation cycle, the more timely information can be obtained. If the performance requirements of image clarity and spatial resolution are not high, the repeated observation cycle of the SAR can be reduced by expanding visual observation. In general, with the orbit altitude of $600 \sim 700$ km and the visible observation zone of $500 \sim 600$ km, the SAR can have a repeated observation cycle of 5 days for a specific area. From this point of view, although the spaceborne SAR is short in time continuity for a certain observation area due to the limitation of the orbit, it can obtain large-scale data continuously, enough to make up for the above shortage. Furthermore, the airborne SAT is not limited by the orbit and can observe a certain area continuously.

The SAR is by far the most effective space sensor because of its unique advantages. One of its contributions to human beings is that it can continuously observe large areas of the sea surface, accordingly obtaining a large range of ocean information and solving problems that cannot be solved through conventional observation.

2.2.2 Related Concepts

2.2.2.1 Backscattering cross-section

The radar backscattering cross-section is the same concept as Radar Cross-Section (RCS) in common use, and its definition is as below:

$$\sigma = \frac{4\pi r^2 S_r}{S_i} \tag{2.2}$$

For a scattering target or for a range of scatterers, in one direction, the ratio of the energy density S_r of the electromagnetic wave returned by scattering to the energy density S_i of the incident electromagnetic wave is called the "backscattering cross-section." A standard backscattering cross-section is also used in some measurements, and its definition is as below:

$$\sigma_0 = 10 log_{10} RCS [db] \tag{2.3}$$

2.2.2.2 Bragg scattering

Electromagnetic wave incidence is divided into vertical incidence and oblique incidence. The echo mechanism of vertical incidence is mirror reflection. For example, the altimeter uses this mechanism to obtain the backscattering cross-section. The echo mechanism of oblique incidence is Bragg resonance, also known as Bragg scattering, and the physical mechanism is actually the optical grating principle. Bragg resonance occurs when the distance difference of the electromagnetic wave is the integral multiples of the wavelength of the electromagnetic wave λ_r . According to the geometric relationship in the schematic diagram, the wavelength of sea surface fluctuation that generates Bragg resonance meets the following relationship with the wavelength of electromagnetic wave:

$$\lambda_s = \frac{n\lambda_r sin\Phi}{2sin\theta} \tag{2.4}$$

n indicates that n-order Bragg scattering will be generated, among which the echo of first-order Bragg scattering is the strongest. Φ refers to the angle between the wave direction and the radar wave plane, and θ is the incident angle of electromagnetic wave. The wavelength of the sea wave that generates the first-order Bragg scattering with the wave direction in parallel to radar wave plane and the wavelength of electromagnetic wave meet the following formula:

$$\lambda_s = \frac{\lambda_r}{2sin\theta} \tag{2.5}$$

2.2.2.3 Polarization

Polarization is an intrinsic property of the electromagnetic wave. As a transverse wave, the electromagnetic wave has the characteristic of transverse wave: polarization phenomenon. The electric field vector E of the electromagnetic wave and the magnetic field vector H are perpendicular to each other, and both E and H are perpendicular to the propagation direction of the electromagnetic wave. The polarization mode is determined by the vertical and parallel relations between the electric field vector E and the incident plane. The polarization of the electromagnetic wave with E in the incident plane is called horizontal (H, H) polarization, and that with E perpendicular to the incident plane is called vertical (V, V) polarization. The electromagnetic waves with different polarization modes have different forms of continuous conditions on the scattering plane, so the backscattering coefficient and phase characteristics at sea are different. They provide additional information except for intensity and frequency of the scatterer. The information content of spatial remote sensing can be increased by different polarization modes of incident waves. For example, Envisat's ASAR sensor has an operating mode with variable polarization, in which more information can be obtained by changing the polarization modes of incident waves.

2.2.2.4 Doppler effect

When there is a relative motion between the wave source and the observer, the frequency received by the observer is different from the frequency emitted by the source. This phenomenon is called "Doppler effect." When they get closer to each other, the frequency received by the observer increases; when they get farther from each other, the frequency received by the observer decreases.

2.2.3 Imaging Principle and Image Features of the SAR

In this section, the imaging principles of the side-looking real and synthetic aperture radars are first described, and then the basic features of SAR imaging are introduced.

2.2.3.1 Side-looking real aperture radar imaging

Figure 2.2 shows the geometric configuration of the earth observation imaging radar. The platforms carrying such radars include aircrafts, satellites and space shuttles. The radar emits an elliptic-cone-shaped microwave pulse beam from a certain side-looking angle θ. The axis of this elliptic-cone is perpendicular to the flight direction of the platform, and the apex angle of elliptic-cone in the plane perpendicular to the orbit, namely the beam height angle ω_v, is related to the width of the radar antenna w, namely,

$$\omega_v = \frac{\lambda}{w} \tag{2.6}$$

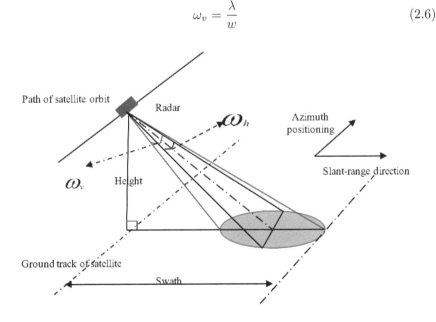

FIGURE 2.2: Geometry of radar imaging.

where, λ refers to the wavelength of the microwave adopted by the radar, and the apex angle of the elliptic-cone ω_k along the orbit is related to the length of radar antenna L, namely,

$$\omega_h = \frac{\lambda}{L} \tag{2.7}$$

The elliptic-cone-shaped microwave pulse beam forms an irradiation belt on the earth surface. The irradiation belt can be considered to be composed of many small spatial surface elements, each of which will scatter the radar pulse backwards, and the pulse will be received and recorded by the radar. Actually, as shown in Figure 2.3, for a row of pixels in the image plane, different radar slant range R is corresponding to different pixels. In this way, during radar flight, a certain width of the earth surface is continuously imaged, and the width W_G can be approximately determined as below:

$$W_G \approx \frac{\lambda R_m}{w cos \eta} \tag{2.8}$$

where, R_m refers to the slant range between the radar center to the center of the elliptic-cone-shaped irradiation belt, and η refers to the incidence angle of the radar at this central point. The minimum distance that can distinguish two adjacent targets is called the "spatial resolution of the radar image." Appearently, the smaller the distance, the higher the resolution. As shown in Figure 2.3, the resolution along the radar flight, namely the azimuth direction, and the radar slant range direction are ΔX and ΔR, respectively. When the slant range resolution ΔR is projected to the horizontal ground, it changes to the slant-range ground resolution ΔY. In combination with Formula 2.9, the azimuth resolution can be determined by the following formula:

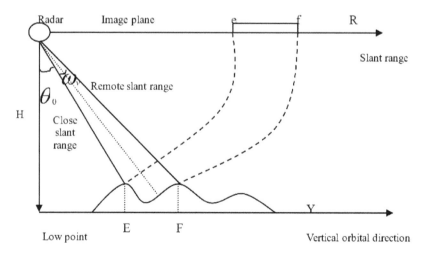

FIGURE 2.3: Projection of the slant range of the radar.

$$\Delta X = R\omega_k = \frac{R\lambda}{L} \tag{2.9}$$

where, R refers to the radar slant range. Actually, ΔX is the along-orbit width of the irradiation belt of the radar. The slant-range resolution and slant-range ground resolution are:

$$\Delta R = \frac{c\tau_p}{2} \tag{2.10}$$

$$\Delta Y = \frac{\Delta R}{cos(90 - \theta_i)} = \frac{c\tau_p}{2cos(90 - \theta_i)} \tag{2.11}$$

where, c refers to the velocity of light, τ_p refers to the pulse width, and θ_i refers to the side-looking angle. In fact, with the side-looking angle θ_i changing within a certain range, as ΔR is a fixed value, ΔY also changes within a certain range. That is to say, the slant-range ground resolution changes. The closer it gets to the bottom point, the lower the slant-range ground resolution, and the farther it is from the bottom point, the higher the slant-range ground resolution. If the side-looking angle of the radar is 0, that is, the radar images exactly opposite to the bottom point, the ground resolution near the bottom point will be very low, which is the main reason why imaging radar must look sideways. It is worth noting that compared with the central projection mode of aerial photogrammetry, the slant-range projection mode of the radar is very special.

Formulas 2.9, 2.10, and 2.11 show that the slant-range or ground resolution of the radar is only related to the characteristics of the radar wave and the side-looking angle, and is independent of the size of radar antenna, but the azimuth resolution mainly depends on the length of radar antenna. For example, if ERS-1/2 satellite radar (using C band, $\lambda = 5166$ cm) operates in RAR imaging mode, a 3 km-long radar antenna will be required in order to achieve the azimuth resolution of 10 meters, which is difficult for normal flight platforms to bear. In other words, it is impossible for a conventional real aperture radar (RAR) system to obtain high-resolution images along the flight direction of the carrying platform. In the synthetic aperture radar (SAR) imaging mode, this problem is solved.

2.2.3.2   Side-looking synthetic aperture radar imaging

When the radar is flying along the orbit, there is relative motion between the imaged ground target and the radar, so the radar pulse frequency reflected by the ground will drift, which is the Doppler frequency shift phenomenon. The synthetic aperture radar (SAR) uses this physical phenomenon to improve the azimuth resolution of radar imaging. Figure 2.7 shows the geometry configuration of side-looking synthetic aperture radar imaging. Given that a real aperture radar with an antenna of length L moves from point a to point b

and then to point c, the radar slant range at the imaging point O will first decrease and then increase again. In this way, the pulse frequency reflected from the ground point and received by the radar will change, that is, the frequency drift decreases. By accurately measuring the phase delay of the radar receiving the pulse and tracking the frequency drift, a pulse can be synthesized accordingly to sharpen the azimuth target, that is, increasing the azimuth resolution (as shown in Figure 2.5). Compared with the azimuth resolution of the RAR, the azimuth resolution of the SAR is greatly improved. At this time, ΔX can be approximately expressed as:

$$\Delta X = \frac{L}{2} \qquad (2.12)$$

Which means that the azimuth resolution only depends on the length of the radar antenna. For example, ERS-1/2 operates in the SAR imaging mode, and can achieve the azimuth resolution of about 5 meters with an antenna of 10 meters in length. The initial data acquired by the airborne/spaceborne radar system based on side-looking imaging geometry can form SAR images only through computer focusing and filtering. In other words, the concept of "synthetic aperture" is realized through data processing.

2.2.3.3 Basic features of SAR images

Each pixel of the SAR image contains not only the so-called gray value that reflects the microwave reflection intensity of the earth surface, but also the phase value related to the radar slant range (generally sampled to the slant range perpendicular to the flight direction of the platform). These two information components can be expressed by a complex number $(a + b * i)$, namely,

$$Pixel \geq a + b * i = \sqrt{a^2 + b^2}e^{i\theta} \qquad (2.13)$$

where, $\sqrt{a^2 + b^2}$ refers to the amplitude (corresponding to greyscale information), and $\theta = tan^{-1}(b/a)$ refers to the phase value. Therefore, the SAR image is also called the "complex image." The larger the amplitude, the higher the reflection intensity of the earth surface. The reflection intensity depends on the side-looking angle, wavelength and polarization mode of the radar, surface orientation, surface roughness, etc. For example, the calm lake is completely black in the radar greyscale image, while the city appears as a bright area. From the perspective of pixel information, SAR images are richer than visible-light remote sensing images, each pixel of which only contains gray information. From the perspective of visual effect, SAR grayscale images are far less clear than visible-light remote sensing images (including aerial photograph), mainly because the so-called speckle noise effect cannot be avoided during radar imaging. Therefore, in order to obtain a good visual effect, multi-look processing, namely smoothing, is generally needed to improve the signal-to-noise ratio at the expense of resolution. The phase of the radar is not only related to the geometric slant range R (distance between the radar platform and the average reflection plane of resolution elements of the earth surface), and various ground objects in resolution elements of the earth surface has a contribution of weighted sum to the overall observation phase, that is, the geometric distance p_i between each ground object in the resolution element and the average reflection plane causes phase delay. Each ground object has different physical backscattering characteristics and thus causes phase delay, which indicates that the phase information recorded by the SAR image pixel contains not only distance information, but also additional phase contributions of ground resolution elements. The latter shows great randomness, so it is generally regarded as noise, bringing inconvenience to interference analysis. Because the radar adopts the side-looking slant-range projection imaging mode and is affected by topographic relief, the SAR image is characterized by geometric distortion, which is reflected in radar foreshortening, radar overlay, radar shadow, etc. Such geometric distortion will bring negative effects on the application and interference analysis of SAR images.

2.2.4   SAR Resolution

2.2.4.1   Range resolution of the SAR

The antenna emits a microwave pulse, which spreads out into a beam, and the beam irradiates on the sea surface at a distance of R. Since the observation angle is slant, there is no direct mirror reflection. It is assumed that Bragg resonance is based on the backscattering mechanism except for very high sea conditions. Thus, for radio-wave radars with the emission wavelength of λ_r, the sea surface disturbances with the maximum contribution to the backscattering mechanism are those with ground spacing close to λ_s. In the formula,

$$\lambda_s = \frac{\lambda_r}{2sin\theta} \tag{2.14}$$

θ refers to the included angle between the slant range direction and the perpendicular direction.

The resolution of a standard antenna is determined by the aperture width D_R, and the standard antenna theory gives the beamwidth of the far field outside the antenna. As shown in Figure 2.4, $\beta = \frac{\lambda_R}{D_R}$. Therefore, in the plane perpendicular to the range direction, the resolution at a distance of R is $\rho = \frac{R}{\lambda RD}$. For slant observation, this resolution becomes larger in the ground direction. In general, the radar operates in a wavelength range of 50 mm to 1 m. Even a low-orbit satellite at a distance of 750 kilometers would require the antenna aperture between 0.375 and 7.5 kilometers to achieve the ground resolution of 100 meters. However, the current technology cannot achieve this, and it is necessary to find a new method. The improvement of the solution in the range direction should be obtained relying on the time that the pulse takes to return. The ordinary side-looking radar can improve the range resolution, but cannot improve the azimuth resolution. In order to improve the azimuth resolution, the principle of synthetic aperture should be adopted. The so-called synthetic aperture is the image, formed through moving along the path, with the resolution equivalent to that obtained by the radar with a very large aperture.

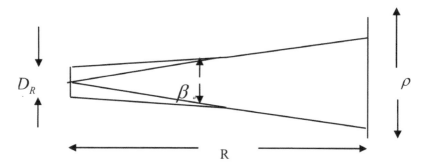

FIGURE 2.4: Geometry of beamwidth.

Here, regardless of the azimuth resolution, only the range resolution of the RAR will be considered, as shown in Figure 2.5.

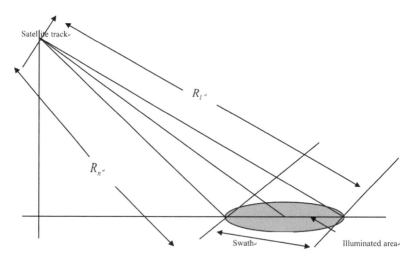

FIGURE 2.5: Geometrical relationship between range resolutions of the side-looking radar.

The radar emits a pulse with the time length of τ, and receives the scattering echo of the target, and the received signal can be transformed to the radar image. Echo signals are usually arranged according to the time that the signals take to return from the target. After the time t_s, the first pulse beam returns from the closest point Rn, and subsequent pulses return successively. Until the time t_f later, the last pulse returns to the satellite from the farthest point R_l. The return signal is actually the convolution of the pulse with the backscattering cross-section σ of the radar. The range resolution of the aperture radar is determined by the pulse width. The value of pulse length τ is the numerical value of σ that is measured independently. σ can be obtained from the swath. If the radar wants to distinguish two adjacent targets, their echoes received by the radar must be at two different times, which cannot be overlapped, otherwise they would be considered as a signal. The radar emits electromagnetic waves, which irradiate two different targets. Due to the distance between the targets, two electromagnetic waves will return with a certain distance difference, and then two independent pulse beams will be returned. On the contrary, the two pulse beams will be confused, and the two targets on the ground will be blurred into one target and cannot be distinguished. Therefore, the ground range resolution is $\rho = \frac{L}{2sin\theta}$, where $L = c\tau$, L refers to the pulse length, c refers to the velocity of light, τ refers to the pulse emission time, and θ is a position function on the swath. Thus, in order to improve the range resolution, the radar pulse should be as short as possible. However, the radar must transmit enough energy to ensure that the return signal has a signal-to-noise ratio (SNR) which is large enough to be recognized. In fact, there is a considerable technical difficulty of the radar system in emitting a very short pulse beam with quite high energy, because the power of the echo signal is proportional to R^{-4} the output power. Therefore, after a pulse is emitted, a long pulse can be effectively compressed into a short pulse using the pulse compression technique. A short pulse with quite high frequency and power will be generated. This process is called signal range compression. The range resolution depends on the fairly frequent sampling ability so that the frequency offset is not confused with the sampling rate.

2.2.4.2 Azimuth resolution of the SAR

For a given pulse, the sea surface reflectors with contributions to signals from a particular distance are within a circle with a satellite radius of that distance. The method for conceptualization of the operating principle of the RAR can be explained in Figure 2.6.

A coherent signal emitted by the radar irradiates targets a, b, c and d. It is assumed that these targets have coherent reflected signals and the reflected signals are radiated from each target individually. The signal returned from Point a, which is corresponding to the antenna center, reaches the antenna center first, while the signals returned from outer points return later. Therefore, for a, there is a small phase change along the aperture of the signal, but the distribution of this phase is symmetric to the center of the aperture. And the sum of the signals on both sides of the aperture is additive. However, this relationship does not apply to points that are much closer to one end of the aperture than the other. Therefore, when the signal is integrated, the phase change of d-point reflected signal at both ends of aperture leads to destructive interference, so there is no contribution to the received signal. The signal

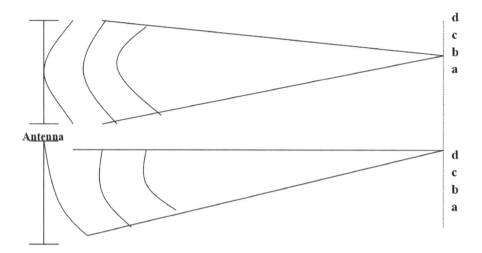

FIGURE 2.6: Phase diagram of the radar beam reflected from the points opposite to the center and edge of the radar antenna.

received by docking of point b and c depends on the phase geometry. This controls the effective beam width and limits the resolution. The larger the aperture, the closer the point at which the reflected signal does not contribute to the received signal is to a. This principle is used for the synthetic aperture. In order to acquire a magnitude based on which the azimuth resolution can be obtained, the following geometrical relationship as shown in Figure 2.7 can be used.

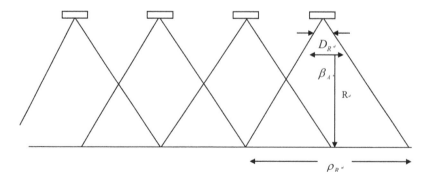

FIGURE 2.7: Geometrical relationship between azimuth resolution.

For the antenna with real aperture D_R and azimuth beam width β_A, its ground coverage of the antenna in the horizontal direction is $\rho_R = \frac{\lambda_R R}{D_R}$. ρ_R refers to the real resolution of this antenna. The length of the aperture that can be synthesized depends on how far the satellite can move when a given point is irradiated. The curvature of the earth and orbit makes things more complicated for satellites, but the principle is the same. The azimuth resolution ρ_A of the synthetic apertures ρ_R is $\rho_A = \frac{\lambda_R R}{2\rho_R} = \frac{D_R}{2}$. So the resolution is independent of the distance and wavelength of the radar, and can be improved with smaller aperture. The wider the real beam, the more information it can transmit about a target. The increase of R indicates that the target is irradiated for a longer time, which offsets the adverse effects of the distant target. The reduction in aperture size is limited by the generation of sufficient signal power. If the Doppler function relation at each point can be clearly identified at the processing stage without being confused, the time sampling is at least twice the maximum Doppler frequency shift in the return signal. That is, the real aperture radar must have at least one pulse within the time it moves half the aperture. On the other hand, with guaranteeing the swath with a suitable width in the range distance, on the premise that the return pulse is not too short and the pulse time is not overlapped, the higher pulse frequency is better. This restrictive relation is shown as below:

$$\frac{2V}{D_R} = \frac{V}{\rho_R} \leq PRF \leq \frac{c}{2W\sin\theta} \tag{2.15}$$

W refers to the swath width, and V refers to the speed of the satellite platform. Both the azimuth resolution and the wider swath should be considered together.

2.2.5 SAR Oil Spill Monitoring

2.2.5.1 Principle of SAR oil spill monitoring

Combined with the foregoing, the most remarkable feature of the oil film on the sea surface is its damping effect on capillary waves and short gravity waves on the sea surface which generate Bragg scattering. As a result of the sea surface wind and gravity, the sea surface will generate small waves, namely capillary waves and short gravity waves. If there is an oil film on the sea surface, the oil film on the surface expands and contracts due to the action of waves, which increases the dynamic elasticity of wave damping. When the sea surface is windy, the presence of the oil film can greatly reduce the growth rate of capillary waves. Sea clutters are generated after capillary waves and short gravity waves reflect radar waves, and they are bright in radar images. When there is an oil film on the sea surface, due to the damping effect on capillary waves and short gravity waves, the sea surface becomes more smooth, and the surface roughness is reduced. As a result, the reflectivity of the oil spill area to radar waves is reduced, that is, the radar backscattering coefficient of the sea surface is reduced. Therefore, the grayscale of the SAR image corresponding to this part of the sea area is reduced, and the image becomes darker. This is why marine oil spills in SAR images appear as darker spots, patches or straps, and this feature can be used to identify the oil spill area in SAR images.

Alpers and Huhnerfuss [35] indicated the damping effect of the oil film on the sea surface on Bragg resonant short waves with the ratio of the radar backscattering coefficient σ_0^S of the background sea surface to the backscattering coefficient σ_0^o of the sea surface covered with the oil film:

$$\frac{\sigma_0^S}{\sigma_0^o} = \frac{\sigma_0^{OSP}(\theta) + \sigma_0^{oB}(\theta)}{\sigma_0^{SSP}(\theta) + \sigma_0^{SB}(\theta)} \tag{2.16}$$

where, σ_0^{SP} and σ_0^B refer to the backscattering coefficient of mirror scattering and Bragg scattering, respectively, and both of them are the functions of the incidence angle θ. When the incidence angle is large, only Bragg scattering is generally considered:

$$\frac{\sigma_0^o}{\sigma_0^S} = \frac{w^o(2k_e sin\theta, 0)}{w^S(2k_e sin\theta, 0)} \tag{2.17}$$

It can be seen from the above formula that the attenuation of radar backscattering coefficient is equal to the attenuation of the sea surface wavenumber spectrum of Bragg resonant short wave. That is to say, the oil film on the sea surface reduces the radar backscattering coefficient by changing the sea surface tension and then reducing the sea surface wavenumber spectrum.

2.2.5.2   Impact of SAR parameters on oil spill monitoring

SAR parameters that affect marine oil film monitoring include wave band, polarization mode and incidence angle. In terms of wave band, studies have shown that compared with other microwave bands, the sea surface backscattering of oil film and seawater in SAR images in C, X and Ku-band are significantly different, which is reflected in the image as large contrast. Although the sea surface wind speed will seriously affect the monitoring of oil films on the sea surface, in C-band, the oil film on the sea surface can be monitored even if the wind speed reaches 13 m/s. In terms of the polarization mode, studies have shown that (H, H) polarization and (V, V) polarization have no essential difference in the monitoring of oil films on the sea surface, but (V, V) polarization mode in C-band is more suitable for the monitoring of oil films. In terms of incidence angle, since the sea surface radar waves have two reflection modes (mirror scattering with an incidence angle of $[0, 15]$ degrees and Bragg scattering with an incidence angle of $[20, 70]$ degrees), and Bragg scattering is also related to wind speed, the variation of incidence angle also affects SAR monitoring of oil films on the sea surface [36].

2.2.6 Analysis Cases of SAR Oil Spill Monitoring Images

[36] collected 27 ERS2 and Envisat-1 SAR images containing oil spills in the waters of the Yellow Sea in China from 2002 to 2005, including 17 full-resolution images and 10 quicklook images. Except one wide-swath image (400 km × 400 km), the coverage area of the other images is all 100 km × 100 km. The coverage area of the images is shown in Figure 2.8. Data processing shall be carried out in the following sequence: image preprocessing, image filtering, image segmentation, feature extraction and distribution statistics. The image processing flow chart is shown in Figure 2.9. First, SAR image filtering is carried out to highlight the features of oil films, and then image segmentation is carried out. Figures 2.10 and 2.11 show the results of image filtering and image segmentation, respectively.

After image segmentation, the information of marine oil spills in each SAR image shall be calculated. The processing results of 27 SAR images for the waters of the Yellow Sea are shown in Figure 2.12. Based on comparison with the sea routes, it can be seen that most of the oil films are caused by illegal pollution discharge ships, and the pollution is mainly concentrated in the routes of coastal cities such as Dalian and Qingdao.

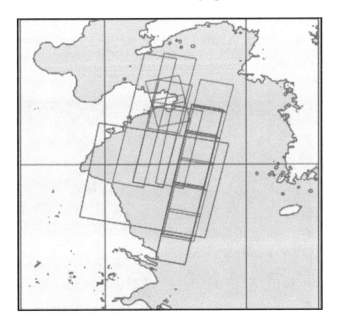

FIGURE 2.8: Map for oil spill statistics.

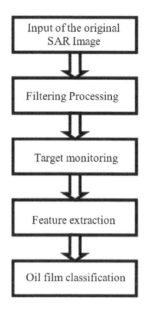

FIGURE 2.9: Block diagram for SAR marine oil spill image processing.

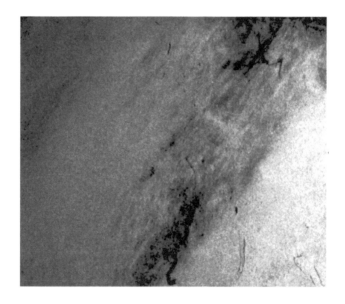

FIGURE 2.10: Effect of image filtering.

FIGURE 2.11: Effect of image segmentation.

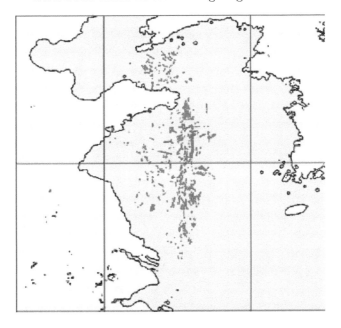

FIGURE 2.12: Statistical result of oil spills in the waters of the Yellow Sea in China from 2002 to 2005.

Chapter 3

Emergency Monitoring of Marine Oil Spill Accidents

Oil spill monitoring includes field investigation, oil spill sample collection, oil spill identification and quantitative analysis of oil spills, and it is an important process of oil spill emergency response. Oil spill monitoring can not only provide an important basis for the formulation of oil spill emergency measures, but also provide necessary physical and chemical parameters for the prediction of oil spill drift trajectory, as well as provide objective evidence for oil spill pollution damage assessment, oil spill clean-up costs or pollution damage claims.

3.1 Field Investigation of Oil Spill Accidents and Oil Spill Sample Collection

3.1.1 Field Investigation of Oil Spill Accidents

After an oil spill accident occurs, emergency personnel must go to the scene of the accident for the first time to investigate, so as to obtain relevant information of the accident. The field investigation area includes waters and coastlines and is close to the sampling site. The investigation conducted by the emergency personnel in the area includes meteorological and hydrological conditions on the day of the accident, the time and place of the oil spill, oil spill situation, pollution sources around the oil spill site, contaminated resources and the information of perpetrators, eyewitnesses or the personnel related to the accident.

After arriving at the site, the investigator should first make the investigation plan according to the site situation, determine the investigation sequence and scope, and then conduct the investigation. During the investigation, the personnel must complete the following tasks: (1) Make investigation notes; (2) Sketch the oil spill site; (3) Make videos on the site; (4) Collect oil spill samples; and (5) Inquire with related personnel on the site. Among them, investigation notes are very important, must be detailed and practical, and should be able to reflect useful information on the site and the technical means used in the investigation process. During the investigation, the personnel should seize the opportunity to collect information related to the oil spill accident in a timely manner, such as initial situation of the oil spill accident, changes in the middle, and ultimate recovery.

After the completion of the field investigation task, an investigation report shall be formed, including investigation time, personnel, technical means adopted, material evidence obtained from the investigation and relevant information. All records made on site shall be listed in the form of annex, and the adverse effects or social reactions caused by oil spill pollution shall be explained. The field investigation report shall be strictly kept in accordance with the provisions of archives management, and relevant information shall be kept confidential to avoid disclosure of relevant information to the perpetrator and the injured party, for later use for pollution claims.

Field investigation should be carried out continuously or intermittently after the oil spill accident, so the content of field investigation at each stage depends on site changes. Therefore, the types of information contained in the field investigations at different stages are not exactly the same.

3.1.2　Oil Spill Sample Collection

During field investigation, it is very important to collect oil spill samples in time. Corresponding plans for oil spill sample collection should be prepared according to the situation of oil spill pollution. The samples collected shall contain oil samples used to identify the source of oil spills or the scope of oil spill pollution, and water samples, sediment samples and aquatic biological samples in the contaminated waters to be used to assess the extent of oil spill pollution. Classified sampling should be carried out according to different monitoring objectives and the specifications for collection of various samples. Oil spill samples are objective evidences of oil spill source and pollution scope. Strict provisions on oil spill sample collection are set out in the Regulations on the Procedures of Oil Sampling During On-water Oil Pollution Accident Investigation issued by Maritime Safety Administration of the People's Republic of China (which took effect on April 1, 2002).

3.1.2.1　Preparation before sampling

Upon receipt of the oil pollution accident report, the oil spill site should be investigated as soon as possible, and relevant preparations should be made before sampling. Relevant documents such as sampling record, monitoring record, sample label and sealing strip as well as necessary tools such as sample bottle, sampling bar, disposable gloves, sealing tape, sample box, oil absorber, wiping rag, adsorbing material or paper, tongue depressor, metal spoon, water sampler and sediment sampler should be prepared before sampling.

3.1.2.2　On-water oil spill sampling

On-water oil spill sampling should be carried out on the oil film before or without dispersing of the oil dispersants, and sampling from fresh oil spill should be done as soon as possible, even if an emulsified mixture of oil and water is formed after the oil spill.

During on-water oil spill sampling, the types of oil spills, such as cargo oil, fuel oil and bilge oil, should be distinguished according to their appearance characteristics such as color and viscosity, and sampling points should be arranged in different types of oil spills. The sampling point should be located in the place where the oil spill is less polluted by other organic substances, the oil film is more concentrated and sampling is more convenient. At least three sampling points should be determined for a large-scale oil spill, and one or two sampling points should be determined for a small-scale oil spill.

In the case of sampling by ship, sampling points shall be located upwind of the ship and away from the exhaust fumes of the ship. When a solid or semisolid sample contains external substances such as algae or sand, the oil and foreign substances should be well stored in the sampling vessel.

According to the morphology of marine oil spills, different methods such as floating oil sampler and oil film sampler are used to collect samples. The specific sampling method can refer to the Guideline for Oil Spill Sampling and Identification issued by the International Maritime Organization (IMO).

3.1.2.3 Sampling of suspected oil spill sources

The suspected oil spill sources shall be sampled in an orderly manner from the most probable ship or part on the basis of administrative investigation and analysis. For the same ship, during collection of relevant material evidences, oil samples should be collected first, in order to prevent oil products from being mixed. For oil samples in different parts of the same ship, those samples from the most probable source of oil spill should be collected first, and then samples in other parts.

Prior to sampling, it is required to initially judge or sketch all probable paths of the oil spill flowing from suspected sources into the water, and sampling points should be arranged accordingly. During the sampling process, the sampling personnel shall be accompanied by the persons receiving sampling. If not accompanied, a written explanation is required. The specific sampling method can refer to the Guideline for Oil Spill Sampling and Identification issued by the IMO. For other suspected oil spill sources not sampled, the sampling personnel shall issue a report to explain the reasons for not sampling.

3.1.2.4 Precautions for sampling

After sampling of on-water oil spill and suspected oil spill sources is completed, the Sample Label and Sealing Tape should be filled immediately and pasted respectively between the bottle body, the bottle cap and the bottle body, and then the bottle cap and the bottle body should be fixed together with an adhesive tape. The Sample Label and Sealing Tape shall be marked with the sample number, sampling time and date, sample name, sampling location and sampling persons (two persons) and shall be signed by the person receiving sampling. For on-water oil spill samples, the words "On-water Oil Spill" shall be indicated in the column of the person receiving sampling.

After on-water oil spill sampling is completed, the Oil Spill Sampling Record shall be prepared. It shall be filled with sampling organization, phone number, address, postcode, sample number, sample name, schematic diagram of oil spill sampling, and sampling date, and shall be signed by the sampling person.

After oil sampling of suspected oil spill sources is completed, the Oil Sampling Record for Suspected Oil Spill Sources shall be prepared. It shall be filled with ship/unit name, ship berthing place, principal of the captain's unit and his/her phone number, sample number and sample description, sampling organization and its phone number, address and postcode, and sampling date, and shall be signed by the sampling persons (two persons) and the person receiving sampling.

3.1.2.5   Sample labeling and storage

In general, at least two samples shall be collected at each sampling point, and each sample shall contain 50 ~ 80 ml of oil. When the oil sample cannot meet the above requirements, it should also be sampled and sent to the laboratory for identification. It is also important to note that the filling volume of the sampling vessel should not exceed 3/4 of its capacity and that relevant measures should be taken to prevent contamination of the sample during the sampling process.

3.2   Oil Spill Identification

3.2.1   Chemical Composition and Physical Properties of Petroleum

3.2.1.1   Chemical composition of petroleum

Petroleum is mainly composed of carbon and hydrogen. The carbon accounts for about 83 ~ 87%, the hydrogen accounts for about 12~14%, and the remaining 1% ~ 3% is mainly sulfur, oxygen and nitrogen, as well as extremely trace elements such as phosphorus, vanadium, potassium, nickel, silicon, iron, calcium, magnesium and sodium. The main component of petroleum is hydrocarbons composed of alkanes, cycloalkanes, aromatic hydrocarbons and cycloalkanes, which account for 50% ~ 98% of the oil and are called petroleum hydrocarbons. The rest are non-hydrocarbon compounds containing oxygen, sulfur, nitrogen and trace elements.

Alkanes are the main components of petroleum and exist in the form of n-alkanes and isoalkanes.

The content of n-alkanes is high, 15% ~ 20% of petroleum. They are non-polar molecules and are composed of C-C single bonds and C-H bonds. The general formula is C_nC_{2n+2}.With the increase of relative molecular mass, the morphology of n-alkanes also changes, and they exist in petroleum in three states: gas, liquid and solid. At normal temperature, $C_1 - C_4$ n-alkanes are gaseous and exist in natural gas; $C_5 - C_{15}$ n-alkanes are liquid (among which, $C_5 - C_{12}$ n-alkanes are gasoline, and $C_9 - C_{22}$ are kerosene, diesel and engine oil); n-alkanes with carbon number above C_{16} are solid (for example, $C_{29} - C_{36}$ n-alkanes are engine oil, heavy fuel oil, lubricating oil and lubricating grease), and they are mostly present in the dissolved state in the petroleum. When the temperature falls, wax will be crystallized.

In petroleum, the carbon number of isoalkanes with branched chains is mainly less than or equal to C10, and a few C11-C25. Isoprenoid alkanes are the most important, and the content of isoalkanes is the highest in pitch-based petroleum [37].

The content of naphthenes is 30% to 60% in the petroleum. Low-molecular-weight cycloalkanes (with carbon number less than or equal to 10), especially cyclopentanes and cyclohexanes and their derivatives, are important components of petroleum, and middle-molecular and high-molecular weight cycloparaffins (with carbon number between 10 and 35) can be monocyclic to hexacyclic. Monocyclic (C_nC_{2n}) and bicyclic (C_nC_{2n-2}) alkanes are the most important, and account for $50\% \sim 55\%$ of cycloalkanes in petroleum, tricyclic alkanes (C_nC_{2n-4}) account for 20%, and tetracyclic alkanes and above account for about 25%. Tetracyclic and pentacyclic alkanes are obviously similar to tetracyclic steroids and pentacyclic triterpenoid compounds in structure, and have optical rotation, which is one of the main evidences of organic origin of petroleum. In petroleum, the content of polycyclic alkanes decreases obviously with the increase of maturity, while in high-maturity petroleum, monocyclic and bicyclic alkanes are dominant. At room temperature and under normal pressure, cyclopropanes (C_3C_6) and methyl cyclopropanes (C_4C_8) are gaseous, monocyclic alkanes are liquid, and alkanes above bicyclic (with carbon number greater than 11) are solid.

The average content of aromatic hydrocarbons and naphthenic-aromatic hydrocarbons in petroleum is $20\% \sim 45\%$. Aromatic hydrocarbons contain only aromatic rings and chain bases, and the common types of aromatic hydrocarbons identified in petroleum include benzene of monocyclic aromatic hydrocarbons (C_nC_{2n-6}), biphenyl of polycyclic aromatic hydrocarbons, naphthalene of polycyclic aromatic hydrocarbons (bicyclic C_nC_{2n-12}), anthracene and phenanthrene (tricyclic C_nC_{2n-12}), and benzanthracene and chrysene (tetracyclic C_nC_{2n-24}) and aromatic hydrocarbons with up to 8 condensed aromatic rings. Among them, the content of benzene, naphthalene and phenanthrene are the highest, and the main types are alkyl derivatives rather than parents. For example, the main component of monocyclic aromatic hydrocarbons is toluene rather than benzene. Naphthenic aromatic hydrocarbons contain one or several condensed aromatic rings and are thickened with saturated rings and chain alkyls. The most abundant in petroleum are bicyclic (an aromatic ring and a saturated ring) indane and tetralin and their methyl derivatives. In petroleum, the most important is tetracyclic and pentacyclic naphthenic aromatic hydrocarbons, most of which are related to steroids and terpenes and are biomarkers. Their content and distribution characteristics are important indicators for studying the origin of petroleum and oil-source correlation.

3.2.1.2 Main physical properties of petroleum

1. Relative density (specific gravity)

In general, relative density, also known as specific gravity, is used to measure the density of petroleum, and refers to the weight ratio of oil to water of the same volume at a given temperature. Specific gravity is related to temperature and the same oil has different specific gravity at different temperatures. The specific gravity of petroleum adopted in China is expressed as d_4^{20}, referring to the density ratio of petroleum at 20 degrees and pure water at 4 degrees under atmospheric pressure of 10^5 pascals. In Europe and America, the specific gravity of oil is expressed as $d_{15.6}^{15.6}$. However, API specific gravity is commonly used internationally, which has the following conversion relationship with $d_{15.6}^{15.6}$:

$$API \ specific \ gravity = \frac{141.5}{d_{15.6}^{15.6}} - 131.5 \qquad (3.1)$$

The relative density of petroleum is generally between 0.75 and 0.98 grams per cubic meter. The petroleum with relative density greater than 0.93 grams per cubic meter is heavy oil, that with relative density between 0.9 and 0.93 grams per cubic meter is medium oil and that with relative density less than 0.9 grams per cubic meter is light oil. Heavy oil has a maximum relative density of more than 1.0 gram per cubic meter and is difficult to exploit with conventional methods.

2. Viscosity

Viscosity refers to the amount of internal friction caused by the relative motion of molecules when oil flows. The greater the viscosity of oil, the lower the fluidity, and vice versa. Viscosity has an important reference value for understanding of oil and gas migration, dynamic analysis of oil wells and oil storage and transportation. Viscosity is divided into kinematic viscosity, dynamic viscosity and relative viscosity.

Kinematic viscosity refers to the product of the time for a certain volume of liquid flowing under gravity, a calibrated glass capillary viscometer under at a constant temperature and the capillary constant of the viscometer. The unit is square meter per second and the international common unit is centistoke (cSt). Dynamic viscosity refers to the product of kinematic viscosity at the temperature of measuring the kinematic viscosity and the density of the liquid at the same temperature, and the commonly used unit is centipoise (cP). In China, the relative viscosity is Engler viscosity, which refers to the ratio of the discharge time for 200 ml of oil flowing through the tuber with diameter of 2.8 mm in Engler viscometer to the discharge time for distilled water of the same volume at 20 degrees flowing through the same tube. In European and American countries, there are Saybolt viscosity and Redwood viscosity, which respectively refer to the time required for oil of a certain volume flowing through the tube of the given size in Saybolt viscometer or Redwood viscometer at a certain temperature t, and the value of viscosity in seconds is called Saybolt seconds or Redwood seconds. According to the value of relative viscosity measured in the laboratory, the kinematic viscosity can be obtained through looking up the viscosity conversion chart, and the dynamic viscosity can be further calculated.

The viscosity depends mainly on the chemical composition of petroleum, molecular weight of each component, temperature and pressure. The smaller the molecular weight of each oil component, the lower the viscosity, and vice versa. Therefore, the viscosity of oil containing high content of colloid and asphaltine is usually very high. The temperature has a great influence on oil viscosity. The viscosity decreases with the temperature increasing. Although pressure has an influence on oil viscosity, the sea surface pressure is about 1 atmosphere pressure and changes little. Therefore, the influence of pressure on oil viscosity is usually not considered.

3. Solubility

Petroleum can be dissolved in a variety of organic solvents, such as benzene, chloroform, carbon tetrachloride, alcohol, etc., but the solubility in water is very low. Generally, the dissolution characteristics of petroleum are used to identify whether the rock contains traces of petroleum or not.

4. Condensation point and pour point

The condensation point is the highest temperature at which the oil no longer flows under the specified experimental conditions. The pour point is the lowest temperature at which the oil can flow under the specified experimental conditions.

5. Distillation range

Since petroleum is composed of multiple components, when heated to boiling, the light components with a low boiling point in petroleum first evaporate, the liquid composition changes, the content of light components decreases, and the boiling point increases. It can be seen that the boiling point of petroleum is different from the fixed boiling point of pure substance under a certain pressure. It has no fixed boiling point and only has a temperature range, which is called "distillation range" in petroleum refining. The lower limit of the distillation range is the initial boiling point, and the upper limit is the final boiling point. The lower the initial boiling point, the lower the temperature at which volatilization begins, and the shorter the time the oil stays on the sea surface. It can quickly evaporate after spilling to the sea surface, and no cleaning measures need to be taken. However, oils that are more volatile also have greater fire and explosion hazards.

3.2.2 Identification of Oil Spills

As mentioned above, the chemical characteristics of various crude oils that are formed under different conditions or environments are significantly different; in addition, in view of the differences in properties of crude oils, refining processes and additives used during crude oil refining as well as subsequent mixing with oil residues in tanks, ships, pipelines, and unloading pipes during transportation, two refined oils of crude oils of the same category are also different. Therefore, all oil spill samples have more or less differences in chemical compositions, and this determines the feasibility of the identification of oil spills [38].

Each oil product has molecular characteristics that are significantly different from other oil products. The molecular characteristics are called oil fingerprints, and oil fingerprint identification is the main technique for the identification of oil spills. At present, the methods for oil fingerprint identifications that are commonly used in laboratories include gas chromatography-flame ionization detector (GC-FID), gas chromatography/mass spectrometry (GC/MS), high performance liquid chromatography (HPLC), infrared spectroscopy (IR), thin layer chromatography (LC), exclusion chromatography, supercritical fluid chromatography (SFC), ultraviolet spectroscopy (UV), fluorescence spectroscopy (FL), gravimetric method, stable isotope mass spectrometry, and the like. According to the characteristics of the detected oil fingerprint information, the oil fingerprint identification can be divided into two categories, namely non-characteristic method and characteristic method.

Traditional non-characteristic methods include gravimetric method, IR, UV, LC, HPLC, exclusion chromatography, SFC, and the infrared fiber optical sensor method that emerged recently. Compared to the characteristic method, the non-characteristic method requires less time for preprocessing and analysis, and the cost is lower. However, the main disadvantage is that the generated data usually lacks detailed characteristic information on individual components and petroleum sources, so the non-characteristic method has some limitations in identifying oil spill characteristics and sources.

The characteristic methods mainly refer to GC/MS and GC-FID. Because these methods make it easier to obtain detailed information on components, especially the diagnostic odds ratios of polycyclic aromatic hydrocarbons and biomarker compounds, they have been widely used for the identification of oil spill sources and the monitoring of oil spill weathering and biodegradation.

Oil fingerprint identification provides a very important scientific basis for the treatment of oil spill accidents through analyzing and comparing various oil fingerprints of suspected oil spill sources and samples.

3.2.2.1   Status of domestic and foreign research on oil spill identification

1. Status of foreign research

A number of foreign countries, including the United States, South Korea, Japan and coastal countries in Europe, have successively established relatively complete oil spill identification systems and oil fingerprint databases, which play an important role in oil spill pollution prevention, oil spill identification and oil spill pollution damage assessment.

(1) The United States

In the mid-1970s, the United States Coast Guard (USCG) began using a variety of analytical methods for oil fingerprint analysis. In 1978, the USCG established the Central Oil Identification Laboratory (COIL), which was responsible for the identification of oil spills. In 1988, the laboratory was renamed the Marine Safety Laboratory (MSL), which strictly followed the oil spill identification procedure, from on-site sample collection of the discharged oil under investigation and evidence (sample) collection to impeccable chain storage of evidences collected, so as to obtain the powerful evidences in support of legal advice [39].

(2) Canada

The Oil Spill Emergency Response Center of Environment Canada has been responsible for the establishment and training of oil spill emergency response systems of provinces in Canada, and carrying out related research and cooperation. The center has established an oil fingerprint identification system based on gas chromatography and gas chromatography/mass spectrometry, explored and determined the oil fingerprint analysis method for more than 100 compounds including n-alkanes, polycyclic aromatic hydrocarbons and biomarkers in oils was determined by internal standard method, analyzed various physical parameters of oils including density, viscosity, water content and sulfur content, carried out simulated weathering research, especially evaporation weathering and microbial degradation weathering, and established the oil fingerprint database for physical parameters and chemical fingerprints of a variety of crude oils [40].

(3) Europe

Based on a large amount of research work, the research institutions of six European countries (Belgium, Denmark, Germany, Norway, Portugal and the United Kingdom) established the European Marine Oil Spill Identification System in 1983. In 1991, the new European Marine Oil Spill Identification System was formed on the basis of revision. The system has developed a series of oil identification procedures including sampling, transportation, storage, analysis and reporting. The step-by-step identification approach is adopted, and the analysis method is gas chromatography and gas chromatography/mass spectrometry for ionization detection, which was accepted by the Bonn Agreement in 1992 as a recommended method for the identification of oil spills within the Bonn Agreement, and is still being revised and improved [41].

(4) South Korea

The Pollution Management Agency of the Republic of Korea Coast Guard is responsible for the management of oil spill accidents and the identification of oil spills. A corresponding oil fingerprint database has been established for all crude oils imported, and the oil fingerprint analysis methods used include GC-FID, IR, FL, and GC/MS [41]. Furthermore, a large number of simulated weathering experiments have been carried out, including weathering effect experiments of different crude oils at different time intervals, experimental comparison between laboratory weathering (including natural weathering and simulated weathering with temperature control using hot plates) and sea weathering, and difference of weathering experiments for different seasons in a year.

(5) Japan

Japan Maritime Security Test and Research Center is the technical support system of Japan Coast Guard, and its chemical analysis department is mainly responsible for the identification of oil spills. The main analysis methods include gas chromatography of the flame ionization detector, gas chromatography of the flame photometric detector, gel permeation chromatography and Fourier transform infrared spectrometry and gas chromatography/mass spectrometry, which are used to analyze alkanes, thiophenols, biomarker compounds and other compounds.

2. Status of domestic research

China is relatively late in carrying out research on oil spill identification technology, and has been carrying out continuous exploration and research. In 1996, on the basis of the preliminary research work, the National Marine Environmental Monitoring Center of the State Oceanic Administration put forward the Specification for the Identification System of Oil Spills on the Sea Surface, which was promulgated and implemented as a national industry standard in 2007 [38]. In order to ensure the legality and effectiveness of sampling and identification procedures in oil spill accidents, the Maritime Safety Administration of the People's Republic of China has compiled the IMO Guidelines for Oil Spill Sampling and Identification, and based on the experience of more than 20 years in oil separation identification, the Regulations for Oil Sampling Procedures During On-water Oil Pollution Accidents has been issued, in order to provide better services for investigation and treatment of oil spill accidents.

In recent years, Both the North China Sea Environmental Monitoring Center of the State Oceanic Administration and Yantai Oil Spill Response Technical Center of Maritime Safety Administration of the People's Republic of China have actively conducted research on oil fingerprint identification, respectively on oil spill accidents occurring during oil exploration, development and transportation in the Bohai Sea and oil sampling and evidence collection for marine traffic accidents and illegal pollution discharge. In addition, according to the latest international oil fingerprint analysis method and identification system, an oil fingerprint identification system that is in line with international standards has been developed, and an oil fingerprint database including some crude oils in the Bohai Sea, land crude oil, refined oil and some foreign oil products has been established. China has made great progress in oil fingerprint identification technology.

3.2.2.2 Oil fingerprint identification method

A few oil fingerprint identification methods that are widely used will be introduced below.

1. Infrared spectroscopy

Different substances have different strength of absorption to infrared light, and different absorption spectra will be produced when infrared light irradiates different substances. The analysis method established on this principle is called infrared spectroscopy (IR).

The IR can be used to detect compounds in petroleum such as aromatic hydrocarbons, olefins, and alkanes, as well as oxygen and sulfur. Similarly, depending on the type and content of these substances in different oils, the infrared spectra produced are different, and thus the oil types can be distinguished. Using the IR for oil analysis is mainly to analyze the polar components in oils, and take infrared vibration spectra of these polar components as the index of oil fingerprint identification, and carry out identification through comparing positions, strength and contours of characteristic peaks of infrared spectra of the two oils. The IR is applicable to unweathered oils and oils with a certain degree of weathering. It can be used to identify crude oil, fuel oil, lubricating oil, film oil on the water surface, oils coated in sandstone or other solid materials, oils from beaches and ships, or emulsified oil in water. This method is the most widely used.

Figure 3.1 shows the absorption curves of the three sample extracts of diesel, gasoline and lubricating oil recorded by IR. Like the infrared spectra of other substances, the infrared spectra of these three oils have no sharp peaks or baseline-resolved peaks. A large number of overlapping peaks and shoulder peaks make difficult the exact attribution of spectral bands, and each infrared band may be a combination of multiple-frequency and combined-frequency bands based on several different fundamental frequencies [42]. It can be seen from the figure that the response modes of diesel, gasoline and lubricating oil in the infrared region are the same, but there are significant differences in the region where the wave number is 5800–6200 per centimeter, and the spectrum outside this region has little effect on characterization samples.

The study of infrared spectrum began in the early 20th century. Since the commercial infrared spectrometer came out in 1940, IR technology has been developing vigorously. In the early days, J. J. Kelly et al. [43] applied infrared spectroscopy to determine the octane number of gasoline. In the follow-up work, J. J. Kelly et al. [43] tried the application of IR in measuring the com-

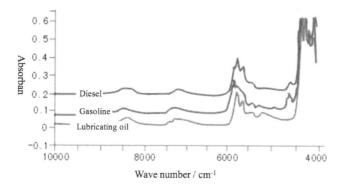

FIGURE 3.1: Near-infrared spectra of extracts of diesel, gasoline and lubricating oil.

position of gasoline family. Later, Bohcs et al. [44] used infrared spectroscopy to study the properties of gasoline such as vapor pressure, density, distillation range, drying point, benzene content, and sulfur content. In the analysis of jet fuels, Lysaght et al. [45] evaluated the feasibility of infrared spectroscopy in determining the content of aromatic hydrocarbons and the freezing point in long-wave and short-wave infrared regions. Cooper et al. [46] studied the use of infrared spectroscopy for determining the content of oxides such as MTBE in gasoline. In the same year, Rest et al. used infrared spectroscopy to determine the composition of light liquid hydrocarbons [47]. In 1999, Chung and Choi [48] successfully applied the method to the identification of types of pure oils.

In China, the IR was first applied to oil spill identification in 1984. Dai and Xu [209] first used IR on the basis of 18 analytical bands selected by Killeen. With the aid of vector models, they found four key bands with representative characteristics to be used as information points for data processing. Through the study on 22 unweathered oils, 58 weathering types and 6342 pieces of data, they introduced the concepts of critical angle, critical line and most achievable area, and established the systematic pattern recognition technology for identifying the sources of marine oil spills with the simple vector model method based on IR. Liu et al. [49] adopted the partial least square method in chemometrics, and used a portable FOX full-automatic infrared fuel analyzer to measure relevant properties of gasoline, including octane number, density, distillation range, aromatics, olefins, and saturated hydrocarbons, and verified the correlation between the results of measurement with IR and the reproducibility requirements of the national standard test method. The study made by [49] showed that it is feasible and reliable to use Fourier transform infrared spectroscopy in combination with multivariate calibration analysis technology in chemometrics for the analysis of quality properties of petroleum fuel. According to the needs of clean diesel fuel process research, production control and product quality inspection, in order to shorten the time for measuring the composition of diesel, with the ASTMD-2425 (mass spectrometry) as the basic measurement method, Guang et al. [50] investigated the feasibility of using partial least square method to establish calibration model for determining the composition of diesel in long-wave infrared region, and compared the effects of different spectrum treatment methods and wavelength range on model quality. The calibration model was tested with the samples of validation set. Given that there is no single way to solve all the problems of oil spill identification, Wang et al. [51] proposed a method for oil spill identification which integrated IR and principal component cluster analysis, and pointed out that this method can be used to correctly and quickly classify marine oil spill samples with the volume fraction of 0.4-0.8 mg/L and can be used as an auxiliary method for oil spill identification. Wang et al. [52] also proposed the method for marine oil spill identification that integrated IR and pattern recognition. They prepared 56 simulated seawater samples of gasoline, diesel and lubricating oil, and used organic solvent to extract oil spills from seawater and recorded its infrared spectrum. Then, after conducting multiplicative scatter correction (MSC) and Norris first-order derivative smoothing preprocessing on the original spectrum, they carried out principal component analysis (PCA) and established the identification model for oil spill samples on the basis of extracting the characteristics of different kinds of oil spill samples. With the IR, Pang Shiping [42] monitored marine petroleum pollutants in real time, analyzed oil spills of different types and concentrations on the sea surface, obtained the intrinsic component information of petroleum pollutants, and acquired the results of qualitative and quantitative analysis with information processing technologies such as pattern recognition. Pang Shiping's research results show that the results of measurement with the three-wavenumber infrared method are in good consistence with the fluorescence method and the ultraviolet method, and this method can be used to measure the concentration of petroleum pollutants in seawater.

To sum up, the IR method has a wide range of application and can be used to identify oil spills in combination with various methods, such as partial least square in chemometrics, multiplication, principal component analysis, pattern recognition, and so on.

The IR method has the advantages of high speed, high efficiency, low cost and good reproducibility. Due to its strong qualitative resolution, the IR technology can be used to not only identify the general geographical origins of oil spills, but also distinguish the oils of the same geographical origin through comparing relative intensity of spectral bands. However, for similar crude oils or refined oils, the use of IR method for oil spill identification has certain limitations.

2. Fluorescence spectroscopy

The so-called fluorescence spectroscopy (FL) is an analysis method established with using ultraviolet light (or short-wave visible light) to irradiate oil samples and detecting the fluorescence phenomena generated by conjugated double-bond organic compound molecules (such as benzene) or some inorganic molecules in oil samples.

Fluorescence spectroscopy can be divided into ordinary fluorescence spectroscopy, synchronous fluorescence spectroscopy, and three-dimensional fluorescence spectroscopy [53]. Figure 3.2 shows the ordinary fluorescence spectrum of a kind of crude oil, which has a wide peak profile, provides little information and cannot reflect the subtle differences between oils. Therefore, its ability to distinguish oil samples is relatively low. Synchrotron FL is the most commonly used synchronous scanning technique. Figure 3.3 shows a synchronous fluorescence spectrum (SFS) of an oil spill. Since the synchronous fluorescence spectrum couples the relationship between excitation and emission wavelengths, it provides a means of selecting features in the three-dimensional space emission spectrum. Compared with the ordinary fluorescence spectrum, this method has the following advantages: simplified spectrogram, narrow spectral band, obvious features, high selectivity and little light scattering interference [54]. Three-dimensional FL is a new fluorescence analysis technology developed in the past 20 years. This technology can identify oils according to the three-dimensional information formed by excitation wavelength, emission wavelength and fluorescence intensity, and provide comprehensive oil fingerprint information. However, this method has a long analysis time, and a variety of chemometric methods are needed for data processing. Currently, it is rarely used in the actual oil spill identification.

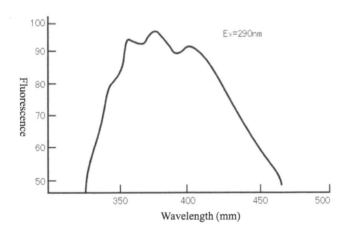

FIGURE 3.2: Ordinary fluorescence spectrum of a kind of crude oil.

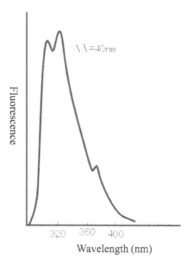

FIGURE 3.3: Synchronous fluorescence spectrum of an oil spill.

The fluorescence spectrometry measures the fluorescence intensity of polycyclic aromatic hydrocarbons and polycyclic compounds. Therefore, it can be used to distinguish different oils according to the difference between their components. The application of fluorescent spectrometry to identify oil spills is not simply to compare spectral profiles, but to process spectra with various methods including visual comparison, fluorescence spectral vector space distance pattern recognition, or derivative fluorescence spectral recognition, and determine the relevance of oil types with related obvious characteristics [42]. When applying the analysis method, first obtain the ordinary fluorescence spectrum, and then perform visual comparison method. After that, perform other treatment as the case may be. Fluorescence spectrometry is only applicable to unweathered oils or oil samples with a certain degree of weathering, and can be used to identify oil types such as crude oil, fuel oil and engine oil.

Foreign studies on FL have been conducted earlier. In the early 1970s, Freegarde et al. [219] proposed a fluorescence contour map, and distinguished various oils by drawing contour plots. Although this method simulates a three-dimensional system to some extent, in actual operation, it is required to record a large amount of data and time, use a computer to help accumulate all necessary data, and adopt the interpolation method that can calculate the same fluorescence intensity points to calculate contours. This method is cumbersome and is not suitable for on-site analysis [56]. Lloyd proposed the synchronous excitation technique for identifying polycyclic aromatic hydrocarbons and refined petroleum products with synchronous fluorescence spectrum, and found that the synchronous fluorescence spectrum had more spectral structures than the ordinary fluorescence spectrum [57].

The application of fluorescence spectrometry in oil spill identification in China began in the early 1980s when ordinary fluorescence spectra were used for oil fingerprint identification for marine oil spills. On this basis, the Marine Environmental Protection Institute of the SOA proposed using five excitation wavelengths (250, 270, 290, 310, and 330 microns) and four spectral characteristics (general shape of the peak, number of peaks, emission wavelength of characteristic peak, and peak profile) as identification indexes for comprehensive analysis, improving the ordinary fluorescence spectrometry [58]. In the 1990s, Zhao and Wang [210] used FL to successively identify seven crude oils, three lubricating oils and five fuel oils in China, and compared the fluorescence spectra of suspected oil samples with actual oil spills on the sea surface to determine the source of pollution. With FL and high-efficiency capillary GC-FID, Li Hong and Jibin [59] analyzed and identified the samples of an oil spill that occurred in the Dongshan Bathing Area in Qinhuangdao, and accurately identified the source of the oil spill. In their paper, they described in detail the oil spill identification method that combines oil spill fingerprint identification technology and the relative peak height ratio of characteristic fingerprint information points. Chen and Zhao [60] studied the influencing factors in the identification process of oil spills on the sea surface with fluorescence spectrometry, and mainly discussed the effects of solvent, concentration, solar radiation, sea weathering, sample collection and sample storage. He et al. [61] used ordinary FL for the platform oil in the South China Sea, adopted synchronous scanning FL, derivative fluorescence technology and three-dimensional fluorescence scanning, and stereoscopically understood the fluorescence spectral characteristics of crude oil samples, providing a scientific basis for identifying marine oil spills in a scientific way. He Xiao-Yuan [61] comprehensively analyzed six platform crude oil samples, made spectral comparison standard crude oil, and found spectral digital characteristics of these samples. They pointed out that ordinary fluorescence spectra and synchronous scanning fluorescence spectra can reflect part of the characteristic information of platform crude oil, but the ability of distinguishing oil sources that are close to crude oil is poor. Derivative fluorescence spectra are favorable for the separation of overlapping bands of ordinary fluorescence spectra, reducing spectral interference and enhancing the fine structure of characteristic spectra and resolution capability. The variation range of the three-dimensional fluorescence spectra is generally not affected by biodegradation and water washing, and the interference factors are few. Therefore, the measurement results can reflect the characteristic information of crude oil samples more accurately. Zhu Lili [54] first selected synchronous fluorescence spectra of high-concentration oil solution (500 mg/L) to approximately differentiate and distinguish crude oils. She proposed a scheme for oil spill identification by applying the synchronous FL of high concentration for approximately differentiating crude oils and then matching oil samples with the gas chromatography (GC)/mass spectrometry (MS). In the process of study, she built a fingerprint database based on SFS and GC/MS of oil samples for approximately differentiating and matching oil samples, applied PCA to digitally identify weathered oil samples. It is reliable and rapid to identify oil spills using this method, which has good application value in oil spill identification.

It can be seen from the above studies that there are many methods for oil fingerprint identification combined with FL, including combination with other oil fingerprint identification methods, such as gas chromatography/mass spectrometry.

Fluorescence spectroscopy has the advantages of high sensitivity, high selectivity, small sample volume, fast analysis results, and suitability for on-site operation. Moreover, both the volatility and water solubility of polycyclic compounds from which fingerprint information is derived are quite low, and natural weathering is slow. Therefore, it is an effective method of identification. However, the high sensitivity also brings a shortcoming, that is, the method is easily affected by various factors. During analysis, every link must be carefully operated to avoid interference from external factors, which may cause the change of spectrum shape and make it difficult to identify correctly. Fluorescence spectrometry is only applicable to unweathered oils or oil samples with a certain degree of weathering. This method is not suitable when the degree of weathering is high or there are some fluorescent substances in samples.

3. Gas Chromatography-Flame Ionization Detector (GC-FID)

The components in the mixture are distributed between two phases, one of which is immobile (stationary phase) and the other (mobile phase) carries the mixture to flow through the stationary phase and interacted with the stationary phase. Under the same driving force, different components stay in the stationary phase for different times, and sequentially flow out from the stationary phase. The components run at different speeds in the chromatographic column. Through a certain length of the column, they are separated from each other, and sequentially leave the column and enter the detector. After amplification of the resulting ionic flow signal, the chromatographic peaks of each component are mapped on the recorder. When the mobile phase is a gas phase, this analysis method is called gas chromatography (GC).

Since petroleum and petroleum products are a mixture with complex components, after the oil enters the chromatographic column, a certain time later, the components are separated out in order, thereby obtaining the chromatogram containing the retention time of each component and the content of each component. The typical hydrogen flame ionization chromatogram of oils has the following characteristics: all n-alkanes of the homologous series are separated, and are evenly spaced according to the order from low carbon number to high carbon number; isoparaffin components are partially separated and completely separated; the distribution range and content of n-alkanes in oils of different types are quite different from the content of pristanes and phytanes; and different types of oils can be distinguished by comparing these differences.

GC appeared in the 1950s, including gas-liquid chromatography, gas-solid chromatography and the like. In some cases, analysis and separation using gas chromatography is often faster, more sensitive, and more convenient. At present, the application of simple gas chromatography is relatively less, and it is generally used in combination with other technologies, mainly for environmental protection, petrochemistry, food, and so on.

GC can provide two types of information about oils: chromatogram and component information. Different oil samples will have different spectra under the same chromatographic conditions. During preliminary identification, the conclusion can be drawn by comparing the contours of gas chromatograms. Although this kind of identification is conducted only through visual comparison, but all chromatographic information is used, so the gas chromatograph is the preferred method for oil spill identification.

The main peak of gas chromatography is an n-alkane. Many smaller peaks form "minute structures" of isoparaffins, cycloparaffins and aromatic hydrocarbons between major n-alkanes. After chromatographic separation of petroleum, since the light-component n-alkanes are easily lost due to weathering, the degree of weathering of oil spills can be determined by gas chromatography fingerprinting. However, because the components measured by this method are susceptible to weathering (even relatively stable pristane and phytane are also affected by biodegradation), it is often necessary to combine it with other methods to determine the source of oil spill.

Ma and Zhao first used GC to discuss the sources of marine oil spills [62]. They used a glass capillary column for chromatographic separation, took n-alkanes as the oil fingerprint feature for oil spill identification, analyzed 9 crude oils, 5 fuel oils and 6 lubricating oils, and preliminarily discussed the influence of weathering oil spill identification, which provided a feasible method for oil spill identification. Xu et al. [63] used GC-FID to detect environmental oil spills, and the result showed that n-alkanes with corresponding carbon number in different types of oils have different peak heights, and different types of oils can be distinguished by n-alkanes, isoprene alkanes, pristanes and phytanes. Du Huaiqin and Sun Hua identified marine oil spills with an HP5890A gas chromatograph, obtained the optimum chromatographic conditions through experiments [64], obtained the GC identification method for marine oil spills through analyzing the ratio of characteristic peak area of each component in crude oil chromatogram, and discussed various factors that affected the identification result. Based on study, Chen and Zhang [65] found GC n-alkane fingerprint technique is a rapid and convenient method for identifying the sources of marine oil spills. It is mainly conducted through analysis and comparison of fingerprint features of n-alkane gas chromatography of different types of oils, including the outline of n-alkane chromatogram, range of peaks, and some characteristic ratios, for discrimination. Gong Jingxia [211] verified the feasibility of gas chromatography through identifying an oil spill in the waters of Xiamen with n-alkane GC fingerprint technique. Ye Liqun et al. identified the sources of oil spills in oceans and port waters with the identification technique combining GC and Fourier infrared spectroscopy. Chen and Zhang [65] took characteristic alkanes such as pristane and phytane as the criteria and identified an oil spill in the waters of Xiamen with n-alkane GC fingerprint technique. By using GC, Bao et al. [66] identified 6 kinds of refined oils with the fuzzy cluster analysis of original fingerprint, relative concentration ratio distribution characteristics, characteristic ratio and relative concentration ratio of n-alkanes. The results showed that the original fingerprints of the oils with large component differences are quite different, and such oils could be directly identified as different oils; the original fingerprints of oils with similar components are more similar, and such oils must be further distinguished according to the fingerprint information of n-alkanes such as relative concentration distribution and characteristic ratios. Malmquist et al. [67] used the PCA model to study the influence of evaporation and decomposition processes on oil spill components, determine the weathering state of oil spills, and distinguish different weathering processes.

GC is one of the most important methods for oil fingerprint analysis and identification in the world. It has the following advantages: high separation efficiency, good selectivity and high sensitivity. It is mainly used to analyze alkanes, pristanes and phytanes in petroleum. Since the light-component n-alkanes are easily lost due to weathering, the degree of weathering of oil spills can be determined by GC fingerprinting. However, this method can only be used to determine the nature with chromatogram fingerprints and carry out quantitative analysis based on peak area, and its measurement range does not include oil samples with serious biodegradation. During oil spill identification, the measured components are vulnerable to the influences of weathering and biodegradation, and even relatively stable pristanes and phytanes may change due to biodegradation. Therefore, this method is applicable to unweathered oil and oil samples with a certain degree of weathering. For oil products seriously affected by weathering, it is necessary to combine this method with other methods to determine the source of oil spill.

4. Gas chromatography/mass spectrometry

Mass spectrometry is an analysis method that measures the mass-to-charge ratio of ions of a sample to be measured. The sample to be analyzed is first ionized, and then the ions are separated as per the mass-to-charge ratio by using the differences between motor behaviors of different ions in the electric or magnetic field to obtain the mass spectrum. According to the mass spectra and related information of the sample, qualitative and quantitative results of the sample can be obtained. In simple terms, chromatography is the separation method of substances, and mass spectrometry is the detection method. Gas chromatography/mass spectrometry (GC/MS) is an analysis method combining separation and qualitative analysis with gas chromatography and qualitative analysis with mass spectrometry, as well as a method that is commonly used for oil spill identification at present.

Both GC/MS and GC can analyze aromatic hydrocarbon components, biomarker compounds and benzene series in oils. However, the content of components such as polycyclic aromatic hydrocarbons and biomarker compounds is low, and they are subjected to interference of other components in gas chromatography and are difficult to be measured accurately. Moreover, the ability of GC to discriminate smaller samples between samples is limited, that is, the ability to discriminate the "minute structure" of isomerides of isoparaffins, cycloalkanes and aromatic hydrocarbons is limited. GC/MS can identify subtle and unique differences between samples only based on the mass spectrometry detector. Thus, most petroleum compounds (and their isomers) can be monitored, and a reasonable method for identifying oil spills is provided. GC/MS distinguishes oil types mainly through analyzing and measuring polycyclic aromatic hydrocarbons and biomarkers in oils.

Pal et al. [212] analyzed the hydrocarbon composition of diesel fractions with supercritical fluid chromatography and GC/MS. On a benchtop GC/MS instrument, Su Huanhua et al. [213] used the ASTM standard method to quantitatively analyze the composition of saturated hydrocarbons and aromatic hydrocarbons of diesel fractions, and saturated hydrocarbons and aromatic hydrocarbons of vacuum gas oil (VGO). Xu Hengzhen et al. [214] analyzed the saturated hydrocarbons in oil spills with GC/MS. The results showed that the parameters of ratios ($Pr/Ph, Pr/nC17, Ph/nC18$) of pristanes and phytanes to their corresponding n-alkanes and the odd-even predominance (OEP) can be used to identify different types of unweathered oils, and the indicator parameter OEP can be used to identify different oils that are heavily weathered. In addition, Xu Hengzhen et al. [214] used capillary GC/MS to detect weathered oil spills in different environments, and selected steranes and terpanes as indicators of severely weathered oil spills, because they were little affected by weathering. Liu et al. [68] used low-resolution GC/MS instead of the traditional magnetic mass spectrometry and equivalently adopted ASTM D2425, D2786 and D3139 methods to analyze saturated hydrocarbons and aromatic fractions of kerosene, diesel and heavy oil. The results can meet the reproducibility requirements of ASTM method and reduce the analysis cost. Mazeas et al. [69] used GC/MS to quantitatively analyze aliphatic and aromatic hydrocarbons in petroleum. Zhao et al. [70] used GC/MS to analyze the samples of oil spills on the water surface and suspected oil spill sources and established a GC/MS identification method for oil spills on the water surface. Wang et al. [71] proposed that GC can be used to initially screen a large number of suspicious oil spill sources, and then a mass spectrometer can be used for further identification. Wen Qiang [53] used a combination of retention time and characteristic ions for qualitative analysis of n-alkanes (containing phytane and pristane), naphthalene, phenanthrene, fluorene, dibenzothiophene and chrysene of oils and their polycyclic aromatic hydrocarbons of alkylation series and steroid and terpenoid biomarkers. Meanwhile, the internal standard quantitative method was used to quantitatively analyze the components, and the oil fingerprint information was used to identify and analyze single-type crude oils and mixed oils, and it was concluded that GC analysis should be conducted for n-alkanes, pristanes and phytanes, and GC/MS analysis should be conducted for polycyclic aromatic hydrocarbons and steroid and terpenoid biomarker compounds. Ni Zhang et al. [72] used GC and GC/MS to analyze and test the heavy fuel oil in a sea area of Zhoushan after two months of weathering, and discussed the changes of its alkane and aromatic hydrocarbon components and oil spill markers. Han Yunli [215] carried out computer analysis on two kinds of crude oils and of seven kinds of refined oils with GC/MS, successfully identifying seven oil samples with combination of original fingerprints and characteristics of concentration distribution of polycyclic aromatic hydrocarbons, established the fingerprint chromatogram on the basis of polycyclic aromatic hydrocarbons, and eventually reached the objective of distinguishing oils and identifying oil spills with polycyclic aromatic hydrocarbon fingerprint chromatogram. For the identification of mild and moderate marine oil spills, Cheng [73] analyzed n-alkanes (containing phytane and pristane), naphthalene, phenanthrene, fluorene, dibenzothiophene and chrysene of oils and their polycyclic aromatic hydrocarbons of alkylation series and steroid and terpenoid biomarkers with GC/MS and in combination with retention time and characteristic ions, obtained chemical fingerprint information, and made clear the rules of weathering of fuel oil, heavy fuel oil and light fuel oil from ships under weathering conditions.

To sum up, GC/MS has the same advantages as GC, such as high separation efficiency, high sensitivity and high selectivity, and makes up for the deficiencies of GC in identifying aromatic hydrocarbon components biomarker compounds and samples of benzene series. However, these relatively stable components are more reliable in identifying oils from different sources, even those oils which are heavy weathered. Therefore, GC/MS has become one of the most important methods for oil fingerprint analysis and identification, and most researchers in China use this method for analysis. However, compared with GC, GC/MS is more expensive and difficult to operate.

5. Stable isotope ratio mass spectrometry (SIRMS)

Biomarker compounds play an obvious role in crude oil identification. However, because the chemical composition may easily change due to environmental processes such as volatilization, leaching and biodegradation, the application of "chemical fingerprints" has certain limitations. In light oils, the content of traditional marker compounds, such as steroid and terpenoid compounds, is low. For example, there are a few low-boiling biomarker compounds in diesel, and there is no biomarker compound available for identification in gasoline. The limited hydrocarbon fingerprints contained in light oils from different sources can be very similar and difficult to identify, and the possible solution to this problem is through stable isotopes, including carbon and hydrogen.

Analysis of oil spills yields two sets of isotope data: total hydrocarbon isotopic composition and single hydrocarbon isotopic composition, and the latter gives useful isotopic fingerprints for petroleum products. The combination of carbon isotopes and hydrogen isotopes provides a powerful fingerprint identification tool for the oils with biomarker compounds missing, and also provides an extremely effective method for the identification of the oils with biomarker compounds still existing.

At present, the stable isotope analysis method used in oil spill identification is mainly carbon stable isotope analysis. Kvenvolden et al. [74] applied the similarity and dissimilarity of carbon isotope component as evidence of the correlation between weathered Exxon Valdez oil and North Slope oil on the shoreline. Mansuy et al. [75] found that the three moderately weathered light fuel oils could be highly matched with their unweathered oil sources through determination of stable carbon isotopes of n-alkanes. Chung et al. [76] used carbon isotope ratios to distinguish four different types of oil based on the age and properties of oil source rocks and environmental conditions of oil production. In China, through analyzing the compositions of A-component carbon isotope and n-alkane single-molecule carbon isotope of typical crude oil and chloroform bitumen in China, Wang et al. [77] discussed their application in oil spill identification. The carbon isotopic composition of crude oil has a maternal inheritance effect. The carbon isotope type curves of A components in crude oils from different sources and their chloroform are different, and this rule can be used for oil spill identification.

3.3   Quantitative Analysis of Oil Spills

In the process of large-scale marine oil spill accidents, most of them involve the assessment of the degree of water pollution in the polluted waters and the degree of pollution of aquatic organisms. In these monitoring projects, oil content analysis is indispensable. The *Marine Monitoring Specification/Water Quality Monitoring and Analysis (Industrial Standard of the People's Republic of China HY003.4-91)* stipulates oil analysis methods, and the commonly used centralized quantitative analysis method for oil spills is introduced below.

3.3.1   Analysis of Oil Spill Content in Water

Petroleum is a complex mixture of multiple components. The petroleum hydrocarbons entering the ocean are affected by wind, light, tide, diffusion, emulsification, evaporation, oxidation, etc., and are mixed with the biogenous hydrocarbons in seawater to form complex petroleum hydrocarbons in seawater. Therefore, different test methods are needed for measurement. According to national industrial standards, fluorescence spectrophotometry, UV spectrophotometry, infrared spectrophotometry and gravimetric method can be used to analyze the oil content in water. The specific analysis method is similar to the oil fingerprint analysis method introduced in the previous section. Here, the applicable scope of various methods in the analysis of oil spill content will be described below.

3.3.1.1 Fluorescence spectrophotometry

The method is sensitive, rapid and highly selective, and the detection limit is $6-13.7 \ g/L$. It is suitable for routine monitoring and investigation in water bodies and can be used for the measurement of oils in various water bodies such as oceans, offshore areas, estuaries, and so on.

3.3.1.2 UV spectrophotometry

This method is suitable for the determination of water bodies with oil content of $0.05-1 \ mg/L$, which is greatly affected by standard oil, and can be used to measure oils in offshore areas and estuaries.

3.3.1.3 Infrared spectrophotometry

The detection range of this method is $0.2-200 \ mg/L$, the test results are less affected by the standard oil than other spectroscopy methods, and it is suitable for environmental monitoring and pollution source analysis.

3.3.1.4 Gravimetric method

The detection range of this method includes water bodies with oil content greater than $4 \ mg/L$. This method is not restricted by oil types, and the measured results can be considered as the total amount of oils. It is suitable for the determination of water samples severely contaminated by crude oil, heavy fuel oil and lubricating oil, and not applicable to the determination of water samples contaminated by kerosene, gasoline and light diesel that contain a large number of low-boiling hydrocarbons.

3.3.2 Analysis of Petroleum Content in Marine Organisms

Another response to oil spill pollution in the sea is the pollution of marine organisms. The analysis of petroleum content in marine organisms is also an important part of the assessment of oil spill pollution. Due to the low content of petroleum hydrocarbons in organisms and the background value range of $3.37-19.7 \ g/L$, fluorescence spectrophotometry can be used to sensitively and rapidly determine the components of aromatic hydrocarbons with high fluorescence in marine organisms such as mussels. According to the experimental results, the status of aromatic hydrocarbons can represent the petroleum hydrocarbons accumulated in organisms. Therefore, fluorescence spectrophotometry is listed as one of the methods for monitoring petroleum pollution in organisms.

3.3.3 Analysis of Petroleum Content in Sediments

After entering the seawater, petroleum hydrocarbons enter sediments after weathering, especially adsorption and sedimentation. Because the suspended particles adsorb a large amount of petroleum hydrocarbons in seawater and sink to the seabed, the amount of petroleum hydrocarbons in the sediment is dozens or even thousands of times higher than that in seawater, so some people compare it to the warehouse of petroleum hydrocarbons. Petroleum hydrocarbons entering the sediment degrade less and slowly due to hypoxia. In addition, the sediment is less affected by the flow of seawater and has the representativeness of the station in a sea area, which can more accurately characterize the situation of petroleum pollution in the sea. The content of petroleum hydrocarbons in marine sediments is about $0.02 - 6 \ mg/L$. It is better to determine the content of petroleum hydrocarbons in sediments with UV spectrophotometry. It should be noted that high-sensitivity fluorescence analysis is not usually used for sediment analysis because fluorescence quenching occurs when the concentration is too high.

3.4 Monitoring of Oil Film Thickness

During treatment of any oil spill accident, the oil spill volume should be determined first. Generally speaking, the oil spill volume can be estimated as per oil spill area and oil spill thickness. The oil spill area is usually estimated through field observation, aerial surveillance, satellite remote sensing monitoring and oil spill model prediction, while the oil film thickness can be calculated, estimated and quickly measured with empirical formulas.

3.4.1 Empirical Formula for Oil Film Thickness

There is a certain quantitative relationship between oil film thickness and oil spill time and oil spill volume. The experts from the Soviet Union developed an empirical formula, which can give satisfactory results for most crude oils.

$$h_t = (\frac{V}{\pi})^{1/3}(\frac{S_w}{3S_0 \cdot (S_w - S_0)k_t t})^{2/3} \qquad (3.2)$$

where
  h_1 refers to oil film thickness related to time t, μm
  V refers to oil spill volume, m^3
  S_0 refers to the density of oil spill, (g/cm^3)
  S_w refers to the density of water, (g/cm^3)
  K_t refers to the constant of a given oil
  t refers to diffusion time, second

3.4.2   Estimation of Oil Film Thickness

On the open sea surface without wind and flow, the oil spill spreads outward with the center of the oil spill point, and forms a roughly circular pollution surface. A cubic meter of crude oil can diffuse into a circular pollution surface with an average oil film thickness of 0.5 *mm* and a diameter of 48 meters within 10 minutes. After 100 minutes, the diameter of the pollution surface can expand to 100 meters and the oil film thickness becomes 0.1 micron.

Experience has shown that the area covered by thick oil layers is relatively small, and the area covered by thin oil layers is relatively large. 90% of the oil covers 10% of the total oil film area, and the other 10% covers 90% of the total oil film area. Oil spill emergency focuses on thick oil layers with small area and large oil volume. The total oil spill volume can be estimated through observing and estimating the thickness of the thick oil layer:

$$Oil\ spill\ volume = [Oil\ film\ area\ (thick\ layer) \times Oil\ film\ thickness \times$$
$$Oil\ density] \div 90\% \tag{3.3}$$

3.4.3   Rapid Measurement Method for Oil Film Thickness

The rapid measurement method is a non-optical measurement method for oil film thickness. Only one sampler and one funnel are needed. The measurement process and principle are as follows: firstly, place the sampler at a certain depth for sampling, and then vertically pull up the sampler to the ship, while the edge of the inner vessel cut off the oil film, and the cylinder part forms an oil column; read the funnel scale, and calculate the actual oil film thickness according to the ratio of the floor area of the cylinder to the cone. This method is suitable for oil films with thickness of 0.2 *m* − 10 *mm*. The measuring time is less than 6 minutes. It is suitable for both dark petroleum products and bright petroleum products.

3.5   Application of Oil Spill Emergency Monitoring

The techniques of physical analysis, spectral analysis and quantitative analysis involved in oil spill monitoring play different roles in the oil spill emergency response. Upon receipt of the report on the occurrence of an oil spill, relevant personnel should immediately understand the general situation of the oil spill for the main contents involved, and quickly prepare a monitoring plan to determine the monitoring content.

TABLE 3.1: The main contents of the oil spill.

Content	Description
Ship name, nationality, gross tonnage, and type of ship	
Date and time when the oil spill occurs	
Place where the oil spill occurs	
Damage to the ship	Floating (fixed or moving with the flow), sunk
Variety of oil spill and its physical and chemical properties	
Type of oil spill	Instantaneous and continuous discharge
Total oil storage and oil spill volume	
Function of the sea area where the oil spill occurs	
Hydrologic and meteorological conditions	At that time, and in next few days

3.5.1   Program for Oil Spill Emergency Monitoring

During emergency monitoring of marine oil spill accidents, the type of oil spill should be determined first, and the physical and chemical properties should be understood. If the variety of oil spill cannot be identified or its physical and chemical properties cannot be found, a plan should be formulated immediately for sampling and measurement. Then, based on the extent of the accident pollution, the regular sampling and distribution plan for oil content in the water should be determined, so as to conduct sampling and monitoring during and after oil spill clean-up.

The general principle of sampling point distribution is that the stations in waters near the oil spill point are dense and those in waters away from the oil spill point are sparse; the stations in sensitive areas are dense, and those in ordinary sea areas are sparse; the stations downstream of oil spill drift are dense, and those upstream of the drift are sparse. The station layout method should be divided into concentric circle, section type and radial type according to the type of accident and the situation of sea area. The first layout type is suitable for sea areas with weak winds and ocean currents, and the latter two types are suitable for sea areas with strong winds and ocean currents. However, some oil spills involve more complicated pollution, and sampling points may increase or decrease based on the specificity of the pollution on the basis of the above-mentioned layout methods.

The monitoring period of oil spill accidents should be determined according to the types of oil spill discharge. In case of instantaneous discharge, sampling should be conducted once every two weeks, and be taken five times consecutively; in case of continuous discharge, sampling should be conducted once a week at the sump oil discharge stage, and after the sump oil discharge ends, monitoring should be carried out once every month and be done three times consecutively. During the process of oil spill clean-up, monitoring should be carried out according to the clean-up plan, and should be done at least before, during and after clean-up operation. The monitoring of water quality recovery should be based on the analysis of the accident scale and the time that the water quality may be affected. Besides, the monitoring cycle should also be determined based on consideration of the measures taken during the water quality recovery and the possible changes in water quality.

3.5.2   Focus of Oil Spill Emergency Monitoring

Key monitoring contents involved in oil spill emergency response include measurement of physical and chemical properties of the oil spill, monitoring of oil spill volume, monitoring of water pollution, monitoring of oil spill pollution scope and confirmation of oil spill source.

3.5.2.1   Measurement of physical and chemical properties of the oil spill

After the oil spill accident, the monitoring personnel or investigators who first arrive at the scene should immediately collect two oil spill samples, each of which should be at least 20 mL, and send them to the laboratory for immediate measurement of density, viscosity, flash point, pour point and coagulation point, and promptly send the measuring results to oil spill model operators and oil spill emergency decision makers immediately.

3.5.2.2   Monitoring of oil spill volume

The monitoring of oil spill is mainly used to estimate the volume of oil spilled, the spill rate and the trend after the oil spill occurred, so as to estimate the total oil spill volume at a certain stage after the oil spill or at the end of the oil spill. On-site continuous monitoring is usually adopted and includes:

(1) Investigate the scene of oil spills, and record the state of the ship, the type of oil spill, and the degree of pollution on the sea surface;

(2) Measure the width and length of the oil zone, the direction and speed of drift, and the thickness of the oil belt;

(3) Record the color and form of the oil zone;

(4) Estimate the oil spill volume.

3.5.2.3 Monitoring of water pollution

This monitoring is mainly intended to effectively monitor the direction of oil migration in water, so as to timely issue forecasts or reports. At the same time, it can also monitor the change of water quality during the clean-up process and water quality recovery period, and confirm the degree of oil pollution in the polluted sea area, so as to conduct a comprehensive assessment of oil spill pollution and provide a basis for taking recovery measures and carrying out oil spill pollution damage assessment.

The degree of oil pollution can be evaluated through comparing the oil content data obtained from water pollution monitoring with the water quality background value before the accident pollution. It should be explained that the degree of water pollution and pollution damage should be evaluated, not only by monitoring and analysis of oil content, but also by monitoring other items related to oil pollution in seawater, such as BOD (biochemical oxygen demand), COD (chemical oxygen demand), etc.

3.5.2.4 Monitoring of pollution scope

In order to identify the scope of pollution caused by oil spill drift, field investigation should be carried out first. The first mode is active investigation, that is, regularly conducting field investigation on the areas where the oil may drift, collecting samples, and analyzing and identifying them. The second mode is passive investigation, that is, investigating the contaminated area at any time according to the report and collecting oil samples for analysis and identification.

3.5.2.5 Oil fingerprint analysis for determining oil spill sources

When the source of the discovered oil spill is unknown or there are a number of suspected oil spill sources and administrative investigation is insufficient for confirmation, oil fingerprint analysis should be used for oil spill identification. Through analyzing and comparing the oil spill samples collected and the suspected oil samples, this application has become one of the most reliable methods for maritime law enforcement in China.

The data from all of the above-mentioned monitoring methods form the overall monitoring results. With respect to the use and treatment of emergency monitoring results, in addition to the data about measurement of physico-chemical properties of oil spills that will be reported to the oil spill emergency headquarters immediately after the measurement, the remaining monitoring data should be analyzed comprehensively once per monitoring cycle and the results of statistical analysis will be reported to relevant authorities at any time. After the monitoring work of each pollution accident is completed, the final overall report should be prepared and comprehensive evaluation should be carried out.

Chapter 4

Model Prediction of Marine Oil Spills

In the unique marine environment, oil spills will eventually disappear under the influence of a series of complex physical and chemical changes and organisms. The behaviors and fates of oil spills in the ocean can be divided into three categories: (1) Spreading process. This process refers to the process that the area of the oil film on the sea surface increases due to its own characteristics. (2) Transport process. This process refers to the migration movement of oil spills under the action of marine environmental dynamic factors, including horizontal drift and diffusion as well as vertical mixing and suspension processes. (3) Weathering process. The process refers to all processes that can cause changes in the composition of the oil spill, including: evaporation, dissolution, emulsification, photooxidation, biodegradation, adsorption and sedimentation, mixing and diffusion of water bodies, and metabolism in organisms [78]. From the perspective of the entire weathering process of the oil spill, evaporation and emulsification are closely related to the emergency response and economic damage assessment of the oil spill in the short term, but the final fate of the oil spill depends on photooxidation and biodegradation. These processes are directly related to people's impact assessment on the global marine environment [79].

With regard to the study of the behavior and fate of oil spills in seawater, numerous scholars have put forward various theories and mathematical models through a large number of experiments, and have obtained fruitful results. With the development of science, many countries have gradually combined these theories and mathematical models with platforms such as the Geographic Information System (GIS), and developed various distinctive oil spill model systems to transform theoretical research results into practical technologies, greatly promoting the development of marine oil spill forecasting and warning technology.

4.1   Numerical Model of Oil Spills

4.1.1   Oil Spill Spreading Model

When the oil spills onto the sea surface, it will spread out under the combined action of surface tension, inertia force, gravity and viscous force. Gravity and surface tension are the driving forces, and inertia and viscous forces are the resistance. At the beginning, gravity plays a major role, so the form of the oil spill largely affects the spreading of the oil. For example, a large amount of oil that spills instantaneously spreads much faster than the oil that spills continuously and slowly. After oil spills for a period of time, as the thickness of oil layer decreases greatly, the effect of surface tension exceeds the effect of gravity and becomes the main factor affecting oil spill spreading, making the oil spill form an oil film on the water surface with the center thicker than the edge. The spreading degree of the oil film is closely related to the time and its own characteristics of the oil, such as variety, viscosity, temperature, etc. For example, a small amount of high-viscosity crude oils and heavy fuel oils are retained on the sea surface because they are not easy to spread. When the ambient temperature is lower than their pour point, these high-viscosity oils do not even spread. The spread oil film is discontinuous under the action of wind, current and wave. As time goes on, its shape and thickness are changing, and it gradually exists in the form of oil belts, fragments or small tar balls.

In the study of oil spill spreading, the spreading range and thickness variation of the oil film has been a common concern of researchers. In the early stage, Blokker used the oil on the free plane as the spreading premise and established the formula for spreading diameter formula based on the conservation of mass of the oil film [80]. This formula ignores surface tension and viscous force and only considers gravity and oil spill volume:

$$D = \left[D_0^3 + \frac{24k_r}{\pi} (d_w - d_o) V_t \right]^{\frac{1}{3}} \tag{4.1}$$

where, D_0 refers to the diameter of the oil film at the beginning, d_o and d_w refer to the specific gravity of oil and water, k_r refers to Blokker constant, and V refers to the total volume of oil spill (m^3). It can be seen from Formula 4.1 that the Blokker formula mainly reflects the inertial spreading of oil under the action of gravity. Later, Fay proposed the three-stage oil film spreading theory, which has achieved groundbreaking results in the research and application of the oil spill spreading model [81]. According to the actual stress condition of oil on the water surface, this theory considers gravity, surface tension, inertia force and viscous force in an all-round way, and according to the roles and size of each force at different stages, the process of oil spill spreading is divided into three stages, namely, gravity-inertia force stage, gravity-viscous force stage and surface tension-viscous force stage. In addition, the theory assumes that under the condition of calm sea surface (i.e., excluding the influence of wind, current, wave, etc.), the property of the massive oil spill remains unchanged and the oil spill is balanced under vertical force during the spreading process, and the oil film spreading always remains circular, and the spreading range can be measured by diameter. Therefore, the diameter of the oil film can be calculated by the pairwise balance of the forces at each stage. The calculation formulas for the diameter of the oil film at the three stages are:

- Inertial spreading stage:

$$D = 2k_1 \left(g\beta V_t^2 \right)^{\frac{1}{4}} \tag{4.2}$$

- Viscous spreading stage:

$$D = 2k_2 \left(\frac{g\beta V^2}{\sqrt{v_w}} \right)^{\frac{1}{6}} t^{\frac{1}{4}} \tag{4.3}$$

- Surface tension spreading stage:

$$D = 2k_3 \left(\frac{\sigma}{\rho_w \sqrt{v_w}} \right)^{\frac{1}{2}} t^{\frac{1}{4}} \tag{4.4}$$

After spreading ends, the diameter of the oil film remains unchanged:

$$D = 356.8V^{\frac{3}{8}} \qquad (4.5)$$

In Formula 4.2 - 4.5, D refers to the diameter of the oil film (m); g refers to the acceleration of gravity $(m/sec.^2)$; V refers to the volume of oil spill (m^3); t refers to the time $(sec.)$; and ρ_w refers to the density of seawater (kg/m^3). $\beta = 1 - \rho_o/\rho_w$, where, ρ_o refers to the density of oil (kg/m^3); v_w refers to kinematic coefficient of viscosity of water $(m^2/sec.)$; σ refers to the difference between air-water surface tension coefficient and oil-air surface tension coefficient and oil-water surface tension coefficient (N/m); k_1, k_2 and k_3 refer to the empirical coefficients at inertial spreading stage, viscous spreading stage and surface tension stage, respectively. The demarcation time of the above stages can be determined according to the condition under which the extension diameters at the two adjacent stages are equal.

Fay's three-stage oil film spreading theory plays an important role in promoting the study of oil spill spreading, but there are still some shortcomings, which are manifested in: (1) The theory assumes that the oil film spreads in a circular shape on the calm water surface (i.e., in a centrally symmetric manner). However, in actual situation, the spreading form of oil film is affected by wind and current and tends to be elliptical or other shapes. Moreover, in the process of oil film spreading, the combination of wind, current and wave will break the oil film, but this theory assumes that the oil film expands outward as a whole, greatly different from the actual situation. (2) This theory assumes that the thickness of the oil film formed by the oil spill is uniform. However, in the actual situation, when the oil film spreads, the thickness is not uniform, but a thick oil film is in the middle, surrounded by a thin oil film. (3) Because the oil film spreading is related to the type of oil spill, a large amount of oil that spills instantaneously spreads faster than the oil that spills continuously and slowly. As a result, the oil that spills initially drifts away quickly from the oil spill point under the action of the current, and the water surface is left blank, and the oil that spills later will fill the gap accordingly, so that a long oil spill belt will be formed in the water flow direction. In Fay's theory, this process cannot be expressed in a formula [82].

After Fay's theory was put forward, many scholars improved the model, and established various spreading models in combination with the spreading process and diffusion process of oil spill, based on its existing foundation. For example, Wang et al. [83] revised the formula for Fay's second stage, i.e., viscous spreading stage, as follows:

$$A_2 = 0.98^2\pi \left(\frac{\Delta\rho gV^2}{\rho_w\sqrt{v_w}} \right)^{\frac{1}{3}} t^{\frac{1}{2}} \qquad (4.6)$$

where, A_2 refers to the area of the oil film (m^2); g refers to the acceleration of gravity $(m/sec.^2)$; V refers to the volume of oil spill m^3; t refers to the time $sec.$; ρ_w refers to the density of seawater (kg/m^3); v_w refers to the kinematic coefficient of viscosity of seawater ($m^2/sec.$); $\Delta\rho$ refers to the density difference between seawater and oil spill (kg/m^3). Mackay [84] considered the influence of wind in Fay's formula for the second stage and established the spreading models for thick oil film and thin oil film, respectively, based on the actual observation results; Lehr et al. [85] modified Fay's theory, took into account the influence of wind field and current field on the asymmetric spreading of the oil film. It is believed that the oil film spreads in the ocean in the form of an ellipse instead of a circle, and its long-axis direction is consistent with the wind direction. Some Chinese scholars have not fully adopted the three-stage division method of Fay's theory to study the oil film spreading process. For example, Liu et al. [86] summarized the unified formula of oil film spreading scale changing with time based on Fay's theory, and the formula for the diameter of the oil film is as follows:

$$D = 0.61 \left[1.3 \left(\beta g V \right)^{\frac{1}{2}} t + 2.1 \left(\frac{\beta g V^2}{\sqrt{v_w}} \right)^{\frac{1}{3}} t^{\frac{1}{2}} + 5.29 \left(\frac{\sigma^2 t^3}{\rho_w^2 v_w} \right)^{\frac{1}{2}} \right]^{\frac{1}{2}} \quad (4.7)$$

where, D refers to the diameter of the oil film (m); g refers to the acceleration of gravity ($m/sec.^2$); V refers to the volume of oil spill (m^3); t refers to the time ($sec.$); and ρ_w refers to the density of seawater (kg/m^3). $\beta = 1 - \rho_o/\rho_w$, where, ρ_o refers to the density of oil (kg/m^3); v_w refers to kinematic coefficient of viscosity of water ($m^2/sec.$); σ refers to the difference between air-water surface tension coefficient and oil-air surface tension coefficient and oil-water surface tension coefficient (N/m).

In addition to Fay's three-stage theoretical model for oil film spreading and the later improved model, according to the analysis of a large amount of observation data, Okubo [87] considered that the approximate normal distribution of the mass of oil film is caused by spreading, and that the diameter of oil film is proportional to the mean square deviation of mass. Under various isotropic conditions, the boundary around the oil film remains circular, and the range of oil film can be measured by diameter. However, in the empirical mode proposed by Okubo [87], only spreading was considered, so the diameter obtained of oil film is only related to time and has major defects. In 1982, Motora from Japan believed that the spreading of oil spill is the result of the potential energy of the oil transforming into viscous dissipation energy and surface tension energy [88]. Motora first assumed that the oil film is always cylindrical and uniform in thickness as it spreads, and the initial radius of oil film is r_0; then, he assumed that it always spreads in a two-dimensional direction, and t seconds later, the radius of the oil film is r [82]. Moreover, the total volume of oil is V_0, and the volume of the oil on the water surface is V. According to the gravity and buoyancy balance of the oil film in vertical direction, the following formula can be obtained:

$$V = V_0 \left(1 - \frac{\rho_0}{\rho_w} \right) \quad (4.8)$$

where, ρ_o and ρ_w refers to the density of oil and seawater, respectively (kg/m^3). Under the condition of calm sea surface (i.e., excluding the influence of wind, current, wave, etc.), it is assumed that the potential energy of oil is transformed into viscous dissipation energy and surface tension energy, and the differential equation of spreading radius obtained is as follows:

$$\frac{dr}{dt} = \frac{-\beta \pm \sqrt{\beta^2 - 4\alpha\gamma}}{2\alpha} \tag{4.9}$$

where,

$$\alpha = \frac{\rho_o}{r} + \frac{4\rho_0 V^3}{\pi^2 r^2} \tag{4.10}$$

$$\beta = \frac{\pi^2 c^2 v_o \rho_o r^4}{2V} \tag{4.11}$$

$$\gamma = -\left(\frac{\rho_o g V^2}{\pi r^3} + 2\pi r\sigma\right) \tag{4.12}$$

In the formula, c refers to the average velocity correction coefficient of oil (taking 0.1); v_o refers to the kinematic coefficient of viscosity of oil $(m^2/sec.)$. The analytical solution is obtained approximately from Formula 4.10, and the spreading radius of the oil just after spilling is obtained:

$$r(t) = \sqrt{2\sqrt{\frac{gV}{\pi}}t + r_0^2} \tag{4.13}$$

After a period of time, the spreading radius of oil film is:

$$r(t) = \left(2\frac{16gV^3}{\pi^3 C^2 v_o}t + r_0^8\right)^{\frac{1}{8}} \tag{4.14}$$

With regard to evaporation and biodegradation of the oil spill, the average thickness of the oil layer is:

$$h(t) = \frac{V_0}{2\pi R^2(t)} \tag{4.15}$$

In addition to the above-mentioned spreading model, Huang Lixian et al. [89] obtained the relationships between the oil spreading speed and the spreading radius and time according to the variation of spreading scale of instantaneous oil spill from a point source on a clear water surface with oil variety and oil spill volume in the experiment. However, this formula itself is an empirical formula, and the function values indicating the nature of the oil and the oil spill volume are also taken from the experimental coefficients [88].

Considering that in the calculation model of oil spill spreading range the main behaviors of oil spill in the early and later stages are spreading and diffusion respectively, the simple spreading model can only simulate the early behavior of oil spill, while the simple diffusion model can only simulate the later behavior of oil spill [90]. It is worth mentioning that the spreading model proposed by [91] can be applied to all spreading and diffusion processes of oil spills.

In the model proposed by [91], s and n refer to the main spreading direction of the oil film and the secondary diffusion direction perpendicular to the main spreading direction, and the long-axis dimension of the oil film in s direction is:

$$D_s = k_{eq} (d_s + d_f) \tag{4.16}$$

The short-axis dimension of the oil film in n direction is:

$$D_n = k_{eq} (d_n + d_f) \tag{4.17}$$

The equivalent circle diameter of the oil film is:

$$D_{eq} = \sqrt{\frac{4A}{\pi}} \tag{4.18}$$

In the formula above, A refers to the area of the oil film (m^2); d_f is calculated as per Fay spreading mode; $d_s(d_s = \omega \sigma_s)$ and $d_n(d_n = \omega \sigma_n)$ refer to the distance increase of the oil film caused by diffusion in s direction and n direction, respectively (m); σ_s and σ_n refer to the standard deviations of mass distribution of the oil film in s direction and n direction: $\sigma_s = a_s t^{1.17}$ and $\sigma_n = a_n t^{1.17}$ (where, $a_s = 2.236 \times 10^{-3}$ and $a_s/a_n = 10^{\frac{1}{2}}$); k_{eq} refers to the attenuation factor, decreasing with time and reflecting the process of the oil film range decreasing and the final extinction time of the oil film, and the expression is:

$$k_{eq} = \frac{d'_s}{\sqrt{a(d_s + d_f)(d_n + d_f)}} \tag{4.19}$$

The value range is from 1 to 0. When $k_{eq} > 1$, make $k_{eq} = 1$, and the expression of d'_s in Formula 4.20 is:

$$d'_s = 2\sqrt{2}\sigma_s \left[\ln \left(\frac{V_e^{-kt}}{2\pi\sigma_s\sigma_n h_c} \right) \right]^{\frac{1}{2}} \tag{4.20}$$

In the formula, k refers to the oil attenuation coefficient that comprehensively reflecting oil spill evaporation and degradation and the value takes $0.5(1/d)$ or is determined based on analysis according to oil variety and sea condition, h_c refers to the minimum thickness of the observable oil film, and the value usually takes $10^{-4} \sim 10^{-3}\, mm$.

An important progress of the oil spill spreading model is the oil particle model proposed by Johansen [216] in the 1980s. The model discretizes oil into a large number of oil particles, and each oil particle represents a certain amount of oil, drifts under the action of surface current field and wind field, and the effect of shear flow and turbulent flow on the movement of oil spill is realized through the random movement of oil particles, that is, particle movement is used to approximate the drift and diffusion process of oil in water, and its discrete path and quality are tracked and recorded as time functions related to the reference grid system [88]. These particles move independently, and all particle clusters must be tracked to the same moment before proceeding to the next step. The spreading of oil film is the migration and spreading of particle swarm, which is controlled by the distribution size of oil particles, shear flow and turbulent flow. It can also be regarded as Lagrange motion of particle swarm under the influence of turbulent flow, which is realized through the random movement of oil particles [88, 92]. The oil particle model will be introduced in detail in Section 4.1.3.

With the improvement and development of oil spill spreading models, a new model, Monte Carlo spreading model, was proposed. In this model, the diffusion of oil particles is regarded as random diffusion under the action of turbulent flow, which is manifested as the chaotic movement of particle swarm along complex trajectories. Its movement direction is random at a certain moment, but this randomness is controlled by the whole playground. When random numbers are given, turbulence intensity, time scale and particle number can also be given, so as to obtain the transport distance of the marked particles [93].

The Monte Carlo spreading model uses the randomness of the diffusion phenomenon. There are two methods for generating random numbers, namely uniform random numbers and regular random numbers.

- if a, b and c are random numbers between -0.5 and 0.5,

$$A = \frac{a}{\sqrt{a^2 + b^2 + c^2}}, B = \frac{b}{\sqrt{a^2 + b^2 + c^2}}, C = \frac{c}{\sqrt{a^2 + b^2 + c^2}} \quad (4.21)$$

The mean value is zero, with a discrete $1/3$ distribution. The relation between the discrete variance σ^2 and the diffusion coefficient is: $k = \sigma^2/2\Delta t$, so the spreading distance of the particle in three-dimensional direction is:

$$l_x = A \cdot \sqrt{6\Delta t \cdot k_x}, l_y = B \cdot \sqrt{6\Delta t \cdot k_y}, l_z = C \cdot \sqrt{6\Delta t \cdot k_z} \quad (4.22)$$

- If a, b and c are normal random numbers between 0 and 1, the mean value of a, b and c is zero, and $(0, 1)$ normal distribution is taken, so the spreading distance of the particle in three-dimensional direction is:

$$l_x = a \cdot \sqrt{2\Delta t \cdot k_x}, l_y = b \cdot \sqrt{2\Delta t \cdot k_y}, l_z = c \cdot \sqrt{2\Delta t \cdot k_z} \quad (4.23)$$

It can be seen that when the Monte Carlo model is used to study the oil spill, the diffusion of particles representing oil particles is regarded as random diffusion under the action of oceanic turbulence, and the transport of particles is a mathematical simulation of a Wiener process.

Assume that the sampling step length is $\Delta t > 0$, and make

$$X_n = X(n\Delta t) \tag{4.24}$$

$$X_n = X_{n-1} + \sigma\sqrt{\Delta t}W_n (\sigma > 0) \tag{4.25}$$

where, W_n refers to mutually independent random numbers of $N(0,1)$.

Increment $X_n - X_{n-k}$ only depends on k variables $(W_{n-k+1}\ldots W_n)$ corresponding to $(n-k,n)(k < n)$, so $X_n - X_{n-k}$ has $N(0, \sigma\sqrt{k\Delta t})$ distribution.

Based on the mathematical process above, the transport distance of particles that represent oil particles can be expressed as below:

$$X_i = X_{i-1} + U \cdot \Delta t + L \tag{4.26}$$

where, $X = (x,y,z), U = (u,v,w)$ and $L = (l_x, l_y, l_z)$.

The horizontal transport speed of oil particles is u, and v includes the speed of the effects of advection, wind-driven surface current and wind on the oil film; the transport speed w of oil particles in vertical direction includes floatation speed and sedimentation speed of oil particles and vertical current velocity in ocean dynamics; turbulent diffusion terms l_x, l_y and l_z can be obtained from Formula 4.22 or 4.23; and turbulent diffusion coefficient is corresponding to the 3D fluid field. Therefore, the transport position of any oil particle can be calculated as per Formula 4.22, and a large number of oil particles can express the behavior process of the oil spill. The behavior process of the oil spill can be described through calculating the positions of a large number of particles.

In summary, in order to study the spreading process of oil spills, the impact of marine environmental dynamic factors must be considered. At the initial stage, the oil spilling into water first experiences the gravity spreading stage, at which, the current and wind have little influence on the spreading of the oil film; as gravity spreading decreases and oil spill thickness reduces, the oil film transits to the shear spreading stage, which is significantly affected by current and wind. The spreading of the oil film is mainly driven by viscous forces between oil film bottom and water and between the top of oil film and wind, and changes in current field and wind field play an important role; when there is no wind, the cyclical change of tide has a decisive influence on the spreading of oil film, and the oil film is often compressed at high tide, but stretched at low tide; when it is windy, the spreading of oil film is affected by current and wind. When both of them are in the same direction, the spreading of oil film slows down, but the drift speed increases; when they are in opposite directions, the spreading of oil film speeds up, but the drift speed decreases. With the weakening of shear spreading and the decrease of oil film thickness, the oil film finally transits to the stage of drift with the flow.

4.1.2   Oil Spill Drift Model

The drift of the oil film at sea is mainly driven by sea surface flow, and its upper boundary is directly affected by wind stress. Drift plays the most important role in the dynamic simulation of oil spills, and the closer it is to the coast, the more important its prediction is. The drift movement depends on the advection condition, and the shear stress of the oil-water interface is closely related to the movement of seawater. The movement of a water body micelle on the sea surface is composed of three parts, including wind-driven current, non-wind-driven current and wave residual-current, without regard to the interaction between them [94]. Wind-driven current and non-wind-driven current form seawater movement under driving forces far greater than the oil film scale, such as tidal force, pressure gradient force and wind stress, and they do not change greatly due to the presence of oil film. The wave residual-current is different. According to Stokes theory, the magnitude of the wave residual-current can reach 2% of the wind speed. However, the oil film will make the surface tension increase and the sea surface tends to be smooth, greatly weakening the non-linear effects of waves. Therefore, in fact, the influence of wave residual current can be ignored [95]. Thus, it can be seen that the essence of the oil film advection is the Lagrange drift process of the oil film under the action of the above driving forces, and it mainly depends on sea surface wind field and current field. The current field can be considered as a vector field composed of density current, wind-driven current, tidal current, gradient current, etc. In offshore waters, tidal and wind-driven currents are the most important factors that determine oil spill drift. Actual observation shows that the drift speed of an oil spill is mainly determined by wind in open waters. In offshore waters or coastal areas, the role of tidal current cannot be neglected; especially in harbors or docks, the role of tidal current is more important. In the model, the influence of wind on oil film is generally calculated as 3%–4% of wind speed, and the influence of ocean current on oil film is calculated as the flow velocity. If the sea area where the oil spill occurred is not greatly affected by the coastal current and abyssal current, better results can be obtained with the above calculation method. To sum up, in order to calculate the oil film drift trajectory, the data of the sea surface current field and wind field should be obtained first.

4.1.2.1   Marine environmental dynamic factors affecting oil spill drift

1. Sea surface wind field

The sea surface wind field plays an important role in oil spill drift, which is shown in the following aspects: (i) The wind-driven current is one of the most important factors affecting the drift of oil film; (ii) the wind itself can act directly on the oil film, driving the movement of the oil film, and also can change the spreading shape and area of the oil film. In addition, the strength of the wind can also affect the weathering process of the oil spill.

The offshore wind field in China is taken as an example. In winter, under the influence of continental high pressure and Aleutian low pressure, the wind in the Bohai Sea and Yellow Sea areas is mostly strong northerly wind, with an average wind speed of 6–7 m/s. The surface of the South Yellow Sea is wide and the average wind speed increases to 8-9 m/s. With the strong northerly wind, there is often cold air or cold waves southward, and the wind speed can reach more than 24.5 m/s. Along the coasts of the Bohai Sea and the northern Yellow Sea, the temperature can drop as much as 10 to 15 degrees Celsius, with occasional heavy snows, making it the main disastrous weather in winter. The cold wave can sometimes trigger a storm surge. As spring comes, monsoons alternate and the southerly wind increases. From June to August, the southerly wind is prevailing, with an average wind speed of 4–6 m/s. However, in case of a cyclone or a northward typhoon, the wind speed can be increased to more than 24.5–28.4 m/s, which is the main disastrous weather in summer. Typhoon No. 3 in 1972 and Typhoon No. 9 in 1985 crossed the Yellow Sea and reached the Bohai Sea, causing severe wind and tidal disasters. The strong wind belts of the Bohai Sea and Yellow Seas are located in Liaodong Gulf, Bohai Strait and the region of Chengshan Cape of Shandong Peninsula as well as in the central and southern areas of the open waters of the South Yellow Sea. The number of days of strong winds (gusts greater than or equal to 8) in the central and western areas of the Bohai Sea averages about 80 days a year, with 110 days in the Bohai Strait and 110 days in the central and south areas of the Yellow Sea.

The East China Sea extends across temperate and subtropical zones. In winter, it is also affected by high pressure from the Asian continent. The wind is mainly strong northerly wind, with an average wind speed of 9–10 m/s. The southern sea area is dominated by northeast wind. Especially in the Taiwan Strait, the wind direction is relatively stable and the wind speed is also high. In winter, when the cold wave goes southward, after the cold front passes, there are often $6 - 8$ northerly winds, accompanied by significant cooling. In winter and spring, the extratropical cyclones formed on the sea surface east and north of Taiwan also have great influence on the East China Sea, and the sudden occurrence of strong northerly winds often causes damage to navigation and fishing operations. In summer, the wind is mainly southerly, with an average wind speed of only 5–6 m/s. However, during this period, tropical cyclones often travel north through the East China Sea. According to statistics, between 1949 and 1969, 154 tropical cyclones passed through the East China Sea, accounting for about a quarter of the total number of tropical cyclones in China's offshore waters. On average, there are also 5 to 6 typhoons and violent typhoons passing through the East China Sea each year, and there are as many as 14 typhoons in some years. Typhoons are most frequent between June and September, and there are also some occasional typhoons in spring and autumn.

The strong wind belts of the East China Sea are located along the coast of Zhejiang, the Zhoushan Islands and the Taiwan Strait. The number of days of strong winds is 120 − 140 days in the northwestern part of the East China Sea, 100 − 120 days in the Taiwan Strait, and only 10–40 days near the Ryukyu Islands.

The South China Sea, located in the tropics, is a typical monsoon climate zone. Around September each year, the northeast monsoon reaches the Taiwan Strait. From November to April, the whole sea area is dominated by the northeast monsoon. In April, the southwest monsoon begins to appear in the Strait of Malacca, which can cover the whole sea area by June, while the peak period is from July to August.

In most areas of the South China Sea, the northeast monsoon is the largest in November, with force 4–5 and sometimes 6–7. The windy areas are in the northern part of the South China Sea, the Bashi Channel and the waters west of the Nansha Islands. Relatively speaking, the southwest monsoon is generally weaker, mostly lower than force 4. However, in the Yingge Sea on the western coast of Hainan Province, the spring wind is relatively strong throughout the year. The average monthly wind speed in April is 5.5 m/s, and the minimum wind speed is in December, just 3.4 m/s.

The average annual number of windy days is the lowest in the South China Sea, 50 days in the offshore waters of Vietnam, about 40 days near the Xisha Islands and less than 40 days near the Nansha Islands. In the east coastal areas of Guangdong near the Taiwan Strait, the number of days with strong winds is as many as 100 days.

Typhoon is the main disastrous weather system in the South China Sea. An average of about 10 typhoons and strong tropical storms occur in the South China Sea every year [96].

To sum up, it can be seen that the sea surface wind field plays an important role among all marine environmental dynamic factors. Therefore, for the simulation of oil spill drift process, the wind field compilation should be carried out first. Usually, the following models and methods are used to compile the wind field:

(1) Regular wind field model

The model is applicable to oil spill process with small time scale and space scale. During calculation, the combined conditions, such as the average value of wind speed and wind direction, the prevailing wind direction and maximum wind speed, the prevailing wind direction and the average wind speed, the prevailing wind direction and the maximum wind speed in this direction, the prevailing wind direction and the average wind speed in this direction, the maximum wind speed and the corresponding wind direction, and the typical wind process, can be selected as the regular wind field conditions.

(2) Random-walk wind field model

In this model, the wind element is regarded as an unsteady and independent random function, that is, the wind speed and direction are regarded as the superposition of average values and random values, respectively. This wind field model is the simplest form, and the main disadvantages are as follows: (i) The cyclical variation of wind elements is not considered; (ii) the autocorrelation and cross-correlation of wind elements are not considered.

(3) Markov wind field model

Markov process is a stochastic process to study the state of an event and the theory of transfer rule between states. It studies the change trend of the state at time $(t_0 + \Delta t)$ through the initial probability of the event at time t_0 and the transition probability relation between the states. The parameters of Markov wind field model can be used to generate the probability distribution from the process from state $t - 1$ to state t by transferring the probability matrix. The "memory" of the first-order Markov wind field model is limited to one-step correlation, while the long-term "memory" is ignored. Such problems are generally dealt with through establishing a higher-order regression model.

(4) Wind field retrieval from satellite scatterometer data

With the development of remote sensing technology, the technology of using satellites to observe sea surface wind has become mature. In the world, the data of sea surface wind field provided by satellite scatterometers has become the main data source of ocean, meteorology, climate research and forecast.

In the existing models, the relationship between sea surface wind field and oil film drift is mostly represented by the transfer matrix of wind field, and the influence of wind stress on the drift velocity of oil film is represented by wind power factor:

$$\vec{U} = \alpha D \vec{W}, \tag{4.27}$$

where, $\vec{U}$ refers to the drift speed of the oil film under wind action (m/s); $\vec{W}$ refers to the wind speed 10 meters above the sea surface (m/s); α refers to the wind power factor, which should be calibrated according to different situations, and is usually $3\% \sim 4\%$; and D refers to the wind field transfer matrix $\left(\begin{bmatrix} \cos\theta & \sin\theta \\ -\sin\theta & \cos\theta \end{bmatrix} \right)$.

2. Wind-driven current

Wind-driven currents are the movement of seawater caused by wind at sea, and is one of the most important factors that influence the oil spill drift process. Among wind-driven currents, when the wind acts on the sea surface at a steady and constant speed for a long time, the steady and constant seawater current generated is drift; when the unevenly distributed steady wind acts on the sea surface, the volume transportation is uneven, resulting in sea surface fluctuation, thus generating pressure gradient force. Such wind-driven currents include drift and slope currents [97].

In the study of drift current, surface seawater is divided into a number of thin layers. In the northern hemisphere, when the first layer of seawater starts to move under the action of the wind, it moves to the right side of the wind direction (to the left side of the wind direction in the southern hemisphere) due to the effect of Coriolis Force. At the same time, momentum is transmitted to the second layer through friction, thus driving the second layer to move together. The second layer of seawater moves on the right side of the movement direction of the first layer. The momentum is transmitted downwards in such a layer and is reduced gradually, and the flow direction is also skewed to the right. Therefore, the flow velocity decreases with the increase of depth. When reaching the friction depth, the flow velocity is only 4.3% of the surface flow velocity, and may be ignored. At this time, the flow direction is also opposite to the surface flow direction. According to Ekman's drift theory, Ekman drift can occur in sea areas with different depths, so it can be divided into drift in infinite deep sea and drift in limited deep sea. The seabed has no effect on drift in infinite deep sea, but for drift in limited deep sea, the friction effect of the seabed must be considered. The actual depth of the ocean is limited, and especially in shallow seas, the friction effect of the seabed is very important. Thus, in the northern hemisphere, the smaller the depth of the ocean, the smaller the angle of rightward deflection of the flow velocity as the depth increases. In quite shallow oceans, drift almost always follows the wind from the sea surface.

3. Tidal current

Tidal currents are the most common hydrological phenomenon in estuarine coastal areas. Sediment, salt, various pollutants including oil spills and heat transport processes are closely related to tidal currents. For oil films on the sea surface, in addition to wind-driven currents, the influence of tidal currents on the drift trajectory is also quite important.

The distribution of tidal currents corresponds to the propagation of tidal waves. The flow velocity is closely related to the tidal amplitude. Generally, in coastal areas, the tidal current is strong, while in the open sea, the tidal current is weak. According to the forms of motions, tidal currents can be divided into reciprocating tidal currents and eddying tidal currents. Reciprocating tidal currents are mostly formed in straits, waterways, estuaries or narrow harbors. Due to the limitation of topographic conditions, seawater changes periodically in both positive and negative directions. Eddying tidal currents mainly appear in relatively open waters. Due to the effect of Coriolis force, the tidal currents have the characteristic of rotation, that is, in a half tidal cycle, both the flow velocity and direction will change. The semi-diurnal tidal current rotates two cycles within a lunar day, while the diurnal tidal current rotates one cycle within a lunar day. The irregular semi-diurnal tidal current usually rotates two cycles every day, but the flow velocity varies significantly. The irregular semi-diurnal tidal current sometimes rotates one cycle at the flow velocity with a large vector, and sometimes rotates for a limited angle at the flow velocity with a small vector. The irregular diurnal tidal current is similar to the irregular semi-diurnal tidal current.

The tidal currents in China's offshore waters are taken as an example. In the Bohai Sea and the Yellow Sea, tidal currents are more important because the water is shallow and the flow is weak. In most areas of the Bohai Sea, the tidal currents are mostly irregular semi-diurnal tidal currents. The maximum flow velocity of semi-diurnal tidal currents and components currents of diurnal tidal currents is 30–40 *cm/s* and 10–20 *cm/s*, respectively. In the Bohai Sea, the flow velocity of tidal current is generally within 2 knots, but in the Bohai Strait and at the head of the Liaodong Bay, the tidal currents are very strong, up to about 5 knots. Because the diurnal tidal current near the Bohai Strait is larger than the semi-diurnal tidal current, it also has the feature of the diurnal tidal current.

Among China's offshore waters, the Yellow Sea has the strongest tidal current. The types of tidal currents are approximately divided by $124°E$, and on the east are regular semi-diurnal tidal currents and on the west are irregular semi-diurnal tidal currents. On the west of the Chengshan Cape-Dalian line, the nodes of the waves of diurnal tidal currents and the antinodes of the waves of semi-diurnal tidal currents are located near Yantai, where the tidal currents are irregular diurnal tidal currents. The tidal currents northeast of Wutiaosha in north Jiangsu are also semi-diurnal tidal currents. The flow velocity of tidal currents in the Yellow Sea is the strongest in the east, the second in the west and the weakest in the center. The maximum flow velocity appears at the head of the bay along the Korean Peninsula. For example, outside Incheon Port, the flow velocity of tidal currents is $2 - 3$ knots, and can even reach 10 knots in some waterways. The area from Chengshan Cape to Changshanchuan is another strong tidal zone in the Yellow Sea. The flow velocity reaches 2 knots near Chengshan Cape, and 5 knots near Changshanchuan. In the west of the Yellow Sea, the flow velocity of tidal currents is generally about 2 knots, and the tidal currents in the center of the Yellow Sea are weak, about 1 knot. In most areas of the Yellow Sea, tidal currents are rotating currents, and the long axis is consistent with the trend of the coastline or the isobath, especially in offshore areas.

The tidal currents in the East China Sea are weaker than those in the Yellow Sea, and the type distribution is relatively simple. Except for regular semi-diurnal tidal currents in the west coastal areas of Zhejiang and Fujian, the other tidal currents are irregular semi-diurnal tidal currents. In offshore areas, tidal currents are mostly reciprocating tidal currents, and in open seas, tidal currents are mostly eddying tidal currents. However, tidal currents in the Yangtze River Estuary and near Sheshan Mountain are mostly eddying tidal currents. Along the coast of Zhejiang and Fujian, the rising tidal current flows to the north and the falling tidal current flows to the south. Especially in the Hangzhou Bay, tidal currents are swift, and the maximum flow velocity reaches 8 knots. The Ryukyu Islands have complex tidal currents. The maximum flow velocity of tidal currents near Amami-Oshima Island reaches 3 knots, and that near Kume-jima Island reaches 3.5 knots. The tidal currents along the west coast of Kyushu are strong, and the flow velocity in some waterways reach 7 knots. In the Taiwan Strait, most of the tidal currents are semi-diurnal tidal currents. However, the tidal currents in the Houlong area along the coast of the strait are irregular semi-diurnal tidal current, and those at the southeast end of Taiwan are irregular diurnal tidal currents. In this strait, the flow velocity of tidal currents is $1 - 2$ knots, and the maximum velocity is 3 knots. The flow velocity in Penghu Channel reaches more than 3 knots. The directions of tidal currents in the south and north of the Taiwan Strait are opposite. In the north, the rising tidal current flows to the south and the falling tidal current flows to the north; in the south, the rising tidal current flows to the north and the falling tidal current flows to the south or southwest. The central part of the strait is a convergence and divergence zone.

Tidal currents in the South China Sea are weak. In most areas, the flow velocity is lower than 1 knot, and only the tidal currents along the continent are stronger. The flow velocity of heavy tidal currents in the Beibu Gulf is not higher than 2 knots, and only the tidal currents in the Qiongzhou Strait are strong, with the flow velocity of up to about 5 knots. Except for irregular semi-diurnal tidal currents along Guangdong, tidal currents in most areas are diurnal tidal currents. In contrast to the Bohai Sea and the Yellow Sea, the maximum flow velocity of component currents of diurnal tidal currents in the South China Sea is much larger than that of component currents of semi-diurnal tidal currents. In the east and south of the Beibu Gulf, to the east of the Leizhou Peninsula, in the north of the Gulf of Siam and at the south end of the Malay Peninsula, strong diurnal tidal currents can be observed. Among them, the maximum flow velocity of component currents of diurnal tidal currents in the east and south of the Beibu Gulf is 30–70 cm/s, 30–40 cm/s in the north of the Gulf of Siam, and 30–70 cm/s east of the Leizhou Peninsula. The semi-diurnal tidal currents are only larger east of the Leizhou Peninsula, near the Mekong River Estuary and near Dashi Bay, and the flow velocity is about 20–30 cm/s. Tidal currents in other areas are weak [96].

The importance of the tidal current lies in the fact that it makes the flow of seawater have obvious three-dimensional spatial characteristics. Therefore, it is necessary to establish a three-dimensional numerical model for tidal currents in shallow seas, in order to accurately predict the drift trajectory of an oil spill. In the mid-1970s, Csanady [98] and Nihoul and Ronday [99] respectively carried out the tidal average of the depth-averaged two-dimensional equations of motion. It was found that in a shallow sea, the astronomical tide motion could not be completely eliminated by the average, and its own nonlinear coupling effect may generate the so-called "tidal stress." Heaps [100] gave a systematic derivation and discussion of this system of equations for the full-flow model of depth integration. Feng Shizuo [101] applied this viewpoint to the equation of three-dimensional spatial motion, and found that there is not only the "tidal stress" of the sea surface, but also the "surface tide source" and "tidal-induced physical force." Therefore, it can be determined that due to the non-linearity of the shallow sea system, the impressed force of the circulation not only has surface wind stress and thermal salt factors, and the non-linear coupling effect caused by tides should also be considered. In China, there have been many achievements in the study of three-dimensional numerical model of tidal currents in shallow seas. For example, Dong Wenjun [102] proposed a step-by-step three-dimensional model of tidal currents and suspended sediment diffusion in coastal areas of estuaries. Xu Jindian and Jiang Yuwu [103] established a tidal current model in three-dimensional σ coordinates and combined it with the difference method of pollutant diffusion equation to build a shallow-sea pollutant diffusion model in three-dimensional σ coordinates. Bao Xianwen et al. [104] constructed a complete system of three-dimensional tidal wave equations with definite solutions by using the tidal current control equation and turbulence closure equation, established a three-dimensional numerical model for tidal currents in the Qinzhou Bay, and simulated the tidal level and tidal current variation of Qinzhou Bay.

In general, under the conditions of incompressible hydrostatic equilibrium and Boussinesq approximation, the system of three-dimensional non-linear control equations of tidal waves in shallow seas has the following expression in the Cartesian coordinate system:

Continuity equation:

$$\frac{\partial u}{\partial x} + \frac{\partial v}{\partial y} + \frac{\partial w}{\partial z} = 0 \tag{4.28}$$

System of momentum equations:

$$
\begin{aligned}
\frac{\partial u}{\partial t} + u\frac{\partial u}{\partial x} + v\frac{\partial u}{\partial y} + w\frac{\partial u}{\partial z} + fv = {} & -\frac{1}{\rho}\frac{\partial p}{\partial x} + \frac{\partial}{\partial x}\left(A_x\frac{\partial u}{\partial x}\right) \\
& + \frac{\partial}{\partial y}\left(A_y\frac{\partial u}{\partial y}\right) + \frac{\partial}{\partial z}\left(A_z\frac{\partial u}{\partial z}\right)
\end{aligned}
\tag{4.29}
$$

$$
\begin{aligned}
\frac{\partial v}{\partial t} + u\frac{\partial v}{\partial x} + v\frac{\partial v}{\partial y} + w\frac{\partial v}{\partial z} + fu = {} & -\frac{1}{\rho}\frac{\partial p}{\partial y} + \frac{\partial}{\partial x}\left(A_x\frac{\partial v}{\partial x}\right) \\
& + \frac{\partial}{\partial y}\left(A_y\frac{\partial v}{\partial y}\right) + \frac{\partial}{\partial z}\left(A_z\frac{\partial v}{\partial z}\right)
\end{aligned}
\tag{4.30}
$$

$$
\begin{aligned}
\frac{\partial w}{\partial t} + u\frac{\partial w}{\partial x} + v\frac{\partial w}{\partial y} + w\frac{\partial w}{\partial z} - fv = {} & -\frac{1}{\rho}\frac{\partial p}{\partial z} + \frac{\partial}{\partial x}\left(A_x\frac{\partial w}{\partial x}\right) \\
& + \frac{\partial}{\partial y}\left(A_y\frac{\partial w}{\partial y}\right) + \frac{\partial}{\partial z}\left(A_z\frac{\partial w}{\partial z}\right) - g
\end{aligned}
\tag{4.31}
$$

In the system of equations, t refers to time; x, y and z refer to the rectangular coordinate system of xoy plane, placed above the undisturbed calm sea surface, with z axis vertically upward; u, v and w refer to the components of flow velocity along x, y and z axes, respectively; ρ refers to the density of seawater, and is a constant; p refers to the pressure intensity of seawater; A_z refers to the vertical eddy viscosity coefficient; A_x and A_y refer to the horizontal eddy viscosity coefficients in x and y directions, respectively; g refers to the acceleration of gravity, and f is the Coriolis parameter.

4. Waves

Waves are one of the most important forms of seawater movement. Among waves, the height of wind waves, surge waves and nearshore waves can range from a few centimeters to more than 20 meters, with a maximum of more than 30 meters. Wind waves are fluctuations of seawater caused by the action of wind. Many waves of different heights and lengths can occur at the same time. The wave surface is steep and the wavelength is short. There are often spindrift or foams near the peaks, and the propagation direction is consistent with the wind direction. In general, the longer the wind with the same state acts on the sea surface and the larger the sea area, the stronger the wind and wave are; when the wind wave is fully developed, it will not continue to develop. The waves formed when the wind waves leave the windswept area are called "surge waves." According to the wave height, the wind waves are usually divided into 10 levels, and the surge waves are divided into 5 levels. Level-0 means that there is no wave or surge and the sea surface is smooth and mirror-like; level-5 larger wave and level-6 high wave correspond to level-4 larger surge, with the wave height of 2–6 m; level-7 violent wave, level-8 violent storm and level-9 furious storm correspond to level-5 high surge, with the wave height from 6.1 m to more than 10 m.

The influence of waves on oil spills is mainly manifested in the following aspects: (i) Wave disturbance will affect the process of crushing, dispersion and emulsification of oil spills (affecting the rate of oil droplets entering water). (ii) The drift process of oil spills will be affected by wave residual-currents generated by non-linear waves such as wind waves and surge waves. Because the second item is related to the drift process of oil spills, it will be discussed in detail. In general, the wave residual-current generated by non-linear waves, such as wind waves and surge waves, is calculated with the second-order Stokes formula for wave residual-currents:

$$U_{wave} = \frac{\omega k a^2 cosh\left[2k(H-z)\right]}{2sinh^2(KH)} \qquad (4.32)$$

In the formula above, ω refers to Pi; k refers to the number of waves; H refers to the depth of water; z refers to the depth of water particle; $a = 0.5H_s$, refers to the significant wave height. It is assumed that the direction of the Stokes wave residual-current is consistent with the wind direction. As mentioned above, although the velocity of the wave residual-current can reach 2% of the wind speed, the presence of the oil film increases the surface tension, and the sea surface tends to be smooth, greatly reducing the non-linear effects of wind waves and surge waves. Accordingly, wave residual-currents are also greatly reduced [95]. Therefore, the effect of wave residual-currents on oil film drift is generally negligible in model calculation.

4.1.2.2 Oil spill drift model

The numerical prediction of marine oil spill drift trajectory began in the 1960s. Many scholars proposed various drift models, most of which combined the spreading and diffusion processes of oil spills. The following is a summary of several representative models:

1. Coast Guard (II) model

The model was developed by Stolzenbach et al. [105] using a hydrodynamics model and a wind model to jointly solve the water flow velocity and predict the oil film movement trajectory. The wind model is established by the balance of pressure gradient force of wind stress and Coriolis force, and considers the influence of wind stress friction near the water surface. The hydrodynamic model considers the vertical average two-dimensional model of discrete pollutants. The wind speed calculated by the wind model is input to the hydrodynamic model to predict the movement trajectory of the oil film [90]. The model emphasizes the dynamic action of wind.

2. SEADOCK model

This model was established by Williams and Hann [106]. The model coordinates the wind velocity vectors in the offshore and open seas through the weight coefficient, and then superimposes them with the surface current vector to predict the drift movement of the oil film under the action of wind and ocean current and obtain the displacement of the center of mass of the oil film. When calculating the oil film drift with this mode, it is considered that the entire oil film is distributed evenly in a circular shape around the center of mass. For each calculation step, the spreading diameter calculated by Fay spreading formula is used as the expansion range of oil film at this time, and the first-order attenuation caused by evaporation and dissolution is taken into account. The amount of sediment is considered as 1% of the volume of oil layer, and the wind field is compiled with Markov model, but the impact of Coriolis force on wind and current is not considered [88].

3. Navy model

The US Navy model established by Webb et al. [107] is mainly used to predict the drift movement of oil tankers and oil spills. The model considers the effects of tidal currents, river inflows, geostrophic currents and wind-driven currents, and establishes the displacement formula for the center of mass of the oil film [90]. One disadvantage of this model is that the self-spreading of the oil film is not considered.

In China, Huang Lixian et al. [89] established a diffusion model through synthesizing the ocean current velocity vector, wind speed vector and velocity vector of oil film spreading in the experiment, and obtained the drift velocity of any point on the edge of the oil film in different directions at different times. Besides, the displacement of the corresponding point at this moment was calculated, and the effect of predicting the drift trajectory of the oil film was achieved. Wu Zhouhu and Zhao Wenqian [108] established an instantaneous oil spill model in which the distribution of the oil film around the center of mass is approximately elliptical, and the long axis coincides with the drift direction at each moment. The movement trajectory of the center of mass of the oil film was obtained firstly with the Euler-Lagrange method, and then the spreading scale of the oil film was obtained. In addition, the irregular shape and development process of the oil film can be described by calculating the trajectory of the boundary particles in the drift process.

4.1.3   Oil Particle Model

In addition to its own surface expansion, the movement of the oil spill on the sea surface is mainly subject to advection and turbulent flow in the ocean. The spreading of the oil film caused by its own gravity and surface tension generally lasts for a short time, while the movement with a longer duration mainly includes advection transport and turbulent diffusion [95].

The oil particle model is a model based on the Lagrange coordinate system. The concept of "particle diffusion" is combined with the Lagrange method to simulate the spatiotemporal behavior of oil spills in the marine environment. The oil spill is regarded as numerous small oil droplets to simulate the transport and diffusion processes of the oil spill in seawater, breaking through the traditional method of solving the convection-diffusion equation to simulate the oil spill. This method can describe the influence of various ocean dynamic factors on the oil spill and avoid the pseudo-numerical diffusion problem caused by the advection-diffusion model [94, 95]. In fact, this method is to track the process of oil particle micelles diffusing and moving along with the advection and turbulent flow of the surrounding water. Because the Lagrange method, which is a deterministic method, is used to simulate the advection process, and the random walk method, which is a stochastic method, is used to simulate the diffusion process, this method is also the combination of a deterministic method and a random method.

4.1.3.1   Concept of the oil particle

Johansen [216], who first proposed the oil particle model, elaborated the concept of the oil particle: an oil film in water is considered as a combination of numerous discrete oil particles, "particle diffusion" is to simulate the concentration field as a "cloud cluster" composed of a large number of particles, each particle represents a certain number of tracers. These small oil particles will move and disperse with the flow of water under the influence of surface flow. The advection process of model particles has Lagrange properties, which can be simulated by Lagrange method. The turbulent diffusion process caused by shear flow and turbulent flow is a random movement, which can be simulated by random walk method, that is, the turbulent flow is regarded as a random current field. The movement of each model particle in the turbulent current field is similar to the Brownian movement of the fluid molecule, and the random movement of each particle causes the diffusion of the entire particle "cloud cluster" in water [82, 95, 109]. The particles on the surface appear as oil films, and those in the water appear as oil droplets. The size of the surface oil film is the superposition of surface particles. The drift and weathering of oil spills on water surface can be simulated by algorithms of surface diffusion, translation, transport, emulsification and evaporation. Therefore, the movement of the oil film depends on the advection and turbulent diffusion of the surrounding water.

Oil particles refer to small spheres with a diameter between 10 and 1000 microns. Considering the diameter range of oil droplets, the actual number of particles required to accurately represent an oil spill film should be quite large. Therefore, it is difficult to calculate too many particle characteristic parameters synchronously in the computer. The maximum possible number of particles should be determined according to the capacity and running time of the computer, and the method of additional volume parameter is used to simulate the properties of oil particles [95]. The volume parameter of some oil particles is defined as:

$$V_i = \frac{\pi}{6}(d_i)^3 \tag{4.33}$$

where, V_i refers to the volume of the i-th oil particle, and d_i is its diameter. The percentage f of this oil particle in the total volume of the oil film is:

$$f_i = \frac{\frac{\pi}{6}(d_i)^3}{\sum_{j=1}^{n}\frac{\pi}{6}(d_j)^3} \tag{4.34}$$

where, n refers to the total number of oil particles. The characteristic volume of each oil particle is: $V_i = f_i \cdot V_0$, where, V_0 refers to the initial volume of the oil spill. Thus, each oil particle represents part of the volume of the oil spill.

4.1.3.2 Oil particle tracking model

The Lagrange particle tracking model can be used to describe the migration trajectory of oil spills. Since oil particles are micelles with a certain volume and a certain mass, in combination with the spreading, drift, turbulent diffusion, evaporation and dissolution of oil spills at sea, the movement trajectory and fate of oil spills can be calculated. In order to simulate the trajectory of oil spill, the oil spill is separated into a large number of oil particles, and each oil particle represents a certain amount of oil, which drifts and diffuses under the combined action of surface water flow and wind [82].

The movement process of oil particles within a period of time is divided into: spreading process, advection process and diffusion process. The deterministic method is used to simulate the advection process of oil particles, and the random walk method is used to simulate the turbulent diffusion process. The spreading process is still calculated with the Fay three-stage spreading formula.

Lagrange particle tracking trajectory is expressed in the following formula:

$$\frac{d\vec{r}}{dt} = \vec{V}(\vec{r}, t) + \vec{V}'(\vec{r}, t) \tag{4.35}$$

In the formula above, $\vec{r} = x\vec{i} + y\vec{j}$ (m); $V(\vec{r}, t)$ refers to the translational velocity (m/s); $V'(\vec{r}, t)$ refers to the horizontal turbulent diffusion velocity of the oil film (m/s); they are expressed in the following formula:

$$\vec{V}(\vec{r}, t) = \alpha D\vec{W} + \beta \vec{U}_W \tag{4.36}$$

The formula can be related to Formula 4.27. In the formula, $\vec{W}$ refers to the wind speed 10 meters above the sea surface (m/s); $\vec{U}_W$ refers to the surface flow velocity (m/s), and can be obtained with a two-dimensional or three-dimensional kinetic equation; α refers to the wind force factor, and is usually $3\% \sim 4\%$; β is the influence factor that makes the oil film drift on the water surface due to the action of water flow, and the value usually is 1.1; D refers to the transfer matrix in consideration of the wind deflection angle:

$$D = \begin{pmatrix} \cos\theta & \sin\theta \\ -\sin\theta & \cos\theta \end{pmatrix} \tag{4.37}$$

When $0 \leq \vec{W} \leq 25m/s$, $\theta = 40° - 8\sqrt{\vec{W}}$; when $\vec{W} \geq 25m/s$; $\theta = 0$.

There are different expressions for the study of horizontal turbulent diffusion velocity of the oil film, including the following expression [110]:

$$\vec{V}'(\vec{r}, t) = V' R_n e^{i\theta'} \tag{4.38}$$

where, $V' = (4E_r/\delta t)^{\frac{1}{2}}$ (m/s); $E_r = 0.4 n_b V_g^{\frac{1}{2}} h^{\frac{5}{6}}$ refers to the turbulent diffusion rate; h refers to the depth of water (m); δt refers to the time step (s); R_n is a random number between 0 and 1; it is assumed that θ' is a random number uniformly distributed between 0 and π. The random walk method for the turbulent diffusion velocity can be described in various forms.

From Formula 4.35, the coordinate of some oil particle can be obtained as below:

$$\begin{aligned} X &= X_0 + (V_x + V_y')\Delta t \\ Y &= Y_0 + (V_y + V_y')\Delta t \end{aligned} \tag{4.39}$$

where, X_0 and Y_0 refers to the initial coordinates of some particle.

In the spreading process in this model, the spreading speed of the oil film calculated with Fay's three-stage spreading formula is still adopted. According to a large number of field observations, Fay proposed the relationship between oil spill area and volume at the end of diffusion:

$$A = 10^5 V^{\frac{3}{4}} \tag{4.40}$$

where, V refers to the oil spill volume (m^3).

According to Formula 4.40, it can be obtained that the thickness of the oil film at the end of spreading is $10^{-5}V^{\frac{1}{4}}$ m.

The spreading and drift trajectory of the oil spill and the affected area can be calculated by tracking the coordinates of oil particles. The affected area of oil spill pollution generally refers to the sea-sweeping area of the oil film, that is, the area of the sea surface that the oil film passes across in a certain period of time. The area of waters that all particles pass across in a certain period of time is taken for calculation.

4.1.4 Oil Spill Weathering Process and Model

4.1.4.1 Oil spill weathering process

The weathering process of oil spills involves multiple processes, such as evaporation, dissolution, emulsification, dispersion, adsorption and sedimentation, photooxidation and biodegradation. These processes are interrelated and occur through changes in the composition of the oil film. For example, photooxidation and biodegradation increase the content of asphaltenes and polar substances in crude oil, thereby affecting the dissolution and the emulsification processes [79]. Emulsification may increase the water content in oil, changes the properties of oil-water and oil-gas interfaces, and thus affects the dissolution and evaporation processes, and evaporation may promote the formation of chemical conditions required to form a stable emulsified mixture. It can be seen that oil spill weathering is a rather complicated process. Therefore, we must first understand the relationships between the composition of oil spills and the physical properties and behavioral properties, in order to carry out certain numerical simulations for the complex non-linear relationships in the weathering process.

The weathering process of oil spills has been studied by many domestic scholars. For example, Yang Qingxiao [111] studied the changes of surface tension, density and viscosity during oil evaporation in China, and established the corresponding weathering formula; Yan Zhiyu [79] simulated the evaporation of crude oil and its emulsified mixture with the shallow plate evaporation method, investigated the evaporation characteristics of different crude oils at different stages and under different emulsifying states, and obtained the evaporation rate equation and the inhibiting factor of emulsification on evaporation. In addition, through observing and analyzing the rules of oil and water movement, he analyzed the kinetic process of emulsification, and established the prediction model of oil spill weathering on the premise that composition and state are the main factors that determine weathering characteristics and the link between different weathering processes; Li Qiong [112] developed the software of the oil spill weathering prediction system through analyzing the weathering characteristics of marine oil spills, based on the existing weathering prediction model, and in combination with Norway SINTEF weathering model and U.S. ADIOS weathering model.

The main processes and related studies of oil spill weathering are described in detail below.

1. Evaporation of oil spills

Evaporation is the process of mass transfer of lighter components of hydrocarbons in crude oils from liquid to gas to the atmosphere, and is one of the main processes of oil spill weathering. Crude oil is mainly composed of hydrocarbons and is a complex mixture of various hydrocarbons; after spilling to the sea surface, most crude oils and their light refined products evaporate fast and the total evaporation is large. Most of the components with the carbon number less than 14 in the crude oil can be evaporated, and this characteristic determines that crude oil is a volatile substance. For light crude oil, diesel and gasoline, the evaporation capacity can reach 50%–90% of the total oil spill volume, while because the content of light components in heavy crude oil and heavy fuel oil is low, they evaporate slowly and the total evaporation is small.

The evaporation rate of oil spills is not only affected by factors such as components, wind speed, sea air temperature, sea conditions and solar radiation intensity, but also by the area of oil spill itself, oil film thickness, oil vapor pressure and mass conversion coefficient. Among them, the vapor pressure and mass conversion coefficient mainly depend on the wind speed. The environmental condition dominated by wind is the main control factor of the evaporation process. The higher the wind speed, the faster the evaporation [113].

In general, the evaporation rate of oil spills decreases with time and the oil spill evaporates quickly in the first few hours. Under normal environmental conditions, most crude oils and their light refined products can evaporate by 25% to 30% within 12 hours and 50% one day. The influence of temperature on the evaporation of oil spills involves the evaporation rate and the total evaporation ratio. The higher the temperature, the faster the oil evaporates. Sea conditions also have a certain impact on evaporation of oil spills. The worse the sea conditions are, the faster the oil evaporates. The oil film thickness also affects the evaporation rate of oil spills. For a certain amount of oil spill, the thinner the oil film is and the larger the area of the oil film exposed to the atmosphere, the faster the oil evaporates. However, the oil film thickness may not affect the total evaporation ratio.

While changing the total oil spill volume and affecting the composition of the oil spill, evaporation also changes the properties of oil, increases the density, viscosity and surface tension of the oil, and increases the pour point. In addition, evaporation also affects other weathering processes, such as diffusion, emulsification, dissolution, etc. [79]. Understanding the evaporation process helps to forecast oil spill residues, make emergency decisions and assess environmental damages. If the type and amount of the oil spill, oil film thickness, environmental temperature and wind speed are given, the evaporation of an oil spill at a certain time can be predicted accurately with the model, and the evaporation curve can be used to estimate the evaporation in the actual oil spill accident.

There are two main methods for calculating the evaporation rate of oil spills, namely the quasi-component method and the analytical method. In the quasi-component method, the oil is considered as a series of hydrocarbons with different molecular weight. The total evaporation is the sum of the evaporation of all components. This method is adopted in the evaporation rate equation proposed by Tkalin [114]. In the analytical method, the oil spill is considered as mixed media, and the empirical relationship between the evaporation rate of the media and the wind speed, vapor pressure, and the area of oil film is established. This method has been adopted in the model proposed by Blokker [80] for approximately simulating the evaporation process of the single-component oil, the oil evaporation model established by Mackay and Matsugu [115] on the basis of the water evaporation process and a large number of experiments, and the equation proposed by Drivas [116] later for calculating the total evaporation rate as per the evaporation rate of single component. At present, many prediction models of oil spill weathering have adopted the method developed by Stiver and Mackay [117], whose calculation formula is as follows:

$$\frac{dF}{dt} = \frac{K_E}{h} e^{6.3 - 10.3(C_1 - C_2 F)/T} \tag{4.41}$$

where, F refers to the volume fraction of evaporation; K_E refers to the mass migration coefficient; h refers to the thickness of the oil film (cm); T refers to ambient temperature ($^\circ C$); C_1 and C_2 refer to the distillation constant of oil. Later, many scholars have improved this formula.

Although the evaporation models have been generally formed, the basic research on oil spill evaporation is only an extension of the research on water evaporation. In the 1990s, through analysis based on shallow plate evaporation experiments, Fingas [118, 119] believed that pure liquids (such as water) and multi-component systems (such as oil) were fundamentally different in evaporation: On the one hand, the evaporation rate of pure liquid is a constant, and the petroleum containing multiple components is exponentially lost with time due to a large number of volatile components, so the amount of evaporation loss is logarithmically related to time; on the other hand, the main control process is different, and water evaporation is mainly controlled by the saturation of the water-gas boundary layer rather than the basic evaporation of water itself. Therefore, water evaporation is more sensitive to wind or turbulent flow, and oil evaporation is not strictly controlled by the boundary layer [120]. The information above means that the simplified evaporation equation will be sufficient to describe the evaporation process without considering the effects of wind speed, turbulent flow, area, thickness, and size. The only important factors are time and temperature. Fingas [118, 119] developed the evaporation equation directly from the calculation of distillation data, and obtained the following formula through combining it with temperature variation:

$$Percentage\ of\ evaporation = [0.165(\%D) + 0.045(T - 15)]\,lnt \qquad (4.42)$$

A few oil types, such as diesel, are conforming to the equation with t as the square root, and the expression is:

$$Percentage\ of\ evaporation = [0.254(\%D) + 0.01(T - 15)]\,\sqrt{t} \qquad (4.43)$$

where, $\%D$ refers to the mass fraction of distillation at $180°C$; t refers to time/min; and T refers to temperature $(°C)$.

According to the study of the evaporation process, most of the equations for calculating evaporation rate are empirical equations based on experiments.

2. Dissolution of oil spills

Dissolution is the process of oil spill forming uniform oil particles under the disturbance of certain energy and entering into seawater. The dissolution amount and dissolution rate depend on the composition and physical properties of crude oil, spreading degree of the oil film, water temperature and turbulivity of water, and emulsification and dispersion degree of oil. Wind speed and sea condition are the most important environmental factors influencing dissolution.

Dissolution is the shortest process after the occurrence of an oil spill. Fan Zhijie and Song Chunyin [121] pointed out that the maximum solubility of oil spills occurs 8-12 hours after the accident, and then the solubility drops exponentially, indicating that this process is affected by evaporation and water mass movement. Solubility is related to oil composition and type. The solubility of low-carbon crude oil hydrocarbons is high, while the solubility of other components in water is generally low. The solubility of lubricating oil and high-boiling fractions is very low. The solubility of alkanes in water varies with the ionic strength of water. The solubility of alkanes in seawater is lower than that in fresh water, and the solubility decreases with the increase of carbon number. For every two more carbon atoms in the same series of hydrocarbon compounds, the solubility decreases by an order of magnitude. The solubility increases with the amount of time that the oil spill floats at sea, because the photochemical process strengthens the dissolution of oil. However, this effect can also be delayed due to the formation of emulsified mixtures [78, 79].

In general, the dissolution amount is much smaller than the evaporation amount. Therefore, during the dynamic simulation of the behavior and fate of the oil spill, dissolution can be ignored in most cases because it has little effect on the mass balance calculation of the entire oil spill.

A number of scholars have conducted in-depth research on the dissolution process from different perspectives. Yang Qingxiao et al. [122] studied the solubility of several petroleum hydrocarbons in different salinities, temperatures, pH values and humic acids and the changes of their components through marine experiments in the waters around Dalian. They established the calculation formula to predict the oil concentration and total dissolution amount as per the vertical diffusion coefficient and dissolution amount of various hydrocarbons [78]. Bobra [123] studied the water solubility of several crude oil hydrocarbon mixtures and found that the dissolved components and concentrations are similar, but in the high-soluble components, the contents of benzene, toluene, ethylbenzene and xylene are high. However, in fuel oil, soluble components are composed of volatile and low-soluble substances such as trimethylbenzene, propylbenzene, naphthalene, alkylated naphthalene, etc. [79].

Although in general the dissolution amount of crude oil differs by two to three orders of magnitude from the evaporation amount, there is still significant competition between evaporation and dissolution in some cases, especially for aromatic hydrocarbons with biotoxicity.

3. Dispersion of oil spills

Dispersion is the phenomenon of small oil droplets entering the water body. Natural dispersion mainly has the following processes: (i) The process of the oil film breaking and forming oil particles under the action of wave; (ii) the process of oil particles entering the water body under the action of wave; (iii) the process of oil particles aggregating in the oil film. The surface solubilizer contained in the oil spill promotes the dispersion process, and the turbulent flow caused by the breaking wave plays the most important role in the dispersion process [113]. The surface tension of oil-water interface, the viscosity of oil and the density of oil spill are also important factors affecting dispersion. The greater the viscosity, the worse the dispersion ability. The smaller the difference between oil and water, the easier the formation of small oil particles and the higher the degree of dispersion.

The natural dispersion rate depends largely on oil properties and sea conditions, and the dispersion process progresses rapidly when breaking waves occur. Low-viscosity oils, such as gasoline and diesel, can be completely dispersed in a few days, while remaining fluid and unaffected by other weathering processes. On the contrary, high-viscosity oils or oils that can form stable water-in-oil emulsified mixtures tend to form thick oil layers that cannot be easily dispersed on the water surface. Such types of oils can remain on the water surface for several weeks, such as heavy crude oil and heavy fuel oil.

At present, the research on dispersion process of oil spills is still in the initial stage. Anderson [217] established an empirical formula for oil spill dispersion based on the wind speed theory, reflecting the disturbance energy required by oil particles to disperse into water. Based on previous research results, Reed [218] added an exponential decay function to the empirical equation to correct the impact of oil spill weathering and the formation of "chocolate mousse" on the model. In the same year, Mackay et al. [84] established a two-stage dispersion model in which the thin oil film and the thick oil film were treated differently, and the results were consistent with the experiment but not confirmed by other studies. Based on the formula proposed by Mackay et al. [84], Reed et al. [124] later proposed the following formula to calculate the dispersion process:

$$D = 0.11(U + 1)^2 \left(1 + 50\mu^{\frac{1}{2}}\delta S_t\right)^{-1} \tag{4.44}$$

where, D refers to the fraction of oil loss at sea per hour due to dispersion; δ refers to the thickness of the oil film (cm); S_t refers to oil-water interface tension ($dyne/cm$); U refers to the wind speed (m/s); and μ refers to the viscosity (cP). Later, Delvigne and Sweeney [125] established a theoretical model based on the data of a series of oil spill dispersion experiments in the laboratory and through analyzing the measured values of dispersion into water, and accordingly calculated the vertical dispersion coefficients of the oil spill and suspended matters to which the oil spill adhered to. In this model, the distribution of oil particles entering water due to the effect of crushing waves is considered to describe the dispersion behavior of oil, which lays a foundation for the development of three-dimensional oil spill model in the future.

4. Emulsification of oil spills

Emulsification of oil spills refers to the process in which crude oil and seawater are mixed together to form an emulsified mixture of oil and water. Emulsification generally begins a few hours after the oil spill occurs. When the oil spill just happens, the thick oil film cannot be damaged in general hydrodynamic conditions. When the oil film spreads to a certain extent, the energy of wind waves is enough to break the oil film, and the emulsified mixture begins to form. It can be said that this process is closely related to the thickness of oil film, the density and viscosity of the oil spill, the strength of wind wave, etc. The water-in-oil emulsified mixtures formed during emulsification appear as dark brown viscous foam-like substances after water droplets are dispersed into the oil. They can float on the sea surface for a long time and are wrapped with secretions and debris of marine organisms [79]. The water-in-oil emulsified mixture is easy to dissipate, and its stability depends on the content of asphaltenes. Oil with an asphaltene content greater than 0.5% is likely to form a stable emulsified mixture, known as "chocolate mousse," and if the asphaltene content is lower than 0.5%, the oil is easy to disperse. When the emulsified mixture of oil is in calm sea conditions or stranded on the shore, it will be separated into oil and water again due to the sun heat. The emulsification speed of oil depends on oil properties and sea conditions, and the water content of emulsified mixture depends only on the oil itself, generally up to $75 - 80\%$. Because of absorbing moisture, the oil often turns to brown, orange or yellow from black. With the development of emulsification, the movement of oil in waves makes the water droplets in the oil smaller and smaller, and the emulsified mixture becomes more viscous. As the water content increases, the density of the emulsified mixture becomes close to that of seawater.

Emulsification of oil spills also affects other weathering processes. After the emulsified mixture is formed, it not only increases the original oil volume (sometimes up to $3 - 4$ times), but also significantly increases the density and viscosity of oil. The further diffusion of the oil spill is hampered, and the evaporation is relatively reduced. Some emulsified mixtures enter deeper water bodies, accordingly changing the later behavior of the oil spill and its possible consequences.

The kinetic mechanism of emulsification has not been well solved. Related studies suggested that when chemical conditions are met and there are certain waves and other turbulent energy, the emulsified mixture will form rapidly. However, due to the lack of sufficient experience and theory, it is not possible to accurately determine the threshold of energy required to form an emulsified mixture. The first-order rate equation first proposed by Mackay et al. [126] is generally used to simulate the emulsification process of oil spills.

$$\Delta W = K_a \left(U + 1\right)^2 \left(1 - K_b W\right) \Delta t, \tag{4.45}$$

where, ΔW refers to water absorption rate; W refers to water content; K_a refers to the empirical constant of absorption rate; U refers to the wind speed (m/s); K_b is a constant approximate to 1.33; and t refers to time $(hr.)$. Some scholars have improved Mackay's equation. For example, Reed et al. [124] modified the constant in Mackay's equation, in order to be more consistent with field observations:

$$\frac{dF_{wc}}{dt} = 2 \times 10^{-5} \left(w + 1\right)^2 \left(1 - F_{wc}/C_3\right), \tag{4.46}$$

where, dF_{wc}/dt refers to water absorption rate; w refers to the wind speed (m/s); F_{wc} refers to the water content in oil; C_3 is a constant of the rate, and the value is 0.7 for crude oil and heavy fuel oil.

Studies have shown that when the total fraction of asphaltenes and resins exceeds 3% and BTEX (benzene, toluene, xylene and hexylbenzene) is less than this value, a semi-stable emulsified mixture will be formed, and when the asphaltene content is greater than 7%, a stable emulsified mixture will be formed. Because evaporation causes BTEX to migrate from the oil film to the atmosphere, the chemical conditions for the formation of the emulsified mixture can be predicted based on evaporation loss and the mass fractions of asphaltenes and resins. Some other studies have shown that the vertical distribution of the emulsified part of the oil in water depends on sea conditions, environmental conditions and oil properties, but the main factor is the role of breaking waves in the ocean.

Yang Qingxiao et al. [127] focused on the external factors of the formation of emulsified mixture and its distribution in the environment. The formation of emulsified mixtures under the action of breaking waves was studied through simulation experiments, and the results show that the oil film thickness is proportional to the amount of emulsification, the oil concentration after emulsification is exponential to the length of stay, the vertical distribution in water changes exponentially, and the change gradient increases with the increase of viscosity. Scholars at home and abroad believe that emulsification is the second most important feature after evaporation and obviously affects the behavior of marine oil spills. The evaporation rate of oil spills after emulsification is relatively reduced by several orders of magnitude; although the diffusion rate of oil spill rates remains at the same magnitude, the speed is also significantly reduced. As the position of oil droplets lowers in the water column, the degree of swaying of oil spills under the action of wind is reduced. In addition, the oil spill after emulsification is difficult to clean-up mechanically or burn.

5. Adsorption and sedimentation of oil spills

After the crude oil enters the water body, some oil particles will adsorb the sediment and other particles in water as well as plankton, microorganisms and bacteria. The oil particles attached to the carriers will move with the carriers, and some oil particles attached to sediment particles and other suspended particles will settle with them. The adsorption process depends on the properties of particulate matters and oil types, and is also affected by temperature and water flow. In addition, due to the influence of weathering processes such as evaporation and emulsification, the density of oil spill will increase continuously. When the density of oil spill exceeds the density of water, sedimentation will occur. Although many achievements have been made in the study of oil attachment to particulate matters, the kinetic process cannot be expressed quantitatively.

6. Photooxidation and biodegradation of oil spills

Photooxidation is the process of the oil spill, exposed to sunlight, undergoing free radical chain oxidation and producing some polar, water-soluble and oxidized hydrocarbons. The oxidation of oil spills in the marine environment is mainly controlled by sunlight and temperature, and the oxidation rate varies with the varieties of oil spills, the intensity of light exposure and the temperature of seawater. Generally, light oil has a high oxidation speed. Under the irradiation with ultraviolet light, the stronger the light source and the higher the temperature, the faster the oxidation. Due to the continuous formation of new substances, physical properties of oxidized oil spills also change, such as color, viscosity, specific gravity, surface tension, and so on. During oxidation, a lot of surfactant substances will be produced. Under the action of such substances and the disturbance of waves, the oxidized oil spill will eventually form what is called a "chocolate mousse." Although the concentration of photooxidation products is not high and the short-term effect is not obvious, the long-term effect of photooxidation is obvious, and it has great influence on the physical process of oil spills and increases the toxicity to organisms. Due to the complexity of weathering process and the limitations of testing methods, it is difficult to determine photooxidation products, and the research on their formation mechanism and their effect on dissolution and emulsification only remains on the qualitative description of speculation. Therefore, although people's understanding of the effects of photooxidation on oil has gradually deepened, it is unlikely to be numerically simulated and used in prediction models in the next few years.

Biodegradation is the most fundamental way for the marine environment to purify oil by itself. At present, more than 200 types of microorganisms that can degrade crude oil have been found, and they generally live on the sea surface and the seabed. The degradation of oil by microorganisms allowed one-third of it to be used in cell synthesis, and two-thirds to be decomposed into water and carbon dioxide. Such degradation causes the oil to be completely eliminated. It has been reported that the biodegradation rate of oil in suitable waters is $0.001 - 0.003$ grams of oil per ton of seawater per day, and $0.5 - 60$ grams of oil per ton of seawater can be eliminated per day in areas that are often contaminated with oil. However, once the oil is mixed with the sediment, the degradation rate is greatly reduced because microorganisms lack nutrients. Therefore, in recent years, oil spill dispersants containing nutrients have been developed, wherein the added nutrient substances can enhance the reproductive ability of microorganisms and strengthen biodegradation. The biodegradation rate does not play a big role in combating oil spills, but it is very meaningful for oil-contaminated marine environments to recover even in months or years.

4.1.4.2 Oil spill weathering model

At present, the study of weathering is gradually transferred to the stage of mechanical research. With modern analytical methods, the composition, property change and internal mechanism of weathering process have been investigated quantitatively. There are mainly the following models:

1. IKU model

The IKU model was developed by the SINTEF Research Institute. The model is based on a large amount of data from the laboratory, and its input section includes experimental data and environmental conditions. The former includes oil distillation curve, properties of the fresh oil, maximum water absorption capacity and viscous limit of dispersants, and the output of the model includes evaporation loss, the total mass balance of oil, change of physical properties, water content, dispersion amount, use of dispersants and field burning time prediction [79].

The information of fresh oil components is given by the distillation curve, and the quasi-component method is adopted for the evaporation process. The calculation formula is as below:

$$\frac{dQ_i(t)}{dt} = -\frac{\alpha(t)Q_i(t)M(t)P_i(t)}{\rho(t)h(t)RT}, \tag{4.47}$$

where, $Q_i(t)$ is the weight of component in unit residual area (kg/m^2); $\alpha(t)$ refers to the mass migration coefficient related to the wind speed (m/s); $M(t)$ refers to the molar mass of liquid mixture $(kg/kilomol)$; $P_i(t)$ refers to the vapor pressure of component (N/m^2); $\rho(t)$ refers to the density of liquid mixture (kg/m^3); $h(t)$ refers to the thickness of the oil film (m); R refers to the gas constant $(J/(mol\dot{K}))$; and T refers to the absolute temperature (K). The dispersion process can be calculated with the empirical equation proposed by Delvigne and Sweeney [125]:

$$Q_r(d_0) = C(o)D_{ba}^{0.57}SFd_0^{0.7}\Delta d, \tag{4.48}$$

where, Q_r is the entry rate of oil droplets of different size with nearby interval Δd per unit surface area $(kg/(m^2 \cdot s))$; $C(o)$ refers to the empirical entry constant depending on oil varieties and weathering state; D_{ba} refers to the energy of breaking waves dispersed on unit surface area (J/m^2); S refers to the coverage rate of oil on the sea surface $(0 \le S \le 1)$; F refers to the number of breaking waves encountered by the sea surface per unit time; d_0 refers to the diameter of oil particle (m); Δd refers to the diameter interval of oil particles (m). The energy of dispersed breaking waves is given according to the semiempirical relationship:

$$D_{ba} = 0.0034\rho_w g H_{rms}^2, \tag{4.49}$$

where, H_{rms} refers to r.m.s value of wave height in the wave field; ρ_w refers to the density of seawater (kg/m^3); g refers to the acceleration of gravity (m/s^2). The empirical equation proposed by Delvigne and Sweeney [125] is established by combining theoretical analysis and experimental results, which is worthy of reference in calculating the dispersion process. The equation $y = f(x)$ is established with nonlinear curve fitting to describe pour point and flash point of the oil, pure oil viscosity and water absorption rate based on experimental data, where, y refers to the property and x refers to the percentage of evaporation. The viscosity of w/o emulsified mixture is calculated with the formula proposed by Stiver and Mackay [117]:

$$\mu = \mu_0 e^{\frac{2.5w}{1-0.654w}}, \tag{4.50}$$

where, μ and μ_0 refers to the viscosity of weather oil and initial oil, respective (cP); and w refers to the water content.

The model is supported by a large database, including properties, distillation curves and laboratory weathering data of more than 200 types of oils. The model can estimate the effective time of mechanical recovery, chemical treatment with dispersant or demulsifier, and field burning through predicting the properties of crude oil. The model can also be combined with the drift mode to track the properties of several oil films simultaneously.

2. ADIOS model

ADIOS is a new weathering model developed by NOAA/HMRAD in the United States, and involves the properties (density, viscosity and water content) and four physical processes (spreading, evaporation, oil-in-water dispersion, water-in-oil emulsification) of three oils.

During model calculation, the spreading process of oil film is calculated with the modified Fay formula, and the evaporation process is calculated with the formula proposed by Stiver and Mackay [117]:

$$\frac{dF}{dt} = \frac{K_E}{h}e^{6.3-\frac{10.3(C_1-C_2F)}{T}} \tag{4.51}$$

The dispersion process is calculated with the modified Delvigne and Sweeney [125] formula.

The water content in the emulsified mixture is calculated with the atomization formula first proposed by Mackay [84]:

$$\frac{dY}{dt} = kU^2 \left(1 - \frac{Y}{Y_f}\right), \tag{4.52}$$

where, Y refers to the volume ratio of water to emulsified mixture; Y_f refers to the volume ratio of water to emulsified mixture after complete emulsification; the value of constant k is 0 before emulsification, and in general, $k = 1.6 \times 10^{-6} s/m^2$.

The calculation of density is based on the assumption that the density is linear with the change of evaporation fraction and temperature in the model applied by Stiver and Mackay [117]:

$$\rho_m = Y\rho_w + (1 - Y)\rho_0 \left[1 - c_1 (T - T_1)\right](1 + c_2 f) \tag{4.53}$$

where, ρ_m refers to the density of the mixture (kg/m^3); ρ_0 refers to the reference density (kg/m^3); T_0 refers to the reference temperature (K); f refers to part of the oil that evaporates; c_1 and c_2 are experimental parameters based on the property of the oil spill.

The viscosity is calculated with Mooney formula:

$$V_m = V_0 exp\left(\frac{c_3 Y}{1 - c_4 Y}\right) exp\left[c_5 \left(\frac{1}{T} - \frac{1}{T_0}\right)\right] exp(c_6 f), \tag{4.54}$$

where, $c_3 = 2.5$, $c_4 = 0.65$, $c_5 = 5.0$ and c_6 is represented by a curve.

This model has an intelligent computer database that stores property information of hundreds of oils for use by the weathering model. Therefore, many oil properties are the result of direct measurements, while others are designed to provide parameters for formulas through experiments. The release of oil can be instantaneous or continuous for a period of time by user selection, and the environmental data is input in the form of chart. The output section shows the mass balance (evaporation, dispersion and residue) and physical properties of a given oil spill in the form of a chart and shows useful reference information for clean-up [79].

3. OSIS

OSIS is an oil spill model system jointly developed by Warren Spring Laboratory (WSL) and BMT Ceemaid Ltd. (BMT) in the UK. The simulation of the oil spill weathering process includes the effects of evaporation, emulsification and dispersion on oil residues and changes in density, viscosity and flash point [112].

In the model, the evaporation process was also calculated with the formula proposed by Stiver and Mackay [117]. The emulsification process allows estimation of vertical movement of oil droplets with a random walk tracking model. The change of oil density and viscosity in the evaporation process is calculated with a formula in a linear relation with the evaporation volume fraction. The viscosity of the emulsified mixture is calculated with the formula proposed by Stiver and Mackay [117]. Since the experiment shows that the parameters of different oils are different, the value in the formula is the actual test result, and the average value of the asphaltene content is used as the default. It can be seen that except for the emulsification process, the equation used by OSIS is basically similar to ADIOS, and only some parameters of the equation are amended.

4. Sebastiao & Guedes-Soares model

In this model, the same formula as OSIS is used to simulate the changes of oil spill evaporation, emulsification and viscosity, etc., and the dispersion process is calculated with the formula proposed by Reed et al. [124]:

$$D = 0.11(U + 1)^2 \left(1 + 50\mu^{\frac{1}{2}}\delta St\right)^{-1} \tag{4.55}$$

where, D refers to the fraction of oil loss at sea per hour due to dispersion; δ refers to the thickness of the oil film (cm); St refers to oil-water interface tension ($dyne/cm$); U refers to the wind speed (m/s); and μ refers to the viscosity (cP).

Given that the amount of residual oil changes as a result of surface evaporation and dispersion into the water body, the model writes the equations into differential forms, forming a system and solving it with the Runge-Kutta method, so that the weathering processes depend on each other, allowing the variables to change simultaneously and allowing environmental conditions (such as wind speed) to change. This method makes predictions more practical. The model predicts well the changes in water content, density and viscosity, but may overestimate the evaporation. The model can also be combined with other prediction components, such as oil spill trajectory prediction, underwater transport prediction, etc. [79, 112].

The biggest improvement of this model is to calculate various weathering processes simultaneously, that is, the equations are jointly built into a system of differential equations and reflect the simultaneous occurrence and mutual interaction of various weathering processes. However, such interaction only reflects the simultaneous effects of evaporation, dispersion and emulsification on the volume of residue oil, but does not reflect the internal influence of changes of components and state. It is required to study the weathering mechanisms and find their inner links, and this cannot be simply solved mathematically.

5. Three-dimensional dynamic forecast model for oil spills

The model was developed by Zhang Cunzhi et al. [128], and the simulation of weathering processes in this model mainly includes evaporation, emulsification and changes in oil density and viscosity. In this model, the same formula as ADIOS is adopted to calculate the evaporation rate and the water content in the emulsified mixture. The evaporation rate is calculated with the formula proposed by Stiver and Mackay [117], and the water content in the emulsified mixture is calculated with the atomization formula proposed by Stiver and Mackay [117]. However, for the calculation of viscosity and density, the model is the same as OSIS. The model is not mainly intended to predict weathering, but is mainly designed to predict the spreading and drift of oil spills. Weathering predictions are calculated with empirical equations that have been publicly published, and there is no database to support them.

The shortcomings of the oil spill weathering model can be summarized as follows: (i) The lack of experimental data makes most oil spill weathering models rely on empirical equations (such as the formula proposed by Stiver and Mackay [117] which is used to simulate the evaporation process). Empirical equations are rarely based on a large number of studies, which will affect the accuracy of predictions. Although some models correct the parameters of equations through experimental data to enhance the accuracy of the simulation, due to the complexity of different types of oil and weathering processes, a set of parameters does not necessarily apply to each type of oil, accordingly affecting the universality and practicality of the model. (ii) The weathering processes of oil spills are quite complicated, and various processes affect and restrict each other. However, the current research on the weathering process basically separates the processes for separate study and it is difficult to integrate all the processes together. The relations among various weathering processes should be based on internal causes, such as composition and state of oil spills. A certain process will affect other processes through changing the composition and state of the oil. Such impact should be based on the study of weathering mechanism, and cannot be solved only by mathematical equations. (iii) In oil spill accidents, there are many internal (such as change of composition caused by weathering of the oil itself) and external factors (such as changes in geographical environment, human intervention, etc.) that affect the mass balance of oil spills. In traditional models, due to the independent calculation of each weathering process, the difference in vertical distribution of the oil spill caused by internal factors is seldom reflected. In order to make the models more practical, factors such as changes in the geographical environment (such as nearshore stranding) and human intervention (such as various clean-up operations) should also be reflected in the mass balance prediction, but not seen in most models [79].

The future development of weathering models should be based on a large number of fundamental studies, in order to find out the essential characteristics and microscopic mechanism of each weathering process of oil spill, reveal the internal mechanism of mutual influence between various processes, reflect the actual situation more truly and comprehensively, and improve the accuracy and practicability of the models.

4.2   Marine Oil Spill Model System

In order to meet the need of emergency decision support for marine oil spill accidents and accurately evaluate oil spills' harms to the environment, studies on the integration of marine oil spill model and Geographic Information System (GIS) technology have been carried out successively at home and abroad, and the integrated system is the oil spill model system. The role of GIS is described as below: (1) GIS can obtain relevant data required by the oil spill model, and feed back the data to the model with different scales, precision and formats. (2) In the model calculation, the flow velocity and water level of the entire current field at the initial moment can be set to zero through GIS. (3) During calculation of the zone boundary, the spatial analysis function of GIS can make some analysis operations in the model simple and easy, making the inspection of model rationality and model correction easier. (4) In GIS, the sea area can be treated with grid discretization according to the resolution requirements, and the digital elevation model can be used to linearly interpolate the water depth or the isobath to obtain the water depth of all the grids matching the model grids [88]. At present, in oil spill model systems at home and abroad, the integration of GIS and the model is mainly divided into three forms: (1) Loose integration. The GIS and the model are two independent systems, and the combination is only reflected in that the GIS provides certain input data for the model and the model operation results can be processed or displayed by the GIS. (2) Tight integration. The GIS and the model are two systems, but have a unified user interface (managing the public data of the two systems and exchanging files). (3) Complete integration. All parts of the GIS and the model are integrated completely, and they have a common user interface and achieve data sharing. These three integration solutions have their own advantages and disadvantages. For example, the loose integration solution cannot carry out human-computer interaction, has a low efficiency, and is error-prone, and the complete integration solution cost is high and has a long cycle. Therefore, based on comprehensive consideration, the current oil spill model system mainly adopts tight integration of the GIS and the model [129]. At present, the marine oil spill model systems at home and abroad are developing towards the tight integration of multiple technologies and models with the oil spill model, such as the integration of remote sensing technology and environmental impact and assessment models, providing strong technical support for the forecasting and early warning of marine oil spill accidents.

4.2.1   Introduction to Oil Spill Model Systems Abroad

The study of oil spill model systems in foreign countries started earlier than in China, so the software is mature and the results are abundant. The Applied Science Associate (ASA) Company in the United States started to study and develop OILMAP system in the 1980s, which can simulate two-dimensional and three-dimensional behavior and fate of oil spills, trace back oil spills, and conduct calculation for emergency response. Now, it has become a powerful and comprehensive oil spill model system with the most extensive application in the world. In addition to OILMAP, the ASA Company has also developed a model system containing biological effect modules, SIMAP, which has also been widely used. Meanwhile, the OSCAR system developed by SINTEF is unique among the oil spill model systems, which can not only be used to simulate the behavior and fate of the oil spill, concentration and depth of the oil spill and the degree of pollution caused to the coastline, but also provide the distribution information of emergency resources and simple emergency response strategies. As part of Australia's oil spill emergency plan, the OSTM system is mainly used to simulate the behavior and fate of oil spills in the waters around Australia, so as to provide the basis for relevant departments to make emergency plans. The UK's OSIS system is a simple and intuitive oil spill model system, which is based on the coastal oil spill diffusion and mixing process of the tidal current research project, and the model will be improved in the future. It is worth mentioning that the OSIS system provides ideal tools for risk assessment, emergency plans and training. In addition, the oil spill model systems RIAM and COMBOS developed by Japanese scientists for the Japan Sea and the Gulf of Tokyo are also advanced. It should be noted that most of current international model systems use the oil particle model to simulate the behavior and fate of oil spills.

The following is a brief introduction to OILMAP, SIMAP, OSCAR, and OSTM systems.

4.2.1.1 OILMAP

1. Brief introduction to OILMAP

OILMAP is commercial oil spill simulation software developed by Applied Science Associate (ASA) of the United States to predict the drift process of oil spills on the water surface and the degree of environmental pollution, including drift trajectory of oil spills, diffusion scope, evaporation amount, coastline adsorption amount and oil spill volume in a specific area at different times and circumstances. Oil spill simulation of this software also relies on environmental background data to improve accuracy, such as data of wind field and current field, natural environmental and geographical data (coastline location, seabed topography, etc.) and properties of oil spills (location of oil spills, total oil spill volume and oil types). These data can be directly set by OILMAP, and through integration of all the set values, the best numerical simulation results are presented. In addition, OILMAP can also be used together with the GIS to present the effect of oil spill simulation, including place names, sensitive resource areas, navigation routes of ships, and timely oil spill observation. OILMAP can not only predict the oil spill after it is reported, but can also trace back the oil spill to determine the location where the oil spill accident occurs. All this shows that the function of OILMAP is quite powerful.

2. Technical characteristics

OILMAP has the following characteristics: (1) Capability of two-dimensional and three-dimensional computation. (2) Capability of tracking the weathering process of the oil spill as well as on-water and underwater movement trajectories of the oil spill. (3) Capability of obtaining actual data through monitoring to update the forecasting result. (4) Capability of predicting the possibility of sensitive area being affected by the oil spill. (5) Inversely inferring the possible source of the oil spill. (6) Risk assessment of important resources. (7) Capability of simulating the interaction between oil spills and coastline, seabed and ice cover. (8) Integration with real-time environment database EDS (forecast, real-time or historical environmental data from around the world).

The system is composed of the following models:

- Oil spill trajectory and weathering model

  The model uses the Lagrange particle model to predict the trajectories of instantaneous or continuous oil spills, and can calculate the diffusion, evaporation, emulsification, water entrainment, and coastline adsorption of oil spills, and the interaction between oil and seabed and ice. In the model, the drift trend of the oil spill in the model is expressed in a dynamic form, and the degree of influence on the coastline can be determined. The change of the oil spill weathering result with time is displayed in the form of a chart, and the GIS shows the extent of the impact of the oil spill on environmental resources.

- Oil spill statistical model

  This model is used for risk assessment and emergency planning of sudden oil spills. This model can be used to determine the most likely oil paths based on the month, season and year and obtain the probability that water surface and coastline near the oil spill site will be affected by the oil spill and the isochrone of oil spill diffusion.

- Traceability statistical model

  This model is designed to analyze the vulnerability due to the effect of oil spills. This model can be used to determine the possibility of pollution damage in a specific area based on the loading/unloading area or the navigation path of the tanker as well as the possible pollution sources according to the oil pollution observed in a specific location.

- Underwater transport model

  This model is intended to simulate the transport and dissolution processes of oil spills carried by the water body.

OILMAP is the world's most advanced computer-based oil spill response system. It is applicable to oil spill emergency response plans and real-time response anywhere in the world and is capable of providing effective prediction at any time. The modeled architecture of OILMAP enables the inclusion of different spill models into the system, and does not increase the complexity of the user interface with advanced data management tool suites. The variation of simulation results with time can be shown in the form of chart, and the GIS can also be used to demonstrate the extent to which oil spills affect environmental resources.

3. Relevant domestic application

The application of OILMAP in China: Sun Jun et al. [130] developed the "The Oil Spill Simulation Information System for Zhoushan Port in China" based on OILMAP. The system can predict the movement trajectory and mass distribution of the oil spill within 48 hours. The specific implementation procedures are described as below: (1) select the scope of Zhoushan Port as the basic map; (2) generate the computing grid according to the coast type (rock type, beach type, swamp type, tidal zone, etc.); (3) provide data on the wind field and current field measured or predicted near the location where the spill occurs, including the speed and time of the wind and ocean current during the simulation of the oil spill trajectory; (4) provide oil spill information: including oil spill location, oil spill volume, oil spill mode and oil property parameters; (5) dynamically display the oil spill trajectory and the mass distribution curve or data table of the oil spill after operation. In addition, Li Jun et al. [131] used the scenario analysis method to apply OILMAP to the oil spill risk study for Qinzhou Bay, and predicted various indicators under various accident scenarios, including the time of the oil spill reaching the shore, sweeping area of the oil film, the length of polluted coastline and amount of the oil adsorbed by the coastline. Moreover, according to the risk analysis of oil spills in Qinzhou Bay and the simulation of accident consequences, they proposed the precautionary measures for oil spill risks in Qinzhou Bay. Based on integration with OILMAP, China Waterborne Transport Research Institute has developed the oil-spill-sensitive resources and emergency resource management system (Oil Spill Emergency Response System 2.0, OSERS 2.0), which will be introduced in detail in Chapter 6.

At present, the user interface of OILMAP has been extended with the Windows framework, and another version uses ESRI's ArcView software. These extensions reflect the usability and linkage of the model to external data. A number of oil spill model systems are developed on the basis of OILMAP.

4.2.1.2 SIMAP

SIMAP (Spill Impact Model Analysis Package) is a model system for predicting the natural fate and ecological impact of oil spills. The system is based on the company's oil spill trajectory model and the oil spill fate and ecological impact submodel in NRDAM/CME model. Therefore, SIMAP contains numerous models, such as two-dimensional oil spill trajectory prediction and oil spill scenario display model, three-dimensional model for natural fate of oil spills, three-dimensional model for ecological impacts of oil spills, and three-dimensional risk assessment and emergency plan preparation and statistical model. In the meantime, the system is coupled with GIS technology, and has a database of necessary data about physical and chemical properties and biodiversity for model input. The system is based on PC operation and operation modes of Windows 95 (+) or Windows NT / XP, and provides friendly graphic input for model operation, data editing and result verification. This system is mainly used for oil spill emergency response planning and risk assessment in marine and freshwater environments.

SIMAP is mainly composed of the following parts:

- Physical fates model: Simulate various processes of oil spills, including spreading, drift, evaporation, dissolution, dispersion, adsorption and sedimentation.

- Hydrodynamics model: Provide the data of current field through independent hydrodynamics models.

- Biological effects model: Simulate the effects on various habitats, fish, shellfish and wildlife under a certain oil spill scenario.

- Three-dimensional stochastic model: Predict the extent of oil spill pollution and count the probability that the oil spill volume will exceed a certain critical point. Be used to risk assessment and emergency plan preparation for oil spills.

- Graphical visualization tool: Provide users with services such as interface input and graphic output.

- Environmental geographic, oil, and biological databases: Provide the model with necessary data about topography, coastline, oil, chemicals and biological effects for the calculation of physical fates and biological effects of oil spills.

Users need to provide relevant oil spill information (time, location, oil type, oil spill amount, etc.) and some environmental constraints (such as temperature and wind speed data) before using the system. Figure 4.1 shows the system structure of SIMAP and the relationships between the models.

1. Physical fates model

This model is intended to simulate the distribution of pollutants (such as mass and concentration) on the water surface, on the coastline, in water bodies and in sediments. The GIS database provides the model with data such as water depth of the entire grid area, sediment type, ecological environment, coastline type, and ice cover value. The oil property and chemicals database provide the model of required physical and chemical parameters. Users need to input information such as the specific time when the oil spill occurs and the wind field where it occurs before using the model.

The model can simulate the processes of oil spill diffusion, drift, evaporation, dissolution, dispersion, adsorption and sedimentation and biodegradation, and provides users with the movement trajectory and weathering information of the oil spill. In the coastline adsorption process, due to the interaction between oil pollution and the shoreline, different types of shorelines will cause oil to produce corresponding material sedimentation and release. In this respect, SIMAP is similar to OILMAP. In general, oil pollution is always present in water bodies and sediments, and the process of entering water bodies and sediments needs to be achieved by simulating the horizontal and vertical substance transport by water. Such substance transport is caused by water advection and turbulent diffusion, respectively. Part of the oil in water adheres to the suspended particles, and the other part is in dissolved and other states. In this model, it is assumed that the ratio between these parts of oil is constant (based on the equilibrium distribution theory), and the amount of oil adsorbed by the suspended particles is distributed according to the proportion of the representative type of sediment. This also includes the amount of oil that is carried into bottom sediments by benthonic animals, which needs to be calculated according to a simple bioturbation algorithm. In addition, the model uses a constant degradation rate to deal with the degradation of oil in water and bottom sediments, on the premise of assuming that such degradation rate is applicable in any environment.

This model is designed primarily for crude oil and refined oil, and is used to calculate the concentrations of oil dissolved in water and sediments as well as the variation of the area of water surface and shoreline covered with oil slick with time and space. The output result is presented to the user with a visualization tool, and the data is transferred to the biological effects model

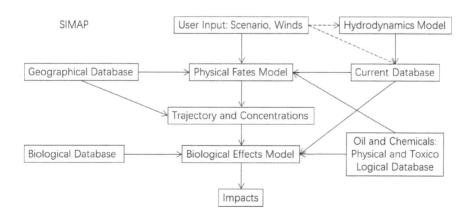

FIGURE 4.1: System structure of SIMAP.

so that it can estimate the effects of the oil with a given oil concentration and coverage area on organisms.

2. Hydrodynamics model

The simulation of the oil spill movement process relies on high-quality current field input data about the relevant sea area. These data can be obtained from observation or model calculation. In this model, users can directly import ASCII code or current field data in other standard formats, and use the visualization tool to draw the current field vector for analysis. In terms of hydrodynamics model, ASA has developed several hydrodynamics models with different complex structures and applicability, including two-dimensional and three-dimensional models. Grids can be rectangular or have the function of shore bridging (this grid is mainly for estuaries and nearshore areas). The model can generate periodic or time-varying datasets. The input forced fields include tides, pressure gradients, wind fields, etc. ASA's hydrodynamics model generally provides standardized current field datasets, which are read by physical fates and biological effects models.

3. Biological effects model

The model is mainly intended to simulate the effects of oil spills on various habitats, fish, shellfish and wild animals.

The habitats in the model include deep water, nearshore area, wetland, and coastal environments. Rectangular grids are laid in areas that may be affected by oil spills, and each grid is marked with a habitat type code. During calculation, this grid needs to match the grid of the physical fates model. In the biological submodule of the model. Adjacent grids with the same habitat type code can be combined into one ecosystem, on the premise of assuming that the number of fish, birds, mammals, and primary productivity are constant and evenly distributed in the system during oil spill simulation, fish, birds and mammals are free to move within their habitats, and plankton (eggs and larvae in water) move with the water.

In the model, oil spills (including crude oil and refined oil) interact with wild animals (including birds, mammals and reptiles), and the concentration of aromatic compounds contained in oil spills and dissolved in water or sediments directly affects the number of such fish, including fish eggs and juvenile fish. It is assumed that in the area covered by oil spills, the death of some wild animals is related to the probability of direct contact with oil spills and the mortality after oil immersion, and then the estimation of this probability will be based on information from field survey, including the mortality rate of organisms in the same environment.

The biological effects model has been validated in about 30 oil spill accidents, and the results of the model can match the actual situation.

4. Three-dimensional stochastic model

The main function of this model is to predict the extent of oil spill pollution and to calculate the probability that the oil spill volume will exceed a certain critical point. When using this model, long-term historical wind and current field data will be used to facilitate the analysis of concerned areas. When using the model, the user needs to operate it multiple times, and choose to randomly specify the initial date and time within a certain month of the year. During each operation, the user may keep the oil spill volume unchanged or randomly choose the oil spill volume from 0 to 100% of the maximum possible volume. The user will select a final result that will provide the probability that the oil spill above is above the critical point.

The output of this model includes the following information:

- Oil distribution on the water surface.

- Oil distribution along the shoreline.

- Concentration distribution of oil dissolved in water.

- Concentration distribution of oil adsorbed to suspended particles in water.

- Oil distribution in sediments.

Meanwhile, the model performs the following five kinds of statistics for each position in the model grid:

- Probability that the oil spill volume will exceed the certain critical point.

- Time when the oil spill volume exceeds the critical point for the first time.

- Expected maximum oil spill volume or maximum oil spill concentration.

- Maximum possible oil spill volume (i.e., the maximum value of all operations), the date and time when the operation involving the maximum possible volume starts.

For each operation of the model, the instantaneous peak values and doses of biological exposure (product of the sum of the concentrations and the time of biological exposure) need to be calculated. The output of the model includes exposure graphs and frequency distribution, and the user can intuitively obtain the standard deviation, minimum and maximum values of the exposure and dose averages.

5. Environmental, chemical and biological databases

These databases are intended to provide necessary environmental, chemical and biological data inputs for physical fates and biological effects models. One of ASA's advantages is the ability to integrate data from a variety of sources, including government and private data services, as well as field survey and research. In addition, the digital model compensates for the observed data to some extent, so that the required data is complete and standardized.

The environmental database is a geographic dataset that includes data about shoreline, water depth, shoreline type, and ecological habitat type. Such information is stored in a simplified GIS system, which facilitates the model's ability to call relevant geographic data in real time, while allowing users to view and edit data through drop-down menus and mouse drivers on a friendly interface.

The chemical database provides various physical and chemical parameters of crude oil and refined oil. ASA edits the source data that exists but is not centralized. The default toxicology database contains $LC50$ (50% of lethal concentration of test organisms) and $EC50$ (effective concentration below 50% control rate) to replace low-molecular-weight aromatic compounds in crude oil and refined oil. This data comes from existing data sources, including EPA's AQUIRE database. The data goes through standardized processing of temperature, pH, and salinity, and then is averaged according to different populations.

The information in the biological database may come from all over the world. In the United States, ASA has developed a corresponding biological database that includes seasonal or monthly abundance and habitat types of species in each biological "subregion" or biogeographic region of the United States. The information on the biological areas of 77 coastal areas and oceans, 11 large lakes and 10 inland waters has been included in the database. Typical biological subregions include estuary, strait, bay, continental shelf, great lake (any of the great lakes in the United States), and some inland biogeographical regions in the country. In a regional or subregional biological database, species, the data about which is edited, include: 96 species of fish or shellfish and 46 species of wild animals (birds, mammals and reptiles). According to the reliability of the information, the abundance is the average of the habitat types, or the total biomass distributed in the habitat that is estimated based on biological history and behavior.

For juvenile fish and shellfish, the database uses the calculation results of the biological population model to supplement the total number of individuals required for the adult population. According to the life history and development process of young creatures, the distribution of habitats is determined to calculate the monthly abundance. The estimation of primary productivity based on seasons and habitat types is also edited. The data is compiled into a standard format with independent (same format) data files for each biological subregion.

4.2.1.3 OSCAR

The OSCAR (Oil Spill Contingency and Response) model system is developed by SINTEF, one of the largest independent research organizations in Europe. Its main role is to provide objective analysis for emergency response to oil spill accidents, in order to ensure that relevant departments can take timely and effective emergency measures. The OSCAR system can maintain a balance between the cost of multiple oil spill response measures and the potential environmental impact of the oil spill. In theory, when an oil spill accident occurs, the emergency department may need to purchase more equipment or require a higher standard of emergency response and assessment, but the OSCAR system can provide a basis for simultaneously reducing the environmental impact caused by the oil spill accident and reducing additional expenses for purchasing equipment.

The structure of the OSCAR system includes:

- Three-dimensional physical fates model: Simulate the drift trajectory and chemical fate of the oil spill.

- Weathering model: Simulate the weathering process of the oil spill based on oil and chemical databases.

- Oil spill emergency response model: The purpose is to conduct a strategic analysis for oil spill emergency response through simulation.

- Biological exposure model: Calculate the short- and long-term effects of oil spills on the sea surface and in water on marine plankton, fish, birds and mammals.

The main components and operating mechanism of the system are shown in Figure 4.2.

The oil and chemical databases can provide the model with required chemical and toxicological parameters. The simulation results will be stored in a computer file as per discrete time steps, facilitating providing data to one or more biological exposure models.

The OSCAR system uses a series of empirical formulas to predict the behavior and fate of oil spills on the sea surface, including diffusion, drift, evaporation, dissolution, emulsification, dispersion, adsorption and sedimentation processes of oil spills. In water, hydrocarbons that drift due to horizontal and vertical convection, dispersion, and dissolution can be simulated through dynamic programming. The division of adsorption and dissolution of the oil spill is the same as the SIMAP system and is also calculated on the basis of the linear equilibrium theory. Broken contaminants that have been adsorbed by the suspended particles can settle together with the particles. In this way, bottom contaminants mix with the previous seabed sediments and may be dissolved into the water. Substances and sediments degraded in water can be presented by the first decay process.

In the oil spill emergency response model, the recovery efficiency of the oil spill is determined by the wave height, and is also related to the wind speed, wind duration and wind area in the model. The recovery efficiency decreases as the wave height (or wind speed) increases, which can be seen from the figure. When the wave height reaches 2.5 meters, or the wind speed reaches 10 meters/second, the efficiency becomes zero.

OSCAR's analysis of oil spill emergency response strategies are applicable to both offshore and coastal areas. With respect to different emergency response strategies, OSCAR can provide comprehensive and quantitative environmental impact assessment of marine environment. The model calculates and records the spatial and temporal distribution of contaminants on the water surface, along the shoreline, and in water bodies and sediments. The model is embedded into the graphical user interface of WINDOWS NT, which makes it easy to apply a variety of standard and customized databases and tools. These databases and tools allow users to create or input wind field time series, current field and grids of arbitrary spatial resolution, and to draw the output results of the model.

4.2.2 Introduction to Oil Spill Model Systems in China

With the continuous maturity of the oil spill model simulation technology, many domestic scholars have begun to develop oil spill information management systems and intelligent systems based on the original oil spill behavior and fate simulation. Most of the domestic development modes are the secondary development based on the original oil spill simulation software. Most of the systems incorporate GIS technology, and the representative system is the oil spill diffusion and drift model system developed by Liu Yancheng et al. [132]. In some oil spill simulation systems, geographic information on geographical environment of oil spill areas and environmentally sensitive areas is added, such as the oil spill forecasting system developed by Xiong Deqi et al. [133].There are also some systems that not only include simulations of movement trajectories and different fates of oil spills, but can also form emergency response decision assistance schemes according to emergency plans and the knowledge and experience of emergency response experts and based on the

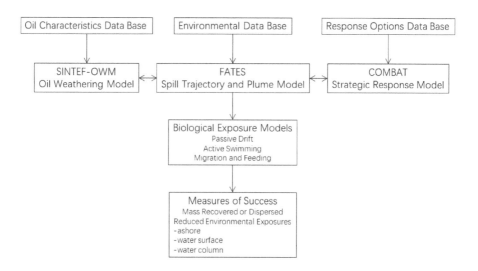

FIGURE 4.2: Schematic diagram of main components of the OSCAR system.

simulation results of simulated calculation parameters, ocean currents and oil spill trajectories, including alternative technologies and appropriate time for the clean-up of oil spill pollution, protection priorities for the retrieval of environmentally sensitive resources, and specific clean-up programs for contaminated shorelines. With the rapid rise of artificial intelligence and its penetration into various fields, the study on the marine oil spill emergency expert system is carried out on the basis of the combination of oil spill simulation and artificial intelligence. Although this is just starting, it is a development direction of the emergency response system. This kind of system mainly uses the Database Management System (DBMS) and GIS technology. It can not only provide decision support for emergency response to oil spill accidents, but also can be used for planning and prioritizing oil spill risk management affairs. Its primary functions include: basic geographic information system; simulation of real-time data input; introduction of oil database and establishment of its connection with the management decision making model; environmental oil spill model; provision of oil spill disaster reduction information; and system performance evaluation.

Some mature model systems in China will be introduced below to provide readers with some references.

4.2.2.1 Oil spill emergency forecasting system of the Bohai Sea

The system is developed by the National Marine Environmental Monitoring Center and is applicable to emergency forecasting, prediction and evaluation and environmental risk assessment of oil spills in the Bohai Sea. The system was developed on the basis of the results of the scientific and technological breakthrough project during the period of the Eighth Five-Year Plan: Microcomputerized Forecasting System for Oil Spills in the Bohai Sea, which is a three-dimensional oil spill forecasting system supported by a microcomputer, and it is designed to predict the dynamic behavior and the change of physical properties of oil spills in three-dimensional space, and to provide decision information for oil spill emergency response. The system consists of three parts: three-dimensional hydrodynamic numerical mode, three-dimensional oil spill stochastic mode and visualization system. The system mainly has the following features:

- In combination with the matching modes and according to the characteristics of different research objects, the most suitable calculation methods can be selected, such as hydrodynamic numerical mode for current field forecasting, and oil particle mode for oil spill behavior and fate forecasting. The matching modes take advantage of both deterministic and stochastic modes.

- The comprehensive matching three-dimensional mode is suitable for sea areas with different water depths. It is suitable for both the Bohai Sea that is relatively shallow and other sea areas.

- The system has a selection function of precision-timeliness that allows for an initial forecast within 5 minutes and an accurate forecast within 15 minutes.

- The system does not depend on the computing environment and can run on the microcomputer of any model.

- The system has the function of visualization, and prompts input through the control menu and output in the form of animation, which is especially convenient for field use.

The key technologies of the system include three-dimensional nonlinear δ stratified tidal current numerical model, oil particle mode for predicting the behavior and fate of oil spills, and output result visualization system.

The Oil Spill Emergency Forecasting System of the Bohai Sea is based on the improvement of the above-mentioned system, and becomes a microcomputer application software that can be used in the Chinese Windows operating environment.

The Oil Spill Emergency Forecasting System of the Bohai Sea consists of two subsystems, i.e., forecasting and result processing, two databases of oil characteristics and environmental conditions, and a model library. Each subsystem contains several models with relatively independent functions. As the core of the whole system, the forecasting subsystem mainly calculates the drift trajectory of the oil spill at sea and the weathering situation and can also trace the oil spill back to the source. This subsystem includes four models of tidal current field calculation, oil spill forecasting, concentration field forecasting and oil source search, which can release information about oil spills hour by hour, such as the central position and coverage area of the oil film on the sea surface, and the concentration distribution of oil in water, specific gravity, viscosity and surface tension of the oil, evaporation amount, dissolution amount, emulsification amount and diffusion amount of the oil, oil residue on the sea surface oil and its average thickness, tidal current field, and so on. The result processing subsystem can visualize the calculation results of oil film trajectory, oil characteristics curve, concentration distribution, and tidal current field, mainly in the forms of graphs and texts, so that users can set control parameters through command selection buttons and dialog boxes under Windows system and obtain better output effects. The oil characteristics database is used to record and maintain the physical and chemical properties of oil products, which is convenient for users to continuously accumulate relevant data of various oil products during the application process, and provides a convenient way for self-expansion system forecasting. The environmental condition database mainly includes two parts: environmental elements and resource distribution. The environmental elements part provides users with various functions including input, maintenance, retrieval and statistical analysis of various meteorological and hydrological parameters, and supports the storage and conversion in various common database formats such as FoxPro and Access. The resource distribution part includes relevant information on environmentally sensitive areas along the coast of the Bohai Sea, such as main protection areas and aquaculture areas, providing reference for hazard analysis and emergency decision making of oil spill accidents [134, 135] .

4.2.2.2 Marine oil spill emergency forecasting information system

This system is developed by DMU Shipping Technology Co., Ltd., and consists of two parts: model calculation and graphical interface. The model calculation section includes an oil spill weathering model based on experimental data, a three-dimensional oil particle model that stimulates oil spill drift and diffusion and chemical behavior, and an oil spill emergency response model that can simulate the clean-up effects of different response schemes (such as mechanical clean-up and use of dispersants).The databases equipped include the database for physical and chemical properties of oil products and chemicals, water depth and topography database, hydrological and meteorological (wind current) data file library, resource distribution database, and oil spill equipment database.

The user can import the parameters of the oil spill accident and the field data into the system within 15 minutes and complete the prediction and simulation of the oil spill trajectory. Due to the adoption of the "oil particle" mode, the system can predict the possible drift trajectory, diffusion range, and landing time and location of the oil spill, and can also visualize the information so that the user can understand the development situation of the oil spill in time. In addition, the system can calculate and predict the residual volume, oil film thickness, density, viscosity, water content, emulsification rate and other characteristics of the oil spill at sea as well as changes of the state according to weather and sea conditions. The system is equipped with GIS technology and built-in electronic chart. It can display the geographical distribution information of environmentally sensitive areas and emergency personnel equipment, which helps to issue pollution warning and carry out emergency deployment in advance [133].

4.2.2.3 Marine oil spill prediction and emergency decision support system for the coastal waters of China

The system is developed by CNOOC Environmental Protection Service Co., Ltd., and mainly includes four functional models: oil spill forecasting and early warning, emergency decision support, database and system operation guarantee.

Here, the oil spill forecasting and early warning model of the system is mainly discussed, which includes the following parts:

- Oil spill drift and spreading model

  In this part, the hydrodynamics model of the system is based on the three-dimensional baroclinic Princeton Ocean Model (POM) mode. The concerned large region is all coastal waters of China, and there are four key areas. The resolutions of the large region and the four areas are $5'$ and $1'$, respectively, and they are combined with nesting technology. The mode adopts the following ten tidal constituents to forecast the tidal current and tidal level: $K_1, O_1, P_1, Q_1, M_2, S_2, N_2, K_2, S_a$ and S_{sa}. The wind field data of the model is mainly from the NCEP wind field data of the US National Centers for Environmental Prediction, and also includes manual input and historical data. Based on the dynamic ring elements provided by the hydrodynamics model, the oil particle model is used to predict the drift trajectory and sea-sweeping area of the oil spill.

- Oil spill weathering model

In this part, the system mainly considers the evaporation and emulsi-fication process and density and viscosity changes of the oil film. The model is based on the previous research results, and the parameters of the evaporation model are corrected according to the experimental data to compensate for the inaccuracy of the model.

This system creates an environmentally sensitive resource map based on GIS technology, and realizes the dynamic rapid coupling of sensi-tive resource map and oil spill model. In the case of oil spill emergency response, the system can give a quick warning to sensitive resources that may be affected and provide corresponding emergency programs, according to the forecasting results of the oil spill model.

Chapter 5

Emergency Prediction and Warning System of Oil Spill in the Bohai Sea

The "emergency prediction and warning system of oil spill in Bohai Sea" is an oil spill model system researched and developed by the National Marine Information Center and its role is to achieve rapid prediction and warning of oil spill pollution in the Bohai Sea.

In order to meet the demand for oil spill emergency response in the Bohai Sea and ensure the successful realization of China's goal of "the 11th Five-Year Plan" for building capacity of emergency response to ship oil spill, the National Marine Information Center undertook one of the seven subprojects of "development of key technologies for rapid emergency response to marine oil spill" of the National Development and Reform Commission in 2007. That is, prediction and warning technology of oil spill pollution in the Bohai Sea. In combination with the "11th Five-Year Plan" of the Ministry of Communications for development of science and technology and the capacity-building plan for marine oil spill emergency response, the National Marine Information Center undertook the major research and development task for the project of "oil spill emergency response capacity building (Yantai Oil Spill Response Technical Center)" and the project of "key laboratory of ship pollution control technology in traffic industry" of Shandong Maritime Safety Administration, and has developed "emergency prediction and warning system of oil spill in Bohai Sea" by learning from the international advanced level in the related fields. The system framework is shown in Figure 5.1.

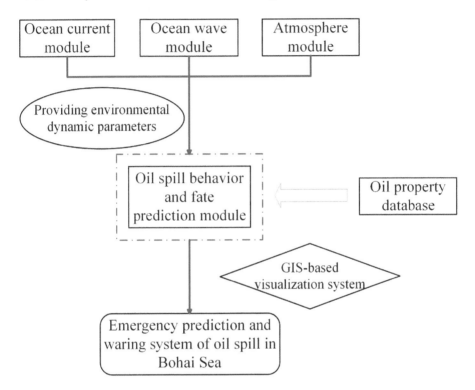

FIGURE 5.1: The framework of "emergency prediction and warning system of oil spill in Bohai Sea."

Based on GIS platform, the system provides high-resolution environmental dynamic parameters to oil particle model through the international advanced current numerical model FVCOM, the wave numerical model SWAN and the atmosphere numerical model WRF, thus realizing accurate prediction of oil spill behavior and fate in the Bohai Sea. Up to now, the system has been operating on the server of Shandong Maritime Safety Administration for one year. It is currently the only business system of Shandong Maritime Safety Administration which runs stably and is used in prediction and emergency decision making of oil spill accidents. The system has the function of unattended automatic forecast and is easy to operate and convenient to use.

5.1   Ocean Current Module

The role of current field is the most important in simulation of oil spill drift trajectory. The main function of the ocean current module of the system is to provide accurate current field parameters to the oil particle model in oil spill behavior and fate prediction module and ensure that the oil particle model can accurately simulate the drift trajectory of spilled oil.

5.1.1   Introduction to the Model

5.1.1.1   FVCOM overview

The FVCOM (Finite-Volume Coastal Ocean Model), a three-dimensional primitive equations ocean numerical model, is adopted in the ocean current module of this system. The model was jointly developed by the research team led by Dr. Chen Changsheng from the School for Marine Science and Technology at the University of Massachusetts and Dr. Robert C. Beardsley from Woods Hole Oceanographic Institution for study on coastal and estuarine tidal circulation. Its greatest feature and advantage is that it combines the finite element method which fits the boundary easily and enables the local refinement, and the finite difference method which facilitates the discrete calculation of the ocean primitive equations. The finite element method uses unstructured triangular grid and gives a linear independent basis function to calculate the undetermined coefficients. This method is characterized in that triangular grid is easy to fit the boundary and is easy to refine locally. While the finite different method directly discretizes the difference ocean primitive equations and is characterized by clear dynamic basis, intuitive difference and high computational efficiency [136]. FVCOM integrates the advantages of the above two methods. The integral form of equation and better calculation format used in the numerical calculation make the momentum, energy and mass have better conservation. The dry-wet method is used to deal with the moving boundary of tidal flat. The applied Mellor-Yamada level-2.5 turbulence closure submodel (1982) makes the whole model closed in physics and mathematics. σ is used vertically to reflect the irregular bottom boundary and splitting of inner and outer modes, so as to save time.

FVCOM contains many modules, such as three-dimensional dry-wet grid processing module, fresh water and groundwater input module, material diffusion and transport module, Lagrange particle-tracing module, sediment transport module, water quality module and ecology module. Such model has been successfully applied to some estuaries in Georgia, South Carolina and Massachusetts and the Great Lakes in the United States and the Bohai Sea in China. It represents a new development direction of numerical model and has a broad application prospect.

5.1.1.2   Unstructured grid

Structured grid means that all internal points in the grid area have the same neighboring cells. In contrast to the definition of the structured grid, unstructured grid means that the internal points in the grid area do not have the same neighboring cells. That is, the number of the grids connected to different internal points in the grid area is different. It can be seen from the definition that the structured grid and unstructured grid have overlapping parts, namely the unstructured grid may contain parts of the structured grid.

5.1.1.3   Design of triangular grid

Similar to the finite element method, the calculation area is divided into some non-coincident triangular cells. Each triangular grid consists of three nodes, one center and three sides. In order to more accurately calculate the water level, velocity, temperature and salinity on the sea surface, the numerical calculation is carried out on a specially designed triangular grid. The variables on a node are calculated by the net flux of the line which connects such point, the center of the triangle and the center of the side, and the variables on the center are calculated by the net flux of the side of the triangle [137].

5.1.1.4   Examples of the Bohai Sea grid

The Bohai Sea is a near-closed shallow sea of continental shelf in the North of China. It communicates with the Yellow Sea through the Bohai Strait in the east. Its north, west and south sides are surrounded by land with complex terrain. With a large population and developed economy in coastal areas, the sustainable utilization of Bohai Sea is one of the strategic concerns of the Chinese people and the distribution and structural characteristics of ocean current, temperature field and salinity field are also one of the important contents of physical ocean studies of this sea area. It can be concluded by using SMS (surface-water modeling system) software to design the grid of the Bohai Sea (Figure 5.2) that the triangular grid can conveniently and correctly fit the complicated shoreline and refine some parts of the grid. The minimum resolution at the open boundary is about 12000 m and the complicated shoreline, bay mouth and key port areas can be refined to make the maximum resolution be 5 m, thus meeting the requirements on resolution in general ocean phenomenon and process.

5.1.1.5   Basic control equations

The FVCOM control equations consist of momentum equation, continuity equation, temperature equation, salinity equation and density equation.

The basic equations describing the hydrodynamic and thermal systems in the ocean include Navier-stokes equation about motion, continuity equation describing mass conservation, state equations of seawater and conservation equation of temperature and salinity. In order to apply these equations to the

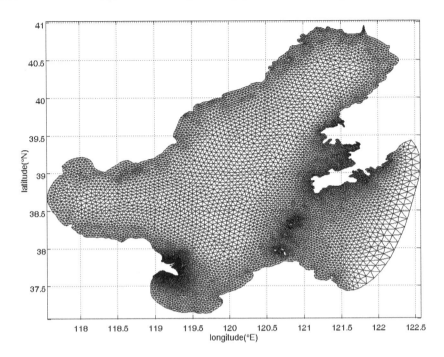

FIGURE 5.2: Example of grid of the calculated sea area.

shallow sea of the continental shelf, some assumptions and approximations are made for these equations according to some characteristics of the shallow sea:

First, in general fluid motion, the compressibility of fluid depends on the square of the ratio of flow velocity to sound velocity. The speed of seawater movement is far slower than the sound velocity, so the seawater can be considered incompressible.

Second, the ratio of depth scale D to horizontal scale L in coastal waters is a small scalar, namely $\delta = D/L \leq 1$. Therefore, the vertical equation of motion can be approximated by static equilibrium. That is, gravity equilibrates to vertical pressure gradient force.

In consideration of the right-hand Cartesian coordinate system, the eastward direction is x-axis positive direction, the northward is the y-axis positive direction and the vertical upward is the positive direction of z-axis. The free sea surface is $z = \eta(x, y, t)$, the bottom topography is $z = -H(x, y)$. $\vec{V}$ is the horizontal velocity vector, $\vec{V} = (U, V)$ and ∇ is the horizontal gradient operator, then the continuity equation can be written as follows:

$$\frac{\partial U}{\partial x} + \frac{\partial V}{\partial y} + \frac{\partial W}{\partial z} = 0 \tag{5.1}$$

The momentum equation is:

$$\frac{\partial U}{\partial t} + \vec{V} \cdot \nabla U + W\frac{\partial U}{\partial z} - fV = -\frac{1}{\rho_0}\frac{\partial P}{\partial x} + \frac{\partial}{\partial z}\left(K_M\frac{\partial U}{\partial z}\right) + F_X \tag{5.2}$$

$$\frac{\partial V}{\partial t} + \vec{V} \cdot \nabla V + W\frac{\partial V}{\partial z} + fU = -\frac{1}{\rho_0}\frac{\partial P}{\partial y} + \frac{\partial}{\partial z}\left(K_M\frac{\partial V}{\partial z}\right) + F_Y \tag{5.3}$$

The vertical static equilibrium equation is:

$$\rho g = -\frac{\partial P}{\partial z} \tag{5.4}$$

The conservation equation of temperature and salinity is:

$$\frac{\partial \theta}{\partial t} + \vec{V} \cdot \nabla \theta + W\frac{\partial \theta}{\partial z} = \frac{\partial}{\partial z}\left(K_H\frac{\partial \theta}{\partial z}\right) + F_\theta \tag{5.5}$$

$$\frac{\partial s}{\partial t} + \vec{V} \cdot \nabla s + W\frac{\partial s}{\partial z} = \frac{\partial}{\partial z}\left(K_H\frac{\partial s}{\partial z}\right) + F_s \tag{5.6}$$

The state equation of seawater is:

$$\rho = \rho(\theta, s) \tag{5.7}$$

Where ρ_0 represents reference density of seawater, ρ represents the field density, g represents the gravity acceleration, p represents the pressure, K_M represents the vertical eddy viscosity coefficient, f represents Coriolis parameter, applying β plane approximation, θ represents potential temperature (in the shallow sea, it is spot temperature), s represents salinity, and K_H represents vertical eddy diffusivity. K_M and K_H are obtained through Mellor-Yamada level-2.5 turbulence closure model. The pressure at the z depth $P(x, y, z, t)$ can be obtained through integral equation.

$$P(x, y, z, t) = P_{atm} + g\rho_0\eta + g\int_z^0 \rho(x, y, z', t)dz' \tag{5.8}$$

It is assumed here that the atmospheric pressure $K_a tm$ on the sea surface is constant.

F_X, F_Y, F_θ and F_s in Equations (5.2), (5.3), (5.5) and (5.6) can be written as follows:

$$F_X = \frac{\partial}{\partial x}\left(2A_M\frac{\partial U}{\partial x}\right) + \frac{\partial}{\partial y}\left[A_M\left(\frac{\partial U}{\partial y} + \frac{\partial V}{\partial x}\right)\right] \tag{5.9}$$

$$F_Y = \frac{\partial}{\partial y}\left(2A_M\frac{\partial V}{\partial y}\right) + \frac{\partial}{\partial x}\left[A_M\left(\frac{\partial U}{\partial y} + \frac{\partial V}{\partial x}\right)\right] \tag{5.10}$$

$$F_{\theta,s} = \frac{\partial}{\partial x}\left[A_H\frac{\partial(\theta s)}{\partial x}\right] + \frac{\partial}{\partial y}\left[A_H\left(\frac{\partial(\theta, s)}{\partial y}\right)\right] \tag{5.11}$$

It should be noted that F_X and F_Y are invariant of coordinate rotation. Where, A_M is the horizontal eddy viscosity coefficient and is calculated with the formula proposed by Smagorinsky [138]:

$$A_M = \alpha \Delta x \Delta y \left[\left(\frac{\partial U}{\partial x} \right)^2 + \left(\frac{\partial V}{\partial y} \right)^2 + \frac{1}{2} \left(\frac{\partial U}{\partial y} + \frac{\partial V}{\partial x} \right)^2 \right]^{\frac{1}{2}} \quad (5.12)$$

Based on needs, the parameter α can be 0.01 to 0.5 and is usually 0.10. By using Prandtl number, the value of horizontal eddy diffusivity A_H can be obtained through A_M. The vertical eddy viscosity coefficient and the vertical eddy diffusivity K_H in Equations (5.2), (5.3), (5.6) and (5.7) can be determined through the Mellor-Yamada level-2.5 turbulence closure model (1982). In this way, the influence of artificial selection of eddy viscosity coefficient on the simulation of physical field can be overcome to some extent. Such model describes two physical quantities: turbulence kinetic energy $q^2/2$ and macroscopic scale of turbulence l. The equations are:

$$\frac{\partial q^2}{\partial t} + \vec{V} \cdot \nabla q^2 + W \frac{\partial q^2}{\partial z} = \frac{\partial}{\partial z} \left(k_q \frac{\partial q^2}{\partial z} \right) + 2(p_s + p_b - \varepsilon) + F_q \quad (5.13)$$

$$\frac{\partial (q^2 l)}{\partial t} + \vec{V} \cdot \nabla (q^2 l) + W \frac{\partial (q^2 l)}{\partial z} = \frac{\partial}{\partial z} \left[k_q \frac{\partial (q^2 l)}{\partial z} \right] + l E_1 \left(p_s + p_b - \frac{\tilde{\omega}}{E_1} \varepsilon \right) + F_l \quad (5.14)$$

Where F_q and F_l represent the horizontal diffusion terms of the turbulence kinetic energy and the macroscopic scale of turbulence respectively and their form is similar to the temperature and salinity diffusion terms in Equation (5.12). Where, $p_s = K_M \left[\left(\frac{\partial U}{\partial z} \right)^2 + \left(\frac{\partial V}{\partial z} \right)^2 \right]$ is the turbulence kinetic energy term caused by shear force; $P_b = \frac{g}{\rho_0} K_H \frac{\partial \rho}{\partial z}$ is the turbulence kinetic energy term caused by buoyancy force; $\varepsilon = \frac{q^3}{B_1 l}$ is the turbulent kinetic energy dissipation rate; $\tilde{\omega} = 1 + E_2 (\frac{l}{kL})^2$ is wall proximity function, and $\kappa = 0.4$ is Von Karman constant.

$$(L)^{-1} = (\eta - z)^{-1} + (H + z)^{-1} \quad (5.15)$$

At a location near the surface, $L \approx l/\kappa$, hence $\tilde{\omega} = 1 + E_2$; at a location far away from the surface, $l << L$, hence $\tilde{\omega} = 1$. K_M, K_H and K_q are determined by the following equations, respectively.

$$K_M = lqS_M \quad (5.16)$$

$$K_H = lqS_H \quad (5.17)$$

$$K_q = lqS_q \quad (5.18)$$

Where, S_M, S_H, and S_q are stable functions and can be written as follows according to the research results of Mellor and Yamada [139] as well as Galperin et al. [140]:

$$S_M = \frac{B_1^{-\frac{1}{3}} - 3A_1 A_2 G_H \left[(B_2 - 3A_2) \left(1 - \frac{6A_1}{B_1} \right) - 3C_1 (B_2 + 6A_1) \right]}{[1 - 3A_2 G_H (6A_1 + B_2)] (1 - 9A_1 A_2 G_H)} \quad (5.19)$$

$$S_H = \frac{A_2 \left(1 - \frac{6A_1}{B_2} \right)}{1 - 3A_2 G_H (6A_1 + B_2)} \quad (5.20)$$

Where $S_q = 0.20$, $G_H = -(\frac{Nl}{q})^2$ is Richardson number, and $N = (-\frac{g}{\rho_0} \frac{\partial \rho}{\partial z})^{\frac{1}{2}}$ is Brunt-Vaisala frequency. A_1, A_2, B_1, B_2, C_1, E_1 and E_2 are empirical constants and $(A_1, A_2, B_1, B_2, C_1, E_1$ and $E_2) = (0.92, 0.74, 16.6, 10.1, 0.08, 1.8, 1.33)$, respectively [139].

In the stable stratified fluid, the macroscopic scale of turbulence l should also satisfy the following [140]:

$$l \leq \frac{0.53q}{N} \quad (5.21)$$

5.1.1.6 Boundary conditions

The above control equations can only be closed by giving appropriate initial conditions and boundary conditions. The boundary conditions include the sea surface boundary conditions, the seafloor boundary conditions, the seafloor solid side boundary conditions and the sea side open boundary conditions.

1. The sea surface boundary conditions

On the free sea surface $z = \eta(x, y, z)$:

$$\rho_0 K_M \left(\frac{\partial U}{\partial z}, \frac{\partial V}{\partial z} \right) = (\tau_{ox}, \tau_{oy}) \quad (5.22)$$

$$\rho_0 K_H \left(\frac{\partial \theta}{\partial z}, \frac{\partial S}{\partial z} \right) = (\dot{H}, \dot{S}) \quad (5.23)$$

$$q^2 = B_1^{\frac{2}{3}} u_{\tau s}^2 \quad (5.24)$$

$$q^2 l = 0 \quad (5.25)$$

$$W = U \frac{\partial \eta}{\partial x} + V \frac{\partial \eta}{\partial y} + \frac{\partial \eta}{\partial t} \quad (5.26)$$

Where, (τ_{ox}, τ_{oy}) is the wind stress vector on the sea surface and its value is represented by $u_{\tau s}$. The calculation equation is as follows:

$$W = U \frac{\partial \eta}{\partial x} + V \frac{\partial \eta}{\partial y} + \frac{\partial \eta}{\partial t} \quad (5.27)$$

Where ρ_a is air density, and C_D is wind force factor. When the wind speed is less than 25 m/s, the wind force factor is determined by the following equation:

$$
C_D = \begin{cases}
1.2 \times 10^{-3} & \left|\vec{W}\right| < 11m/s \\
(0.49 + 0.065\left|\vec{W}\right|) \times 10^{-3} & 11m/s \leq \left|\vec{W}\right| < 25m/s \\
2.115 \times 10^{-3} & \left|\vec{W}\right| \geq 25m/s
\end{cases}
\tag{5.28}
$$

$\left|\vec{W}\right|$ is the sum of wind speed vector. $\dot{H}$ is the net heat flux on the sea surface which includes four parts: solar radiation, long-wave radiation, sensible heat and latent heat. $\dot{S} = \frac{S(0)(\dot{E}-\dot{p})}{\rho_0}$ is virtual salt flux, $(\dot{E}-\dot{p})$ is freshwater flux of sea surface evaporation / precipitation and $S(0)$ is the sea surface salinity.

2. Seafloor boundary conditions

At the seafloor, $z = -H(x,y)$:

$$
\rho_0 K_M \left(\frac{\partial U}{\partial z}, \frac{\partial V}{\partial z} \right) = (\tau_{bx}, \tau_{by})
\tag{5.29}
$$

$$
\frac{\partial \theta}{\partial z} = \frac{\partial S}{\partial z} = 0
\tag{5.30}
$$

$$
q^2 = B_1^{2/3} u_{\tau b}^2
\tag{5.31}
$$

$$
q^2 l = 0
\tag{5.32}
$$

$$
W_b = -U_b \frac{\partial H}{\partial x} - V_b \frac{\partial H}{\partial y}
\tag{5.33}
$$

Where, $H(x,y)$ is the bottom topography, (τ_{bx}, τ_{by}) is bottom friction stress vector at seafloor and its value is represented by $u_{\tau s}$. It is calculated with the following equation:

$$
(\tau_{bx}\tau_{by}) = \rho_0 C_Z \left|\vec{V}_b\right| \vec{V}_b = \rho_0 C_Z \sqrt{U^2 + V^2}(UV)
\tag{5.34}
$$

Where, $C_Z = MAX\{\frac{\kappa^2}{[ln(z/z_0)]^2}, 0.0025\}$ is the bottom friction coefficient, $\kappa = 0.4$ which is Von Karman constant and Z_0 is the roughness parameter of seafloor.

3. The solid side boundary conditions

It is generally assumed that the normal velocity vertical to the solid coast at the solid side of the researched area is zero. Namely, $\vec{U} \cdot \vec{n} = 0$. In addition, it is assumed that there is no heat and salt exchange between seawater and the solid boundary. Namely, $\frac{\partial \theta}{\partial \vec{n}} = \frac{\partial s}{\partial \vec{n}} = 0$.

4. The open boundary conditions

On the open boundary of the sea area, the boundary conditions of the sea level are calculated from the harmonic constant of the main constituents at the boundary with the following equation:

$$\eta = E_{mean} + \sum_{i=1}^{6} a_i \cos(w_i t - \varphi_i) \tag{5.35}$$

Where,

a_i is the vibration amplitude of the constituent i;

w_i is the frequency of the constituent i;

ϕ_i is the angle of delay of the constituent i;

E_{mean} is the residual water level at this point relative to the mean sea level.

In addition, the open boundary of current velocity is given by radiation condition or zero-gradient boundary condition, and the open boundary condition of temperature and salinity is given by zero-gradient boundary condition or upwind convection scheme in radiation inflow region of outflow region.

5.1.1.7 Vertical coordinate transformation

To better fit the bottom topography, σ coordinate transformation is adopted in the model in the vertical direction. Namely:

$$x^* = x, y^* = y, \sigma = \frac{z - \eta}{H + \eta} t^* = t \tag{5.36}$$

$H(x, y)$ represents the bottom topography and $\eta(x, y)$ represents the sea surface relief. At the point $Z = \eta$, $\sigma = 0$; at the point $z = -H$, $\sigma = -1$.

Assuming that $D = H + \eta$, the transformation of new and old coordinates is:

$$\frac{\partial G}{\partial x} = \frac{\partial G}{\partial x^*} - \frac{\partial G}{\partial \sigma} \left(\frac{\sigma}{D} \frac{\partial D}{\partial x^*} + \frac{1}{D} \frac{\partial \eta}{\partial x^*} \right) \tag{5.37}$$

$$\frac{\partial G}{\partial y} = \frac{\partial G}{\partial y^*} - \frac{\partial G}{\partial \sigma} \left(\frac{\sigma}{D} \frac{\partial D}{\partial y^*} + \frac{1}{D} \frac{\partial \eta}{\partial y^*} \right) \tag{5.38}$$

$$\frac{\partial G}{\partial z} = \frac{1}{D} \frac{\partial G}{\partial \sigma} \tag{5.39}$$

$$\frac{\partial G}{\partial t} = \frac{\partial G}{\partial t^*} - \frac{\partial G}{\partial \sigma} \left(\frac{\sigma}{D} \frac{\partial D}{\partial t^*} + \frac{1}{D} \frac{\partial \eta}{\partial t^*} \right) \tag{5.40}$$

$$w = W - U \left(\sigma \frac{\partial D}{\partial x^*} + \frac{\partial \eta}{\partial y^*} \right) - V \left(\sigma \frac{\partial D}{\partial y^*} + \frac{\partial \eta}{\partial x^*} \right) - \left(\sigma \frac{\partial D}{\partial t^*} + \frac{\partial \eta}{\partial t^*} \right) \tag{5.41}$$

So Equation (5.26) can be transformed into

$$w(x^*, y^*, 0, t^*) = 0 \tag{5.42}$$

$$w(x^*, y^*, -1, t^*) = 0 \tag{5.43}$$

In addition, any variable G which is integrated from the seafloor to the sea surface can be expressed as follows:

$$\bar{G} = \int_{-1}^{0} G d\sigma \tag{5.44}$$

Therefore, Equations (5.1), (5.2), (5.3), (5.4), (5.5), (5.6) and (5.7) can be written as (for ease of representation, the asterisk is omitted):

$$\frac{\partial \eta}{\partial t} + \frac{\partial U D}{\partial x} + \frac{\partial V D}{\partial y} + \frac{\partial \omega}{\partial \sigma} = 0 \tag{5.45}$$

$$\frac{\partial U D}{\partial t} + \frac{\partial U^2 D}{\partial x} + \frac{\partial U V D}{\partial y} + \frac{\partial U \omega}{\partial \sigma} - fVD + gD\frac{\partial \eta}{\partial x}$$
$$= \frac{\partial}{\partial \sigma}\left(\frac{K_M}{D}\frac{\partial U}{\partial \sigma}\right) - \frac{gD^2}{\rho_0}\frac{\partial}{\partial x}\int_{\sigma}^{0} \rho d\sigma + \frac{gD}{\rho_0}\frac{\partial D}{\partial x}\int_{\sigma}^{0}\sigma\frac{\partial \rho}{\partial \sigma}d\sigma + F_X \tag{5.46}$$

$$\frac{\partial V D}{\partial t} + \frac{\partial U V D}{\partial x} + \frac{\partial V^2 D}{\partial y} + \frac{\partial V \omega}{\partial \sigma} + fUD + gD\frac{\partial \eta}{\partial y}$$
$$= \frac{\partial}{\partial \sigma}\left(\frac{K_M}{D}\frac{\partial V}{\partial \sigma}\right) - \frac{gD^2}{\rho_0}\frac{\partial}{\partial y}\int_{\sigma}^{0} \rho d\sigma + \frac{gD}{\rho_0}\frac{\partial D}{\partial y}\int_{\sigma}^{0}\sigma\frac{\partial \rho}{\partial \sigma}d\sigma + F_Y \tag{5.47}$$

$$\frac{\partial \theta D}{\partial t} + \frac{\partial \theta U D}{\partial x} + \frac{\partial \theta V D}{\partial y} + \frac{\partial \theta \omega}{\partial \sigma} = \frac{\partial}{\partial \sigma}\left(\frac{K_H}{D}\frac{\partial \theta}{\partial \sigma}\right) + F_\theta \tag{5.48}$$

$$\frac{\partial SD}{\partial t} + \frac{\partial SU D}{\partial x} + \frac{\partial SV D}{\partial y} + \frac{\partial S\omega}{\partial \sigma} = \frac{\partial}{\partial \sigma}\left(\frac{K_H}{D}\frac{\partial S}{\partial \sigma}\right) + F_S \tag{5.49}$$

$$\frac{\partial q^2 D}{\partial t} + \frac{\partial U q^2 D}{\partial x} + \frac{\partial V q^2 D}{\partial y} + \frac{\partial \omega q^2}{\partial \sigma}$$
$$= \frac{\partial}{\partial \sigma}\left(\frac{K_q}{D}\frac{\partial q^2}{\partial \sigma}\right) + \frac{2K_M}{D}\left[\left(\frac{\partial U}{\partial \sigma}\right)^2 + \left(\frac{\partial V}{\partial \sigma}\right)^2\right] + \frac{2g}{\rho_0}K_H\frac{\partial \rho}{\partial \sigma} - 2\frac{Dq^3}{B_1 l} + F_q \tag{5.50}$$

$$\frac{\partial q^2 l D}{\partial t} + \frac{\partial U q^2 l D}{\partial x} + \frac{\partial V q^2 l D}{\partial y} + \frac{\partial \omega q^2 l}{\partial \sigma}$$
$$= \frac{\partial}{\partial \sigma}\left(\frac{K_q}{D}\frac{\partial q^2 l}{\partial \sigma}\right) + E_1 l\left\{\frac{K_M}{D}\left[\left(\frac{\partial U}{\partial \sigma}\right)^2 + \left(\frac{\partial V}{\partial \sigma}\right)^2\right] + \frac{qD^3}{\rho_0}K_H\frac{\partial \rho}{\partial \sigma}\right\} - \frac{Dq^3}{B_1}\bar{W} + DF_l \tag{5.51}$$

Where, ω is a vertical velocity vertical to σ layer after the coordinate transformation.

$$F_X = \frac{\partial D\hat{\tau}_{xx}}{\partial x} + \frac{\partial}{\partial y}(D\hat{\tau}_{yx}) \tag{5.52}$$

$$F_Y = \frac{\partial D\hat{\tau}_{yy}}{\partial y} + \frac{\partial}{\partial x}(D\hat{\tau}_{xy}) \tag{5.53}$$

$$\hat{\tau}_{xx} = 2A_M\left(\frac{\partial U}{\partial x}\right) \tag{5.54}$$

$$\hat{\tau}_{xy} = \hat{\tau}_{yx} = A_M \left(\frac{\partial U}{\partial y} + \frac{\partial V}{\partial x} \right) \tag{5.55}$$

$$\hat{\tau}_{yy} = 2A_M \left(\frac{\partial V}{\partial y} \right) \tag{5.56}$$

$$F_{\theta i} = \frac{\partial D \hat{q}_x}{\partial x} + \frac{\partial D \hat{q}_y}{\partial y} \tag{5.57}$$

$$\hat{q}_x = A_H \left(\frac{\partial \theta_i}{\partial x} \right) \tag{5.58}$$

$$\hat{q}_y = A_H \left(\frac{\partial \theta_i}{\partial y} \right) \tag{5.59}$$

Where, θ_i represents θ, S, q^2 and $q^2 l$, respectively.

5.1.1.8 Inner and outer mode splitting algorithm

In order to save computer time and improve calculation efficiency, FV-COM model adopts the method of inner and outer mode alternate calculation. The outer mode is two-dimensional and the time step is shorter when calculating the water level and vertical average velocity. The inner mode is three-dimensional and the time step is longer when calculating the physical quantities like current velocity, temperature, salinity and turbulence kinetic energy.

The equations for outer mode calculation can be obtained by integrating Equations (5.45), (5.46) and (5.47) vertically from the seafloor to the sea surface:

$$\frac{\partial \eta}{\partial t} + \frac{\partial \bar{U} D}{\partial x} + \frac{\partial \bar{V} D}{\partial y} = 0 \tag{5.60}$$

$$\begin{aligned}
&\frac{\partial \bar{U} D}{\partial t} + \frac{\partial \bar{U}^2 D}{\partial x} + \frac{\partial \bar{U} \bar{V} D}{\partial y} - f\bar{V}D + gD\frac{\partial \eta}{\partial x} - D\bar{F}_X \\
&= -\overline{WU}(0) + \overline{WU}(-1) - \frac{\partial \overline{DU'^2}}{\partial x} - \frac{\partial \overline{DU'V'}}{\partial y} \\
&\quad - \frac{gD^2}{\rho_0} \frac{\partial}{\partial x} \int_{-1}^{0} \int_{\sigma}^{0} \rho d\sigma' d\sigma + \frac{gD}{\rho_0} \frac{\partial D}{\partial x} \int_{-1}^{0} \int_{\sigma}^{0} \sigma' \frac{\partial \rho}{\partial \sigma} d\sigma' d\sigma
\end{aligned} \tag{5.61}$$

$$\begin{aligned}
&\frac{\partial \bar{V} D}{\partial t} + \frac{\partial \bar{U} \bar{V} D}{\partial x} + \frac{\partial \bar{V}^2 D}{\partial y} + f\bar{U}D + gD\frac{\partial \eta}{\partial y} - D\bar{F}_Y \\
&= -\overline{WV}(0) + \overline{WV}(-1) - \frac{\partial \overline{DU'V'}}{\partial x} - \frac{\partial \overline{DV'^2}}{\partial y} \\
&\quad - \frac{gD^2}{\rho_0} \frac{\partial}{\partial y} \int_{-1}^{0} \int_{\sigma}^{0} \rho d\sigma' d\sigma + \frac{gD}{\rho_0} \frac{\partial D}{\partial y} \int_{-1}^{0} \int_{\sigma}^{0} \sigma' \frac{\partial \rho}{\partial \sigma} d\sigma' d\sigma
\end{aligned} \tag{5.62}$$

Where,

$$(\bar{U}\bar{V}) = \int_{-1}^{0} (UV) d\sigma \tag{5.63}$$

$$(\overline{U'^2}, \overline{V'^2}, \overline{U'V'}) = \int_{-1}^{0} (U'^2, V'^2, U'V') d\sigma \qquad (5.64)$$

$$(U'V') = (U - \bar{U}V - \bar{V}) \qquad (5.65)$$

$$D\overline{F_X} = \frac{\partial}{\partial x}\left(2A_M \frac{\partial \bar{U}D}{\partial y}\right) + \frac{\partial}{\partial y}A_M\left(\frac{\partial \bar{U}D}{\partial y} + \frac{\partial \bar{V}D}{\partial x}\right) \qquad (5.66)$$

$$D\overline{F_Y} = \frac{\partial}{\partial y}\left(2A_M \frac{\partial \bar{V}D}{\partial y}\right) + \frac{\partial}{\partial x}A_M\left(\frac{\partial \bar{U}D}{\partial y} + \frac{\partial \bar{V}D}{\partial x}\right) \qquad (5.67)$$

where, $-\overline{WU}(0)$, $-\overline{WV}(0)$ and $-\overline{WU}(-1)$, $-\overline{WV}(-1)$ are the sea surface wind stress and bottom stress component, respectively.

It is firstly assumed that all variables at the time points t^{n-1} and t^n have already been obtained, then the right-hand terms of Equations (5.61) and (5.62) can be obtained and it is considered that they remain unchanged from the time point t^n to t^{n+1}. Equations (5.60), (5.61) and (5.62) and the time step of outer mode DTE are used to show the several steps of integral to the time point of t^{n+1}, thus the water level and vertical average velocity can be obtained for calculation of the inner mode.

5.1.1.9   Lagrange residual current calculation module

Sun Wenxin and Feng Shizuo (1992) pointed out that although Euler residual current is a solenoidal vector field [141], it does not meet the condition of conservation of material surface composed of fluid micro-aggregates in the current field. The mass transport speed is not only a solenoidal vector field, but can also be proved to satisfy the conservation equation of material surface in the current field in the most general sense. It is of practical significance in the lowest level to describe the circulation issues in the shallow sea by using Lagrange residual current which represents the velocity field constituted by mass transport velocity.

Euler is usually used to represent the shallow sea circulation, i.e., Euler residual circulation. It is defined as low-pass filtering at any selected point in the current field space, or simply time averaging over a tidal period, which is called Euler average velocity or Euler residual current.

Sun Wenxin and Feng Shizuo [141] pointed out that the mass transport velocity was derived as Lagrange residual current of the lowest level. The calculation process is as follows: calculate the Euler average velocity first and then add the Stokes drift velocity to get the Lagrange average velocity.

The calculation equations are:

Lagrange average velocity:

$$U_L = U_E + U_S \qquad (5.68)$$

Euler average velocity:

$$U_E = \frac{1}{nT} \int\limits_{t_0}^{t_0+nT} U(x_0 t) dt \tag{5.69}$$

Stokes drift velocity:

$$U_S = \langle N\xi_0 \cdot (\nabla U_0) \rangle \tag{5.70}$$

According to the above definition, the part for solving the Euler and Stokes average velocity is added to the model, and the Lagrange average velocity can be obtained by simply adding the simulation results of Euler and Stokes average velocity.

5.1.1.10 Dry and wet grid processing technology

One of the most difficult problems in the numerical model of estuaries and coasts is to provide accurate simulation of the volume transport into or out of the tidal flat areas. This kind of volume transport may be several times larger than that of the main rivers. In the past 30 years, two methods have been used to solve this problem: one is variable boundary method and the other is dry and wet point judgment method. In the first method, the calculation area is limited to a boundary between land and water. On such boundary, the total water depth and vertical transport are equal to zero. Since this boundary moves in dry and wet areas, the model grid needs to be regenerated in each step. This method is suitable for ideal estuaries or coasts with simple shapes, but not for real estuaries or harbors with complex shapes, including small ports, islands, obstacles and water bay. In the second method, the calculation area covers the maximum wet area. The numerical grid consists of dry and wet points and has a boundary defined by the boundary between land and water. The dry and wet points are distinguished with the local total water depth $D = H(x,y) + \zeta(x,y,t)$ (of which, H is the reference water depth, ζ is the surface relief, as shown in Figure 5.3). When $D > 0$, it is wet point; when $D = 0$, it is dry point. At the dry point, the velocity is automatically defined as zero, but the salinity value of the previous step is maintained at such point. Because this method is relatively simple, it is widely used in model of estuaries and coasts to simulate the volume transport of waterbody in intertidal zone.

Whichever method is used to simulate the drying and wetting processes in estuaries or coastal intertidal zones, it must be ensured that the mass conservation is observed. The dry-wet point in all these methods is determined by some empirical criteria and the estimation of water transport in the dry-wet conversion area depends on: (1) criteria used to define the dry-wet point; (2) time step of numerical integration; (3) the horizontal and vertical resolution of the model grid; (4) amplitude of surface water level; and (5) bottom topography. In the σ coordinate transformation model, it is also related to the thickness of bottom viscous boundary layer.

A new dry and wet treatment method is used in FVCOM. This method has been proved to be effective in a series of tidal simulations in an ideal semi-closed harbor with intertidal zone. The relationship between time step and grid resolution, externally forced amplitude, slope of dry and wet zone and thickness of bottom viscous boundary layer are discussed, and the criterion of selecting time step is obtained. The criterion used in validation is mass conservation which is a prerequisite for the objective assignment of dry-wet points in a bay.

The criteria for dry or wet judgment of nodes are as follows:

$$\begin{cases} Wet, if D = H + \zeta + h_B > D_{\min} \\ Dry, if D = H + \zeta + h_B \le D_{\min} \end{cases} \tag{5.71}$$

For triangular grid cells:

$$\begin{cases} Wet, if D = \min(h_{B,\hat{i}}, h_{B,\hat{i}}, h_{B,\hat{i}}) + \max(\xi_{\hat{i}}, \xi_{\hat{j}}, \xi_{\hat{k}}) > D_{\min} \\ Dry, if D = \min(h_{B,\hat{i}}, h_{B,\hat{i}}, h_{B,\hat{i}}) + \max(\xi_{\hat{i}}, \xi_{\hat{j}}, \xi_{\hat{k}}) \le D_{\min} \end{cases} \tag{5.72}$$

In the above equations, D_{min} represents thickness of bottom viscous boundary layer, h_B represents the height of the terrain connected to the main channel boundary of a river and $\hat{i}, \hat{j}, \hat{k}$ are integers used to determine the three nodes of a triangular grid cell.

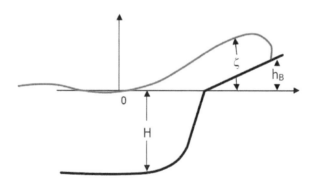

FIGURE 5.3: Definition of reference water depth (H), sea level (ζ) and terrain height (h_B).

When a triangular cell is dried, the velocity at the mass center of the triangular grid cell is defined as zero and there is no flux on the three boundaries of the triangular grid cell. This triangular grid cell will be removed from the flux calculation. For example, the integral form of the continuity equation in FVCOM is written as:

$$\iint\limits_{\text{TCE}} \frac{\partial \zeta}{\partial t} dx dy = -\iint\limits_{TCE} [\frac{\partial(\bar{u}D)}{\partial x} + \frac{\partial \bar{v}D}{\partial y}] dx dy = -\oint\limits_{l} \bar{V}_N dl \qquad (5.73)$$

Where, $\bar{u}$ and $\bar{v}$ are the x and y component of vertical average velocity. In the dry-wet point system, only wet triangles are considered in the calculation, for the boundary flux of dry triangles is zero. In this way, the conservation of volume is ensured.

5.1.2 Model Configuration and Result Verification

5.1.2.1 Model configuration

The initial temperature and salinity field of the model is the annual mean temperature and salinity field obtained from the historical observation data; the open boundary is forced by the water level predicted by tidal harmonic constant; the six main harmonic constants of semi-diurnal constituent and diurnal constituent M_2, S_2 , N_2 , K_1 , O_1 and P_1 are used for prediction. The tidal harmonic constants used are obtained from the tidal model simulation results of the Bohai Sea, Yellow Sea and East China Sea with adjoint variational method. The model is vertically divided into eleven σ layers and the upper surface layer has relatively higher resolution. The time step of the outer mode of the model is 2 seconds and that of the inner mode is 12 seconds. The model automatically forecasts the conditions in 48 hours from zero in the morning and outputs the calculation results every hour.

The terrain used in the model comes from the charts of different resolution provided by Shandong Maritime Safety Administration and the charts splicing is completed by using ArcGIS and other professional software. In addition, the related terrain and shoreline information extracted from the basic geographic information database of National Marine Information Center is supplemented and the method of nearest point interpolation of set thresholds is used to complete the production of high-precision terrain data products for the forecast area (as shown in Figure 5.4 to Figure 5.7).

The sea areas calculated by the model are the Bohai Sea and Yellow Sea ($37° \sim 41°N$, $117.42° \sim 126.25°E$). Grid refinement was carried out in key areas like Kiaochow Bay, coastal waters of Qingdao, Chengshan Cape routeing waters and Bohai Strait. The refined areas are divided into three levels: the primary refined area has a grid precision of 100 m and includes the following three sea areas: Kiaochow Bay, coastal waters of Qingdao, Chengshan Cape routeing water and the Bohai Strait; the secondary refined area has a grid precision of 300 m and includes the following sea areas: the whole Bohai Sea and the sea areas west to $124°E$; the rest of the sea areas belong to the tertiary refined area which has a grid resolution of 1-bit (as shown in Figure 5.8 to Figure 5.12). The resolution near the islands in the Bohai Strait can reach the order of 100 m and basically all the islands can be preserved and distinguished,

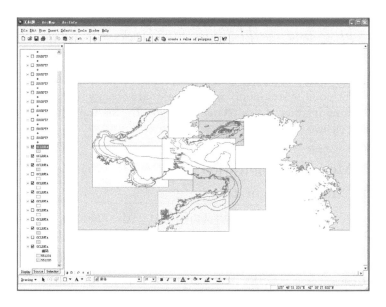

FIGURE 5.4: Block division in charts with different resolution.

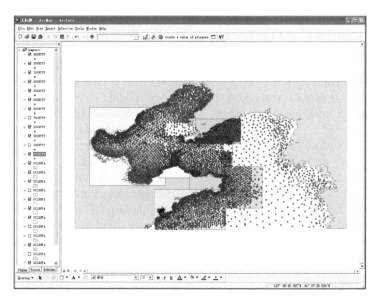

FIGURE 5.5: Distribution of deep water data points in charts with different resolution.

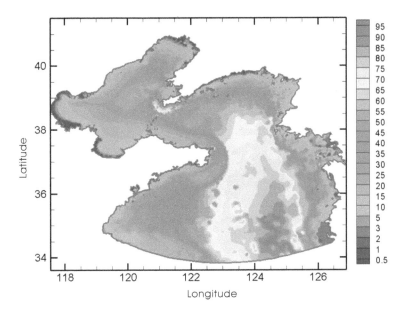

FIGURE 5.6: High-precision Bohai Sea and Yellow Sea terrain after processing.

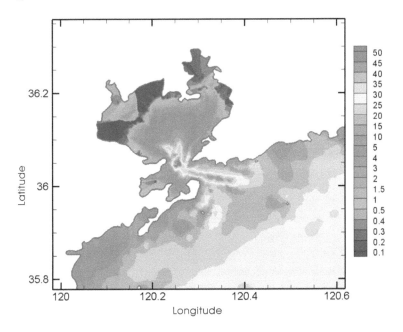

FIGURE 5.7: High-precision Kiaochow Bay terrain after processing.

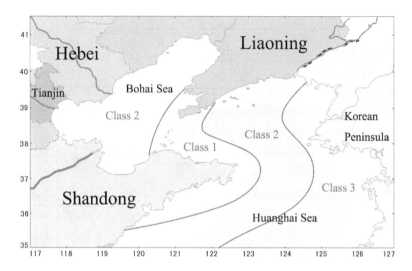

FIGURE 5.8: Schematic diagram of researched sea area.

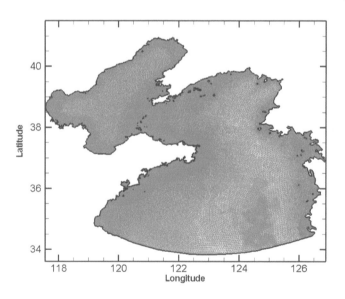

FIGURE 5.9: Grid distribution of the Bohai Sea and Yellow Sea; the shading represents the water depth.

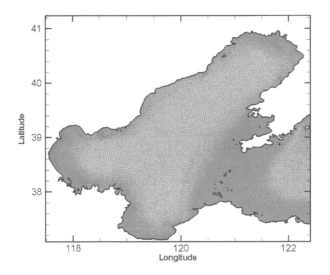

FIGURE 5.10: Grid distribution of the Bohai Sea.

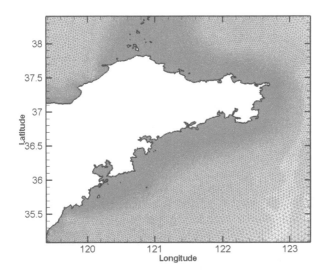

FIGURE 5.11: High-resolution grid distribution of Shandongtou.

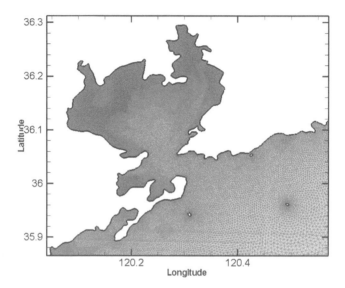

FIGURE 5.12: Grid distribution of Kiaochow Bay.

avoiding the situation in the previous forecast that the ocean current passes through the islands. The grid resolution in the non-key areas is reduced, which not only ensures the high-resolution simulation of the key areas, controls the calculation cost, and further optimizes the forecast efficiency of ocean current module.

5.1.2.2   Result verification

1. Tidal harmonic constant verification

Table 5.1 shows the comparison between the harmonic constant of the constituent M_2 simulated by the model and the harmonic constant observed at the tidal stations in the Circum-Bohai Sea Region and Yellow Sea. The average amplitude error simulated at the 19 tidal stations is only 2.8 cm, the average error of delay angle is only 3.7 degrees. The harmonic constant obtained after stabilized calculation is well consistent with the observation results. The reasons for the difference may be the following: (1) the difference in location of the grids and the tidal stations, for the model's horizontal grid spacing is thousands of meters, such a grid still has poor resolution of coastlines; (2) changes in the shoreline and errors in topography and water depth which are especially obvious in the Yellow River Estuary and Laizhou Bay; (3) the harmonic constant precision of the constituents at the boundaries.

TABLE 5.1: The comparison between harmonic constant of the constituent M_2 simulated by the model and the harmonic constant observed at various tidal stations.

Station	Amplitude (cm)			Phase (degree)		
	Observed value	Calculated value	Error	Observed value	Calculated value	Error
Dengsha River	116	116.6	0.6	275	267	-1.3
Lvshun	82	79.5	-2.5	299	305.6	6.6
Yangtouwa	61	61.0	0	321	320.4	-0.6
Beihuangcheng	61	54.4	-6.6	299	301.2	2.2
Caofeidian	61	57.3	-3.7	60	66.5	6.5
Qiansuowai	13	16.9	3.9	176	172.2	-3.8
Huangdi	40	39.1	-0.9	150	153	3.0
Changshansi	61	62.6	-1.6	159	150.2	-8.8
Yingkou	130	137	-7	129	127.1	-1.9
Bayuquan	119	113.3	-5.7	125	126.8	1.8
Changxing Island	54	54.4	0.4	83	82	-1
Hulutao	61	52.5	1.5	11	14.3	3.3
Yingchengzi Bay	55	55.4	0.4	348	341.1	-6.9
Terrace 28	20	19.8	-0.2	334	347.3	13.3
Yantai	61	64.9	3.9	288	285.8	-2.2
Longkou	40	29.1	-10.9	329	328.8	-0.2
Tanggu	113	109	-4	91	91	0
Hulu Island	94	92.4	-1.6	151	145.4	-5.6
Dalian	94	91	-3	287	285.3	-1.7
Average error			2.8			3.7

Figure 5.13 to Figure 5.16 are the diagrams of co-tidal hours of the tidal constituents M_2, S_2, K_1 and O_1 in the Bohai Sea and Yellow Sea obtained from the model simulation. The co-tidal lines and iso-amplitude lines are well consistent with the distribution characteristics as set forth in the Atlas of Bohai Sea, Yellow Sea and East China Sea [142].

Since there are many islands in the Bohai Strait, its current field structure is very complicated. When the tidal current is introduced into the Bohai Sea from the open sea, the islands distributed in the Bohai Straits divide the Straits into six main waterways, so that the current rate will inevitably increase due to the sudden narrowing of the waterways, of which the Laotieshan Waterway which is located in the northernmost part is the deepest one and has the largest current velocity. At low tide, the residual current of a small amount of ebb tide still has a relatively large velocity when it passes through the narrow waterways between islands and presents a right-hand helicity, and the tide will turn in the next moment. Due to the characteristics of free refinement of unstructured grid in FVCOM model, the grid refinement in Bohai Strait can finely distinguish many islands in the Strait and more accurately simulate the current field in the Bohai Strait (as shown in Figure 5.17 to Figure 5.19).

One-year tidal level and ocean current observations were made at Changxing Island Observatory (121(16.320'E), 39(30.433'N)) from March 15, 2008 to March 20, 2009. The tide and tidal current were harmonized with the observed data and the harmonized results were then compared with the numerical

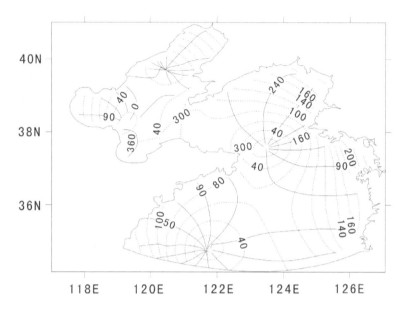

FIGURE 5.13: The calculated iso-amplitude lines (cm) and co-tidal lines (degree) of the constituent M_2 in the Bohai Sea and Yellow Sea. (The dotted line is iso-amplitude line (cm); the solid line is co-tidal line (degree).)

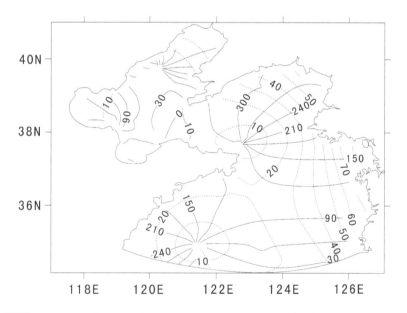

FIGURE 5.14: The calculated iso-amplitude lines (cm) and co-tidal lines (degree) of the constituent S_2 in the Bohai Sea and Yellow Sea. (The dotted line is iso-amplitude line (cm); the solid line is co-tidal line (degree).)

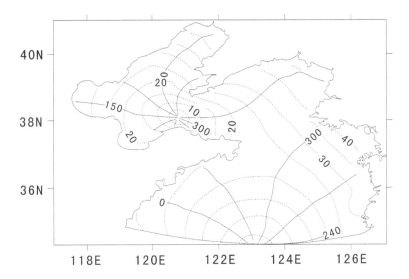

FIGURE 5.15: The calculated iso-amplitude lines (cm) and co-tidal lines (degree) of the constituent K_1 in the Bohai Sea. and Yellow Sea. (The dotted line is iso-amplitude line (cm); the solid line is co-tidal line (degree).)

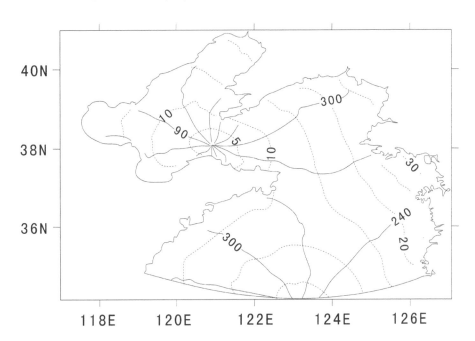

FIGURE 5.16: The calculated iso-amplitude lines (cm) and co-tidal lines (degree) of the constituent O_1 in the Bohai Sea and Yellow Sea. (The dotted line is iso-amplitude line (cm); the solid line is co-tidal line (degree).)

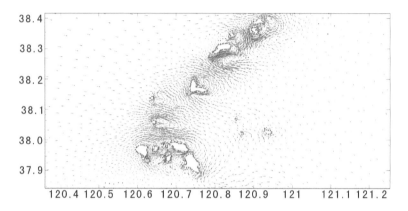

FIGURE 5.17: Current field at low tide in the Bohai Strait.

simulation results, as shown in Table 5.2. It is found from the comparison that the average delay angle error of tidal current is 9.3 degrees and the average error of amplitude of tidal current is 5.7 cm; the average delay angle error of tide level is 9.4 degree and the average error of amplitude of tidal level is 2.4 cm. The numerical simulation results are in good agreement with the observed results.

2. Tidal current verification

A total of 13 comparisons of the data at point A $(120°43'E,\ 39°52'N)$, point B $(121°21'E,\ 40°30'N)$, point C $(117°57'E,\ 38°39'N)$, point D $(121°24E,\ 40°38'N)$, point E $(121°24'E,\ 40°38'N)$ and point F $(119°25'E,\ 37°59'N)$ in the Bohai Sea and Yellow Sea were carried, out and the results are shown in Table 5.3 and Figure 5.20 to Figure 5.32. It can be seen that the simulated tidal current is close to the actually observed tidal current, and the

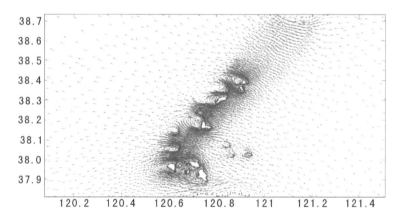

FIGURE 5.18: Current field in the middle of high tide in the Bohai Strait.

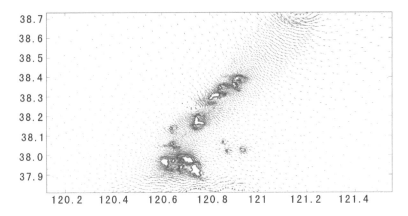

FIGURE 5.19: Current field in the middle of ebb tide in the Bohai Strait.

current velocity, current direction, high and low tidal current and the turn of tidal current are basically the same.

The ocean current observation data of Liaodong Bay in March 2009 were selected for comparison. The points compared were No.1 ($120°35.827'E$, $40°18.619'N$) and No.2 ($120°38.35'E$, $40°17.178'N$). There are 4 comparison processes, as shown in Figure 5.33 to Figure 5.36, respectively. It is found through the comparison that the simulation results of ocean current are in good agreement with the observed results.

The observed ocean current at five points, namely (H1: $120°15.5E$, $34°21.5'N$; H2: $120°17.5'E$, $34°24'N$; H3: $120°17'E$, $34°19'N$; H4: $120°21'E$, $34°21'N$; H5: $120°20'E$, $34°15'N$), in the Yellow Sea area near Lianyungang on September 7 and 8, 2006 were selected for comparison. The results are shown in Figure 5.37 to Figure 5.41. The current velocity and direction of

TABLE 5.2: Comparison of harmonic constant of tidal current and tide level observed at Changxing Island Observatory.

Element	Tidal constituent	Harmonic constant					
		North component (observed)		North component (simulated)		Error	
		Delay angle	Amplitude	Delay angle	Amplitude	Delay angle	Amplitude
Tidal current	O_1	306	8.3	293.96	6.57	12.04	1.73
	K_1	359	12.8	345.63	10.49	13.37	2.31
	M_2	19	58.1	4.07	46.31	14.93	11.79
	S_2	80	17.3	75.04	10.41	4.96	6.89
Tide level	O_1	28.4	22.3	23	18.91	5.4	3.39
	K_1	71.5	27.9	62.57	22.81	8.93	5.09
	M_2	35	42.8	36.28	43.65	-1.28	-0.85
	S_2	84.9	14.1	56.27	12.3	28.63	1.8

TABLE 5.3: Comparison between observed and simulated tidal current in Bohai Sea and Yellow Sea.

Point	Time	Current velocity		Current direction	
		Relative error(%)	Correlation coefficient	Relative error(%)	Correlation coefficient
A	August 16 and 17, 1997	8.8	0.97	9.8	0.90
A	August 23 and 24, 1997	9.8	0.90	5.2	0.98
A	July 15 and 16, 1997	9.8	0.95	9.8	0.91
A	July 25 and 26, 1997	7.1	0.96	9.7	0.95
A	May 22 and 23, 1998	5.8	0.99	9.6	0.94
B	January 19 and 20, 1991	9.6	0.90	4.7	0.97
C	December 9 and 10, 1991	9.8	0.91	10	0.92
C	December 14 and 15, 1991	9.9	0.93	7.1	0.96
D	October 24 and 25, 1988	8.3	0.94	9.3	0.92
E	September 3 and 4, 1989	9.5	0.91	6.5	0.97
E	November 2 and 3, 1989	9.3	0.89	9.7	0.96
F	July 19 and 20, 1988	6.6	0.99	9.3	0.93
F	August 3 and 4, 1988	9.7	0.88	9.6	0.87

simulated tidal current are highly consistent with that of the observed tidal current, indicating that the model results can basically describe the right tidal current process.

The observed data at two points in Laizhou Bay, namely (L1: $118°55.650'E$, $38°11.600'N$; L2: $118°53.553'E$, $38°8.547'N$), on April 28 and 29, 2004 were selected for comparison and the results are shown in Figure 5.42 and Figure 5.43, respectively. The comparison shows that the simulation results are basically consistent with the observation results and can represent the actual tidal current process.

3. Sea level verification

A total 12 comparisons of the data at point BH1 ($119°25'E$, $37°59'N$), point BH2 ($117°57'E$, $38°39'N$), point BH3 ($121°24E$, $40°38'N$) and point

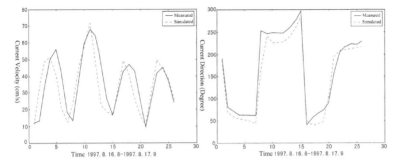

FIGURE 5.20: Comparison between observed and simulated tidal current on August 16 and 17, 1997 at point A. Left: Comparison of current velocity, the relative error is 8.8% and the correlation coefficient is 0.97. Right: Comparison of current direction, relative error is 9.8% and the correlation coefficient is 0.90.

BH4 ($120°43'E$, $39°52'N$) were carried out (as shown in Figure 5.44 to Figure 5.55). The results are as follows:

- Point BH1: the relative error between the simulated value and the observed value is 5.4%, the root mean square error is 0.0333 and the correlation coefficient is 0.91.

- Point BH2: the relative error between the simulated value and the observed value is 5.1%, the root mean square error is 0.0301 and the correlation coefficient is 0.93.

- Point BH3: the relative error between the simulated value and the observed value is 10.1%, the root mean square error is 0.025 and the correlation coefficient is 0.90.

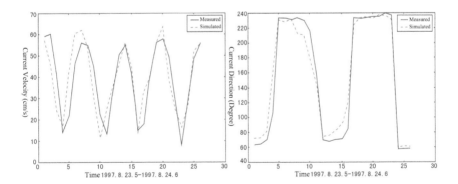

FIGURE 5.21: Comparison between observed and simulated tidal current on August 23 and 24, 1997 at point A. Left: Comparison of current velocity, the relative error is 9.9% and the correlation coefficient is 0.90. Right: Comparison of current direction, relative error is 5.2% and the correlation coefficient is 0.98.

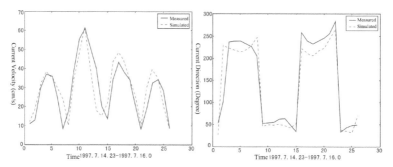

FIGURE 5.22: Comparison between observed and simulated tidal current on July 15 and 16, 1997 at point A. Left: Comparison of current velocity, the relative error is 9.8% and the correlation coefficient is 0.95. Right: Comparison of current direction, relative error is 9.8% and the correlation coefficient is 0.91.

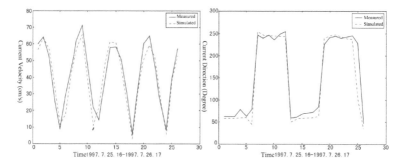

FIGURE 5.23: Comparison between observed and simulated tidal current on July 25 and 26, 1997 at point A. Left: Comparison of current velocity, the relative error is 7.1% and the correlation coefficient is 0.96. Right: Comparison of current direction, relative error is 9.7% and the correlation coefficient is 0.95.

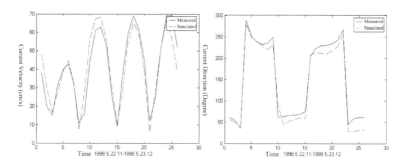

FIGURE 5.24: Comparison between observed and simulated tidal current on May 22 and 23, 1998 at point A. Left: Comparison of current velocity, the relative error is 5.8% and the correlation coefficient is 0.99. Right: Comparison of current direction, relative error is 9.6% and the correlation coefficient is 0.94.

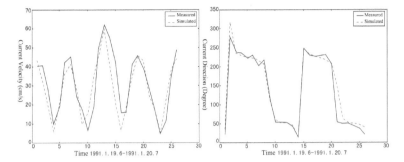

FIGURE 5.25: Comparison between observed and simulated tidal current on January 19 and 20, 1991 at point B. Left: Comparison of current velocity, the relative error is 9.6% and the correlation coefficient is 0.90. Right: Comparison of current direction, relative error is 4.7% and the correlation coefficient is 0.97.

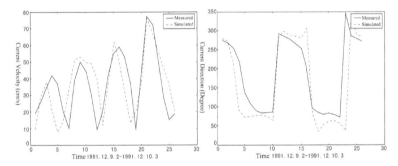

FIGURE 5.26: Comparison between observed and simulated tidal current on December 9 and 10, 1991 at point C. Left: Comparison of current velocity, the relative error is 9.8% and the correlation coefficient is 0.91. Right: Comparison of current direction, relative error is 10% and the correlation coefficient is 0.92.

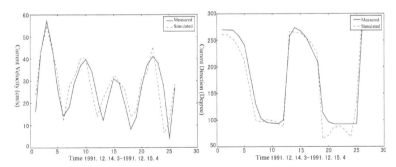

FIGURE 5.27: Comparison between observed and simulated tidal current on December 14 and 15, 1991 at point C. Left: Comparison of current velocity, the relative error is 9% and the correlation coefficient is 0.93. Right: Comparison of current direction, relative error is 7.1% and the correlation coefficient is 0.96.

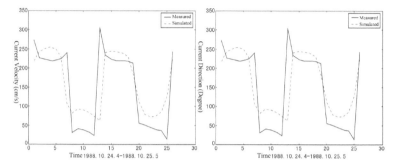

FIGURE 5.28: Comparison between observed and simulated tidal current on October 24 and 25, 1988 at point D. Left: Comparison of current velocity, the relative error is 8.3% and the correlation coefficient is 0.94. Right: Comparison of current direction, relative error is 9.3% and the correlation coefficient is 0.92.

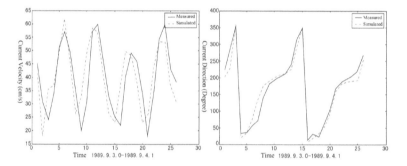

FIGURE 5.29: Comparison between observed and simulated tidal current on September 3 and 4, 1989 at point E. Left: Comparison of current velocity, the relative error is 9.5% and the correlation coefficient is 0.91. Right: Comparison of current direction, relative error is 6.5% and the correlation coefficient is 0.98.

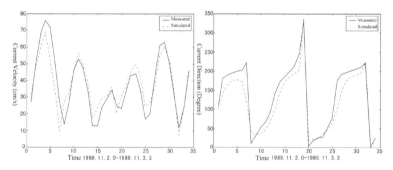

FIGURE 5.30: Comparison between observed and simulated tidal current on November 2 and 3, 1989 at point E. Left: Comparison of current velocity, the relative error is 9.3% and the correlation coefficient is 0.89. Right: Comparison of current direction, relative error is 9.7% and the correlation coefficient is 0.96.

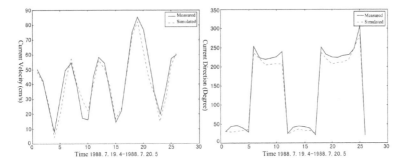

FIGURE 5.31: Comparison between observed and simulated tidal current on July 19 and 20, 1988 at point F. Left: Comparison of current velocity, the relative error is 6.6% and the correlation coefficient is 0.99. Right: Comparison of current direction, relative error is 9.3% and the correlation coefficient is 0.93.

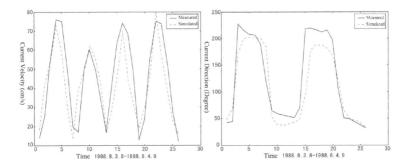

FIGURE 5.32: Comparison between observed and simulated tidal current on August 3 and 4, 1988 at point F. Left: Comparison of current velocity, the relative error is 9.7% and the correlation coefficient is 0.88. Right: Comparison of current direction, relative error is 9.6% and the correlation coefficient is 0.87.

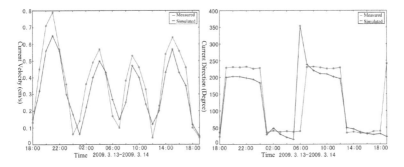

FIGURE 5.33: Comparison between observed and simulated tidal current on March 13 and 14, 2009 at point No.1 (left: current velocity; right: current direction).

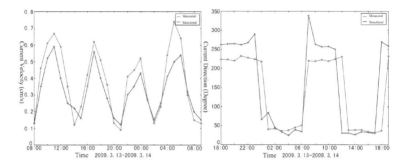

FIGURE 5.34: Comparison between observed and simulated tidal current on March 13 and 14, 2009 at point No.2 (left: current velocity; right: current direction).

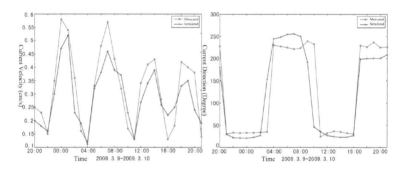

FIGURE 5.35: Comparison between observed and simulated tidal current on March 9 and 10, 2009 at point No.1 (left: current velocity; right: current direction).

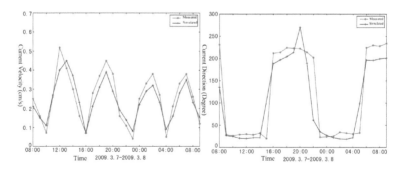

FIGURE 5.36: Comparison between observed and simulated tidal current on March 7 and 8, 2009 at point No.2 (left: current velocity; right: current direction).

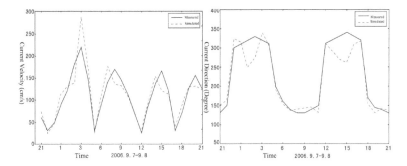

FIGURE 5.37: Comparison between observed and simulated tidal current on September 7 and 8, 2006 at point H1 (left: current velocity; right: current direction).

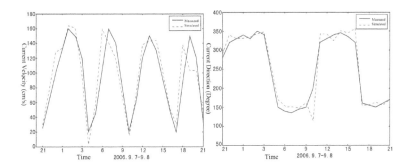

FIGURE 5.38: Comparison between observed and simulated tidal current on September 7 and 8, 2009 at point H2 (left: current velocity; right: current direction).

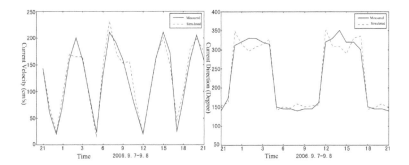

FIGURE 5.39: Comparison between observed and simulated tidal current on September 7 and 8, 2006 at point H3 (left: current velocity; right: current direction).

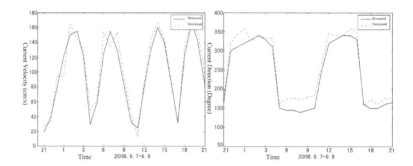

FIGURE 5.40: Comparison between observed and simulated tidal current on September 7 and 8, 2006 at point H4 (left: current velocity; right: current direction).

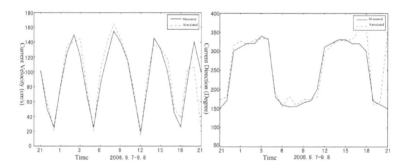

FIGURE 5.41: Comparison between observed and simulated tidal current on September 7 and 8, 2006 at point H5 (left: current velocity; right: current direction).

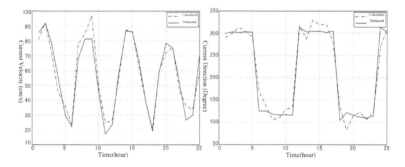

FIGURE 5.42: Comparison between observed and simulated tidal current on April 28 and 29, 2004 at point L1 (left: current velocity; right: current direction).

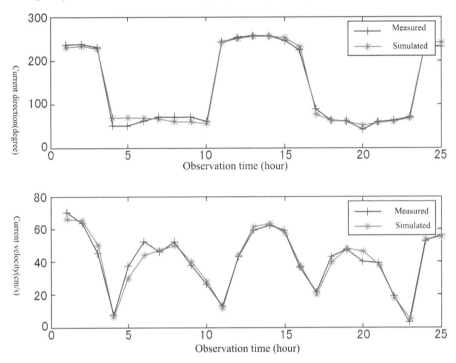

FIGURE 5.43: Comparison between observed and simulated tidal current on April 28 and 29, 2004 at point L2 (upper: current velocity; lower: current direction).

- Point BH4: the relative error between the simulated value and the observed value is 5.8%, the root mean square error is 0.013 and the correlation coefficient is 0.90.

It can be seen from the comparison results that the simulated sea level agrees well with the actually observed sea level, the root mean square error is basically within 0.1 cm, the ratio of the observed deviation to the mean value is all less than 10% and the correlation coefficient is greater than 0.9.

It can be found through the above comparison of harmonic constant, tidal current and sea level that the FVCOM model can better simulate the characteristics of tide and tidal current in the Bohai Sea and Yellow Sea, and it can be affirmed that such model can be reliably applied in forecast of the conditions in the Bohai Sea and Yellow Sea and the forecast results truly reflect the current field characteristics of the forecasted sea areas.

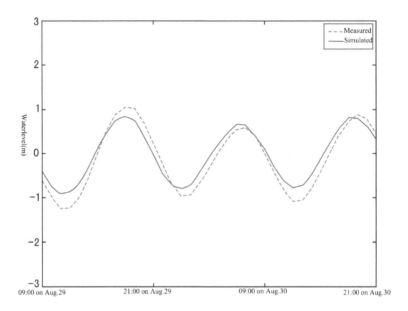

FIGURE 5.44: Comparison of sea level at point BH1 on August 29 and 30, 1988.

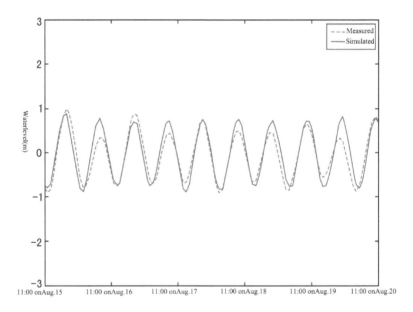

FIGURE 5.45: Comparison of sea level at point BH1 from August 15 to August 20, 1988.

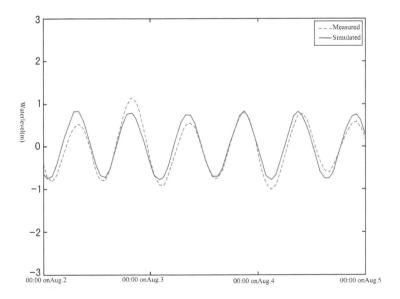

FIGURE 5.46: Comparison of sea level at point BH1 on from August 2 to August 5, 1988.

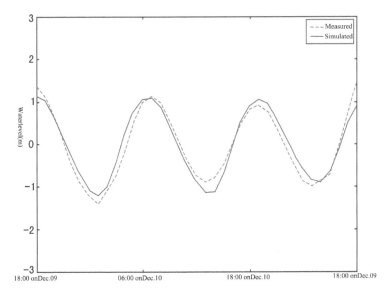

FIGURE 5.47: Comparison of sea level at point BH2 from December 9 to December 11, 1992.

Information Engineering for Marine Oil Spills

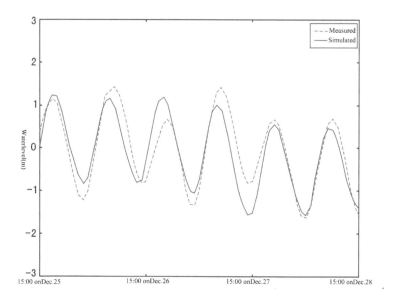

FIGURE 5.48: Comparison of sea level at point BH2 from December 25 to December 28, 1991.

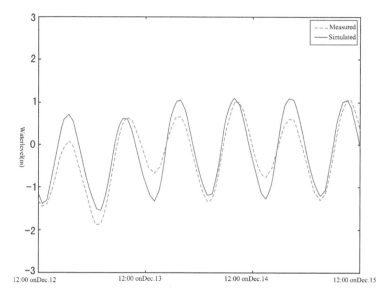

FIGURE 5.49: Comparison of sea level at point BH2 from December 12 to December 15, 1991.

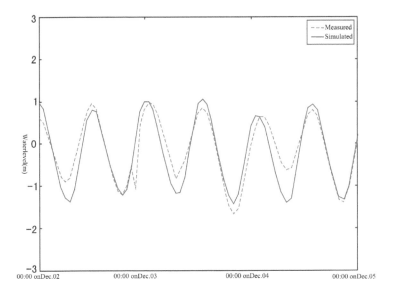

FIGURE 5.50: Comparison of sea level at point BH2 from December 2 to December 5, 1991.

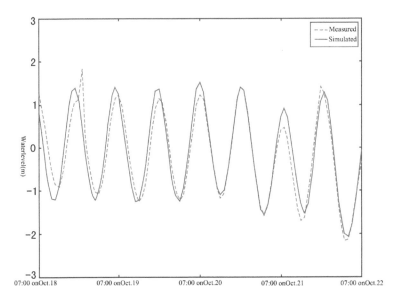

FIGURE 5.51: Comparison of sea level at point BH3 from October 18 to October 22, 1988.

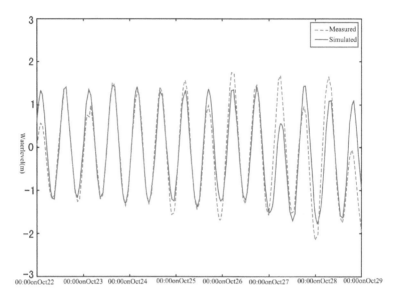

FIGURE 5.52: Comparison of sea level at point BH3 from October 22 to October 29, 1988.

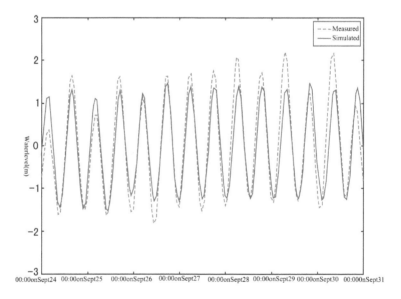

FIGURE 5.53: Comparison of sea level at point BH3 from September 24 to September 31, 1988.

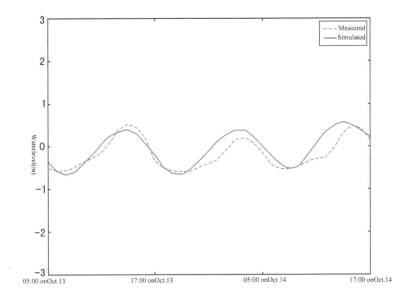

FIGURE 5.54: Comparison of sea level at point BH4 from October 13 to October 14, 1997.

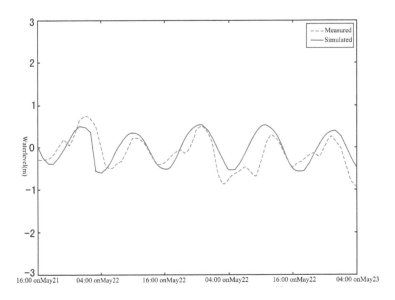

FIGURE 5.55: Comparison of sea level at point BH4 from May 21 to 23, 1998.

5.2   Ocean Wave Module

The impact of waves on oil spill is mainly reflected in wave disturbance (especially broken waves). Wave disturbance affects the process of breaking, dispersing and emulsifying of spilled oil, and makes spilled oil enter the water body in the form of oil particles. Therefore, wave is the control factor of the rate that oil particle enters into the water and one of the environmental forces causing oil slick rupture. In addition, the wave residual current generated by non-linear waves such as wind waves and surges also has a certain influence on the drift of the oil slick. However, the oil slick will make the surface tension increase and the sea surface tend to be smooth, greatly weakening the non-linear effects of wind waves and surges. So the influence of wave residual current on drift of oil slick can generally be ignored in model calculation [143].

In this system, the ocean wave module provides necessary wave element field for the oil spill behavior and fate prediction model, so as to accurately simulate the process that the spilled oil enters into the water.

5.2.1   Introduction to the Model

The model adopted by the ocean wave module is SWAN, the most advanced third generation shallow sea wave model in the world. This model was developed and maintained by the Department of Civil Engineering of Delft University in the Netherlands. From the first publicly released version (SWAN 30.51) to the current version (SWAN 40.81), it has been continuously improved, with gradually improved performance and gradually enhanced functions.

SWAN model takes more physical processes into account, including the latest results of current ocean wave forecast research. The factors considered are:

1. Wave propagation process

- Refraction caused by current and unsteady water depth changes;

- The shallowing effect caused by water bottom and current changes;

- Obstacles and reflections in counter current propagation;

- Propagation of waves in geometric space;

- The obstruction to waves caused by sub-grid obstacles and the propagation of waves through sub-grid obstacles;

- Wave-induced setup.

2. Generation and dissipation of waves

- Wind input;

- White-cap breaking;

- Breakage caused by shallowing;

- Bottom friction;

- Three-wave and four-wave non-linear interaction.

The SWAN model can be used in calculation of the wind waves and surges of laboratory scale and continental shelf sea scale. In addition to nesting calculation, the WAM and WAVEWATCH III models can also be easily embedded into such model. The SWAN of version 40.20 and above have been added a parallel calculation module. In addition, SWAN's grid has been changed from rectangular grid to triangular finite element grid to facilitate easy coupling with the FVCOM ocean current model with triangular grid.

The specific construction of SWAN is detailed below.

5.2.1.1 Representation of waves

In the theory of linear surface wave, it is assumed that the wave height is relatively small, compared with the wavelength, and the local surface of water can be regarded as the superposition of a series of sine or cosine waves with different wavelengths (or frequencies) and propagation directions. The waves with small amplitude can be expressed as (for convenience, the spatial coordinates and wavenumber in the following equations are one-dimensional):

$$\eta(x,t) = a\cos(\sigma\,t - kx + \varphi) \tag{5.74}$$

Where η is the functions of displacement (x) and time (t) and represents the undulating surface; a is the amplitude; σ is the relative frequency; k is the wavenumber; and ϕ is the initial phase.

In the presence of current, there is a Doppler frequency shift relationship between velocity U, wavenumber k, relative frequency σ and absolute frequency ω.

$$\omega = \sigma + kU \tag{5.75}$$

It is assumed that the vertical profile of the current does not vary with depth. And the relative frequency σ can be expressed as:

$$\sigma = gkth(kd) \tag{5.76}$$

Where, d represents the water depth (can be dependent on time). The relative velocity C and group velocity C_g can be expressed as:

$$C = \frac{\sigma}{k}$$
$$C_g = \frac{\partial \sigma}{\partial k} = \frac{C}{2}\left(1 + \frac{2kd}{sh(2kd)}\right) \tag{5.77}$$

The density of wave energy is:

$$Etot = \frac{1}{2}\rho g a^2 \tag{5.78}$$

Where ρ represents the density of seawater. If the change of water surface is a steady process in time and space, the spectral density function of the wave can be obtained by Fourier transform of the covariance of the wave surface. Spectral density functions usually can be represented by the following: $E(\omega,\theta)$, $E(\sigma,\theta)$, $E(k,\theta)$. Obviously, the following relational expression is true:

$$\int_0^{2\pi}\int_0^\infty E(\sigma,\theta)d\sigma d\theta = <\eta^2> \tag{5.79}$$

Under non-steady circumstances, the spectral density functions of wave become the functions of coordinates and time and expressed as $E(\sigma,\theta;x,t)$. In the actual process, it is the action spectral density function expressed as $N(\sigma,\theta) = \frac{E(\sigma,\theta)}{\sigma}$ instead of energy spectral density function that has to be considered. This is because the action spectral density is a better conserved quantity in the presence of current and when the water depth varies with location.

5.2.1.2 Wave propagation

The following change rate in the geometric space and spectral space can be obtained from the wave packet theory of linear wave and crest conservation law [144]

$$\frac{d\vec{x}}{dt} = \vec{C}g + \vec{U} = \frac{1}{2}\left[1 + \frac{2kd}{sh(2kd)}\right]\frac{\sigma\vec{k}}{k^2} + \vec{U} \tag{5.80}$$

$$\frac{d\sigma}{dt} = C_\sigma = \frac{\partial\sigma}{\partial k}\left[\frac{\partial d}{\partial t} + \vec{U}\cdot\nabla d\right] - Cg\vec{k}\cdot\frac{\partial\vec{U}}{\partial s} \tag{5.81}$$

$$\frac{d\theta}{dt} = C_\theta = -\frac{1}{k}\left[\frac{\partial\sigma}{\partial d}\frac{\partial d}{\partial m} + \vec{k}\cdot\frac{\partial\vec{U}}{\partial k}\right]$$
$$\frac{d}{dt} = \frac{\partial}{\partial t} + (\vec{C}_g + \vec{U})\cdot\nabla\vec{x} \tag{5.82}$$
$$\nabla\vec{x} = \left(\frac{\partial}{\partial x}\frac{\partial}{\partial y}\right)$$

Where, $\frac{\partial}{\partial s}$ and $\frac{\partial}{\partial m}$ represent the directional derivative of σ and θ direction, $\vec{k}$ represents wavenumber unit vector. The change rate of action spectral density can be expressed by action spectral equilibrium equation [145, 146, 147]:

$$\frac{\partial}{\partial t}N + \frac{\partial}{\partial x}C_x N + \frac{\partial}{\partial y}C_y N + \frac{\partial}{\partial\sigma}C_\sigma N + \frac{\partial}{\partial\theta}C_\theta N = \frac{S}{\sigma} \tag{5.83}$$

The first term on the left of Equation (5.83) represents the rate of change of action spectral density along with time; the second term and third term represent the propagation of action spectral density (at the speed of C_x and C_y) in geometric space, respectively; the fourth term represents the frequency shift (at the propagation speed of C_σ) caused by current and changing water depth; the fifth term represents the refraction and shallowing effect (at the propagation speed of C_θ) caused by current and changing water depth. S on the right of Equation (5.83) represents the energy source and can be written as a sum of source terms of different types.

$$S = S_{in} + S_{ds} + S_{nl} \tag{5.84}$$

Where S_{in} is the wind input; S_{ds} represents the dissipation effect caused by white-cap, bottom friction, shallowing and breakage; and S_{nl} is the four-wave interaction and three-wave interaction.

5.2.1.3 Physical process

1. Wind input

Up to now, there are two different types of mechanisms to describe the transmission of energy and momentum from the wind to the waves. The first mechanism is Phillips [148] resonance mechanism which considers the linear growth of waves over time. The second mechanism is Miles [149] instability mechanism which considers the exponential growth of waves over time. Based on these two mechanisms, the wind input term is the sum of linear growth and exponential growth:

$$S_{in} = A + BE(\sigma\theta) \tag{5.85}$$

A represents the linear growth:

$$
\begin{aligned}
A &= \tfrac{1.5 \times 10^{-3}}{2\pi g^2} \{U_* \max\left[0 \cos(\theta - \theta_w)\right]\}^4 H \\
H &= \exp\left[-(\sigma/\sigma_{PM}^*)^{-4}\right] \\
\sigma_{PM}^* &= 2\pi \tfrac{0.13g}{28U_*} \\
C_D &= \begin{cases} 1.2875 \times 10^{-3} & U_{10} < 7.5m/s \\ (0.8 + 0.065 \times U_{10}) \times 10^{-3} & U_{10} \geq 7.5m/s \end{cases}
\end{aligned} \tag{5.86}
$$

B represents the exponential growth. The following are the empirical equations proposed by Komen [150] and Janssen [151, 152, 153], respectively:

$$B = \max\left\{00.25\frac{\rho_{air}}{\rho_{water}}\left\{28\frac{U_*}{C_{phase}}\cos(\theta - \theta_w) - 1\right\}\right\}\sigma$$

$$B = \beta\frac{\rho_{air}}{\rho_{water}}\left\{\frac{U_*}{C_{phase}}\right\}^2 \max\left\{0\cos(\theta - \theta_w)\right\}^2\sigma$$

$$\begin{cases} \beta = \frac{1.2}{k^2}\lambda\ln^4\lambda & \lambda \le 1 \\ \lambda = \frac{gz_e}{C_{phase}^2}e^r & r = kc/\left|U_*\cos(\theta - \theta_w)\right| \end{cases}$$

$$U(z) = \frac{U_*}{k}\ln\left(\frac{z+z_e-z_0}{z_e}\right) \tag{5.87}$$

$$z_e = \frac{z_0}{\sqrt{1-\tau_w/\tau}}$$

$$z_0 = \hat{\alpha}\frac{U_*^2}{g}$$

$$\vec{\tau}_w = \rho_{water}\int_0^{2\pi}\int_0^\infty \sigma BE(\sigma,\theta)\frac{\vec{k}}{k}d\sigma d\theta$$

2. Energy dissipation term

Three types of dissipation mechanisms were considered in the model. In deep water, the white-cap breaking of wind and waves dominates and controls the saturation of the high frequency part of the spectrum. In water of medium depth and shallow water, the bottom friction becomes important. When the waves spread to the vicinity of the broken zone of shallow water, the wave breaking caused by shallowing predominates.

(1) White-cap breaking

The white-cap dissipation term of wind and waves describes the energy loss caused by wave breaking in deep water and the wave steepness controls the dissipation degree:

$$S_{ds,w} = -\Gamma\bar{\sigma}\frac{k}{k}E(\sigma,\theta)$$
$$\Gamma = C_{ds}\left\{(1-\delta) + \delta\frac{k}{k}\right\}\left\{\frac{\bar{s}}{\bar{s}_{PM}}\right\} \tag{5.88}$$
$$\bar{s}_{PM} = (3.02 \times 10^{-3})$$
$$\bar{s} = \bar{k}\sqrt{E_{tot}}$$

Where, Γ is the coefficient depending on wave steepness, $\bar{\sigma}$ and $\bar{k}$ are average frequency and average wavenumber, respectively.

(2) Bottom friction

When waves travel from deep water to the water of limited depth, their interaction with the water bottom becomes important. This energy dissipation is controlled by a variety of different mechanisms, such as bottom friction, percolation, and muddy movement at the water bottom. Due to possible actions under different conditions at the water bottom, different empirical equations of bottom friction were used in the model [145, 154, 155].

$$S_{ds,bo}(\sigma,\theta) = -C\frac{\sigma^2}{g^2\sinh^2(kd)}E(\sigma,\theta) \tag{5.89}$$

Where, C is the bottom friction coefficient; g is the gravity acceleration; and d is the water depth. In SWAN, there are three options of energy dissipation source function of bottom friction according to the method of determining C:

- The empirical constant determined according to JONSWAP [145]:

$$C = \begin{cases} 0.038 \ m^2/s, \ \text{Surge} \\ 0.067 \ m^2/s, \ \text{Wind waves} \end{cases} \tag{5.90}$$

- The form simplified by Collins [154] based on the non-linear expression proposed by Hasselmann [156] according to the bottom stress is:

$$C = C_f g U_{rms} \tag{5.91}$$

Where C_f is the bottom friction coefficient; and U_{rms} is the root mean square of wave-induced horizontal velocity of water particle at water bottom.

- Expression of eddy viscosity model proposed by Madsen [155]:

$$C = f_W g U_{rms}/\sqrt{2} \tag{5.92}$$

Where, f_W is obtained from Jonsson [157].

In SWAN, the first case, namely the bottom friction dissipation source function corresponding to JONSWAP plan, is the default bottom friction dissipation source function of the model.

(3) Breakage caused by shallowing

When the wave travels from deep water to the water of limited depth and the ratio of wave height to water depth becomes large, the waver energy is dissipated due to breakage. The equation adopted in the SWAN model is:

$$S_{ds,br} = -\frac{Dtot}{Etot} E \tag{5.93}$$

Where, D_{tot} is average dissipation rate per unit horizontal area [158].

3. Non-linear interaction

Non-linear interaction refers to the energy exchange between resonant wave components to redistribute the energy. In deep water, the four-wave interaction is relatively important while the three-wave interaction becomes important in shallow water.

The energy is transferred from high-frequency part to low-frequency part through the four-wave interaction, which plays an important role in maintaining the spectral shape and determining the directional distribution of energy. The four-wave interaction is given by Boltzmann integral [159, 160, 161]. But its numerical calculation is very time-consuming and hence difficult to be used in practice. An approximate calculation method called Discrete Interaction Approximation (DIA) was proposed by Hasselmann et al. [162]. DIA approximately reflects the growth characteristics of wind and waves and is used in many models. The four-wave transmission rate is small when the water depth becomes shallow, so a shallow water factor R should be multiplied in case of shallow water.

In shallow water, when the wave steepness is large, the energy is transferred from the low-frequency part to the high-frequency part through the three-wave interaction. Near the shore, the spectrum of a single crest changes into that of multiple crests. The three-wave interaction adopted in SWAN model is the equation LTA (Lumped Tried Approximation) proposed by Eldeberky and Battjes [163].

5.2.1.4   Numerical methods

1. Advection term

According to the properties of action spectral equilibrium Equation (5.83), the state of each grid point is determined by the state of grid point of head waves. So the most effective difference method is implicit upwind difference method which is unconditionally stable. SWAN adopts simple backward difference in the time term and three difference schemes in the geometric space. The first type is simple first-order explicit backward difference scheme which was used in previous versions of SWAN 40.11. The second type is S&L difference scheme (second-order upwind difference) under stable conditions [164]. The third type is SORDUP difference scheme (second-order upwind difference) under stable conditions. The latter two are both second-order difference schemes which are newly added to AWAN 40.11. The second-order difference scheme is used in frequency and direction space, including center difference, upwind difference and any scheme in between.

In the calculation of advection term, 4-SWEEP technique was also used to divide the geometric space into four quadrants, and in each quadrant, parts other than refraction and non-linear interaction were calculated independently.

2. Source function item

It is easy to calculate the linear growth term of wind input. The other terms are divided into positive source terms and negative source terms. For the positive source terms, an explicit scheme is adopted:

$$S^n = \varphi^{n-1} E^{n-1} \tag{5.94}$$

For the negative source terms, an implicit scheme is adopted:

$$S^n = \varphi^{n-1} E^{n-1} + \left(\frac{\partial S}{\partial E} \right)^{n-1} (E^n - E^{n-1}) \tag{5.95}$$

3. Matrix solution

In order to obtain the action spectral density value at each grid point, the algebraic equation after discretization of the action spectral equilibrium equation must be solved.

$$A \cdot N = B \tag{5.96}$$

Where A and B are known. In the absence of current and when the water depth does not vary over time, A is a common tridiagonal band matrix and easy to be solved. In the presence of current or when the water depth varies over time, A is not a tridiagonal band matrix. The previous version of SWAN40.20 used ILU-CGSTAB method. The SIP method [165] was added to SWAN40.20 on the original basis. Such method is three to five times faster than ILU-CGSTAB. But ILU-CGSTAB is still useful, especially when calculating wave-induced setup.

5.2.2 Model Configuration and Result Verification

This module adopts the latest version of SWAN model. That is, this module introduces SWAN model of unstructured computing grid. The calculation area and grid of this model is the same as that of FVCOM model, realizing the coupling with FVCOM. Both are coupled by means of marine environment element exchange. FVCOM provides current field to SWAN and SWAN provides wave element field to FVCOM. Both exchange the data on each time step for coupling and finally provide more accurate current field and wave element field for oil spill behavior and fate prediction module.

The SWAN model was used to have numerical simulation of the big waves near Qinhuangdao on October 11 and 12, 2009 and the comparison between the simulation results and the value observed at Qinhuangdao Observatory on October 11, 2009 is shown in Figure 5.56. It can be seen that the simulation results are in good agreement with the actually observed data.

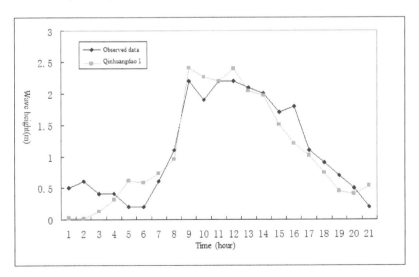

FIGURE 5.56: Comparison between data observed at Qinhuangdao Observatory on October 11, 2009 and the results simulated by the model.

5.3   Atmosphere Module

The influence of the sea surface wind field on the oil spill behavior and fate is mainly shown in the following aspects:

- The wind-induced current is one of the most important factors affecting the drift of oil slick;

- The wind itself can directly act on the oil slick, drive the oil slick to move and can change the shape and area of oil slick spread;

- The wind will affect the sea waves which are an important environmental dynamic factor in the dispersion process of spilled oil;

- The wind itself may also affect the weathering process of the spilled oil, such as the evaporation rate of the spilled oil.

The atmosphere module adopts an advanced Weather Research Forecast (WRF) model which provides high-precision wind field forecast products for the system.

5.3.1   Introduction to the Model

The high-resolution forecast system of wind field on the surface of the Bohai Sea adopts the atmospheric mesoscale model-WRF model, which is very mature in dynamic module.

The WRF model is a new generation of mesoscale forecast model and assimilation system developed by scientists from many US research departments and universities. The model is completely compressible and non-static and the control equations are written in flux form. The terrain-following hydrostatic barometric coordinates are used as the vertical coordinates. The use of Arakawa C-shaped horizontal and vertical staggered grid is conducive to improving the accuracy in high-resolution simulation. Complete time-split scheme is used as the time integral. The outer loop Runge-Kutta technique has a larger time step. The inner loop is an acoustic time integral, which can allow a larger time step and shorten the calculation time while ensuring the stability of the integral. In order to meet the needs of simulating the actual weather, the model has a set of physical processes and parameterization processes, including cloud microphysical processes, cumulus convective parameterization scheme, long-wave radiation, short-wave radiation, boundary layer turbulence, near-surface layer, land surface parameterization, and subgrid turbulence diffusion. The model uses inherited software design, multi-level parallel decomposition algorithm, selective software management tools and intermediate software package structure. At present, the horizontal grid precision of the WRF model can be as accurate as 1 km or higher. The

WRF model has become a tool to improve the forecast accuracy of important weather features at different scales from cloud to weather. Meanwhile, the WRF model has advanced 3D variational assimilation technique (3DVAR) which can fully and effectively assimilate all kinds of information and data into the initial field of the model, providing higher quality initial values for the model and achieving the goal of significantly improving the quality of numerical forecast.

The WRF version adopted in this system is WRF-ARW Version 3.1.1 released on July 31, 2009. This version adopts fully compressible non-static Euler equations and Arakawa C grid as the horizontal grid and mass-based terrain-following η coordinate [166, 167] as the vertical coordinate. The η layers can be changed according to needs. The model framework contains dynamic and physical processes of atmospheric change. The physical model includes the microphysical model describing the phase transformation of water vapor and cloud physical process, cumulus parameterization scheme considering cumulus effects on subgrid scale, multi-layer land surface physical model including the simple thermodynamic process on the ground and the soil and ground vegetation model considering the impacts of snow and sea ice, the planetary boundary layer model using 2-order turbulence closure or non-local K closure schemes, and atmospheric radiation model considering multi-spectral long-wave radiation scheme and simple short-wave radiation scheme about cloud and ground surface radiation. New radiation scheme–RRTMG [168, 169] was also added to the version V3.1.1. Such scheme includes long-wave and short-wave radiation scheme of which the long-wave radiation scheme was developed from Rapid Radiative Transfer Model (RRTM) in MM5. RRTMG also uses Monte Carlo Independent Cloud Approximation (MCICA) technique to effectively describe the role of subcolumn cloud. In addition, the terrain drag effect of the model and the high-order boundary layer turbulence closure scheme are added, and variable time steps can be used to shorten the integral time in single zone simulation.

The WRF-ARW model consists of the following modules:

- WRF Preprocessing System (WPS)

  In this module, the simulated area is set, topographic data (topographic data with 10-bit and 2-bit resolution was applied in the system) are interpolated into the simulated area and meteorological data from other models (such as global model) are interpolated into the simulated area. The purpose of this module is to provide a background field for simulation.

- WRF Data Assimilation (WRFDA)

  The assimilation scheme (in this system, 3DVAR was applied) is used to assimilate the objective data such as observations at stations, satellite data and radar data to improve the initial field and boundary conditions required for simulation. This module is optional.

- Numerical Simulation Master Program Module (ARW)

  Generate the initial background field and time-varying side boundary conditions for the simulation and obtain the numerical integral equations.

- Post-processing Module

  Analyze and process the output of the model (files in NetCDF format) and display the graphics.

  At present, the dynamic framework and calculation scheme of WRF model have been quite perfect and such model has great advantages for the simulation of mesoscale weather phenomena.

5.3.2   Model Configuration

The system focuses on the Bohai Sea and the North Yellow Sea, but the actual simulation area is not confined to the Bohai Sea and the North Yellow Sea since the weather system is mobile. Considering the zones and resolution comprehensively, the double nesting technique was adopted in the simulation. In the coarser resolution area (30 km), the entire weather system affecting the coastal waters of East China can be captured. In the higher resolution area (10 km), the high resolution simulation and reproduction of weather system of the Bohai Sea and North Yellow Sea can be realized. The zone settings are shown in Figure 5.57.

In order to match with the horizontal resolution, the vertical resolution of the model has been adjusted to a large extent and altogether $44n$ layers were designed (generally, 27 layers are applied in simulation). The specific values are: $n = 1.0000, 0.9975, 0.9925, 0.9850, 0.9775, 0.9700, 0.9540, 0.9340, 0.9090, 0.8800, 0.8506, 0.8212, 0.7918, 0.7625, 0.7084, 0.6573, 0.6090, 0.5634, 0.5204, 0.4798, 0.4415, 0.4055, 0.3716, 0.3397, 0.3097, 0.2815, 0.2551, 0.2303, 0.2071, 0.1854, 0.1651, 0.1461, 0.1284, 0.1118, 0.0965, 0.0822, 0.0689, 0.0566, 0.0452, 0.0346, 0.0249, 0.0159, 0.0076, 0.0000$. The altitude of each layer below 850 kpa is about 0 m, 32 m, 61 m, 110 m, 169 m, 228 m, 322 m, 469 m, 658 m, 890 m, 1149 m and 1417 m, respectively.

The WRF model itself provides many physical options, including atmospheric boundary layer turbulence scheme, cumulus parameterization scheme, cloud microphysical scheme, long-wave and short-wave radiation scheme and land/sea surface process. The appropriate scheme was determined for the sea area under study according to the past experience, as shown in Table 5.4. For the sea surface wind, the atmospheric boundary layer turbulence scheme is the most important. The boundary layer scheme adopted here is the one developed by Yonsei University (YSU). After a large number of numerical experiments, it is proved to be very suitable for numerical simulation of the sea surface. The cumulus parameterization scheme is mainly related to the model resolution, and Kain-Fritsch scheme is more suitable for 10 km resolution.

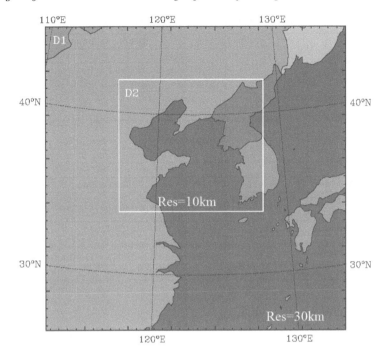

FIGURE 5.57: WRF simulation area settings (D1: 30 km; D2:10 km).

5.3.3   System Framework

The real-time sea surface wind forecast system is completely modular and mainly includes the following modules:

- Data collection and processing module

  (1) Have real-time download of the forecast field data of global atmospheric model to provide initial field and time-varying boundary conditions for the model.

  (2) Have real-time download of the conventional radiosonde data, ground and ship observation global telecommunications system (GTS) data.

  (3) Have real-time download of China's offshore unconventional satellite radiation and observation data.

  (4) Conduct decoding, quality control and format conversion of the observation data.

- Real-time forecast module

  (1) Assimilate the conventional and unconventional observation data on the basis of the forecast field data of global atmospheric model to provide initial field and time-varying boundary conditions for the model.

  (2) Run the numerical forecast system.

TABLE 5.4: WRF model settings.

Zone and options	Specific settings	
	D1	D2
Zone and resolution	Double nesting, Lambert conformal projection Center point (36.0°N, 123.0°E)	
	Number of grid points 70×70	Number of grid points 100×91
	The horizontal resolution: 30km	The horizontal resolution: 10km
	Vertical resolution: 44 η layers	
Time step	120 seconds, 40 seconds	
Boundary layer scheme	YSU scheme	
Cumulus parameterization scheme	Kain-Fritsch scheme	
Microphysical scheme	Lin scheme	
Radiation scheme	Long-wave radiation: RRTM scheme Short-wave radiation: Dudhia scheme	
Land surface process	NOAH land surface model	

- Post-processing Module

(1) Extract the sea surface wind data from WRF forecast results.

(2) Convert the sea surface wind data from Lambert project coordinates to regular latitude and longitude coordinates.

(3) Use GrADS grid point analysis and display system software to draw the sea surface wind at the grid points of latitude and longitude into an image.

(4) Process the sea surface wind at the grid points of latitude and longitude into NetCDF format.

(5) The interconnection between the modules is controlled by Linux/ UNIX-shell script program, which is fully automatic and does not require manual intervention. The system framework is shown in Figure 5.58.

5.4 Operation of Forecast System

5.4.1 Data Acquisition

All the data required by the forecast system are from the Internet, and they are automatically downloaded with Linux/UNIX-SHELL scripts at fixed time and the download tools such as lftp and wget. After the data is downloaded, format conversion and quality control of the data are carried out first and then data assimilation with 3DVAR is carried out to provide initial field and time-varying boundary conditions for the forecast system. These data

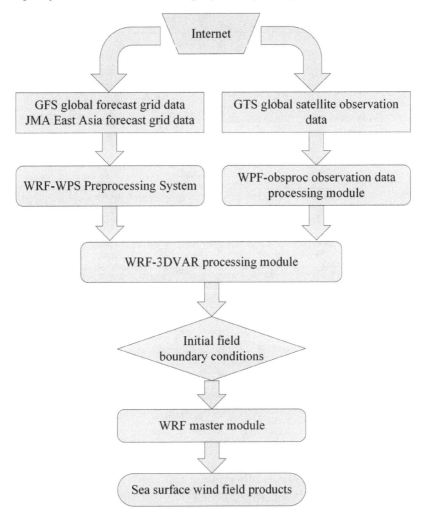

FIGURE 5.58: Framework of WRF real-time sea surface wind forecast system.

include conventional ground observation data, conventional radiosonde data, ship observation data, satellite sea surface wind observation data and satellite radiation observation data. Their distribution is shown in Figure 5.59.

5.4.2 Data Assimilation

High-quality forecast results cannot be achieved without high-quality initial field. Currently, single-time 3DVAR is used for data assimilation to form the initial field of real-time forecast. If the network speed is very fast and

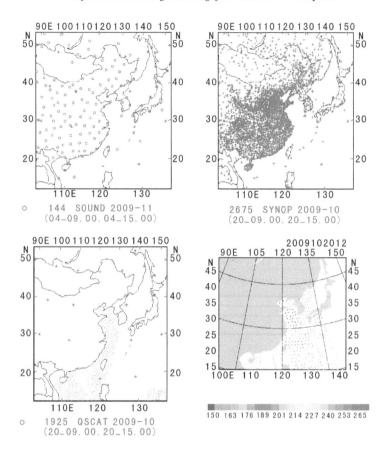

FIGURE 5.59: Types of observation data used by the forecast system. (Top left: ground and radiosonde observation; top right: ground automatic observation; Lower left: satellite sea surface wind observation; lower right: satellite radiation observation.)

enough observation data can be obtained, multi-step 3DVAR (i.e., cyclic 3DVAR) can be further adopted in order to form a higher quality initial field. The data assimilation process is shown in Figure 5.60.

WRF Preprocessing System (WPS) and real.exe form the initial field required for the first 3DVAR and wrf.exe drives WRF forward integration. In the figure, bg, obs and be represent the background field, observation and the background field error covariance; 0, Δt and $2\Delta t$ are the assimilation time. For instance, the data assimilation window at Δt is $0.5\Delta t - 1.5\Delta t$. That is, all observation data in this period are assimilated at Δt.

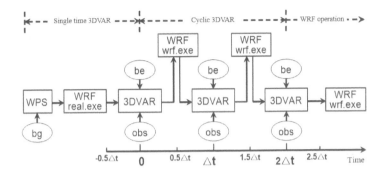

FIGURE 5.60: Cyclic 3DVAR process.

5.4.3 Operation Process

GFS grid point data, GTS observation data (conventional radiosonde and ground observation data), satellite observation data and GOOS sea surface temperature (SST) are automatically acquired from the Internet and decoded, and WRF-3DVAR data assimilation is carried out to form the initial and boundary values of WRFM5 forecast system. Then run the WRF master program (completed by Linux workstation) and the final result is graphically processed. The whole process is fully automated, and Linux/UNIX cron tools and shell script files control the whole automation process. The process topology diagram is shown in Figure 5.61.

The forecast system runs once a day, with forecast starting at 08 Beijing and ending at 08 Beijing the day after tomorrow. The forecast period is 48 hours. The system starts automatically at 00 o'clock every night and ends around 2 o'clock in the morning. The whole operation takes about two hours. See Figure 5.62 for the diagram of operation time control of the forecast system.

5.4.4 Installation of the Forecast System

5.4.4.1 Installation environment

The system operation environment is 24CPU Dawning Server; the operating system is SuSE-Linux; the FORTRAN compiler used is pgf77, pgf90 (software of Portland company, USA); openMPI-1.4.1 is used for parallel operation; the graphics drawing software is the GrADS grid analysis and display system developed by the University of Maryland, USA.

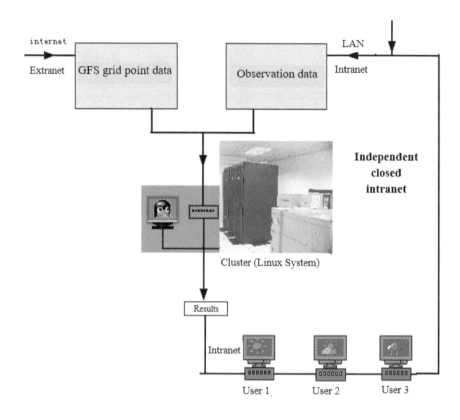

FIGURE 5.61: Process topology diagram.

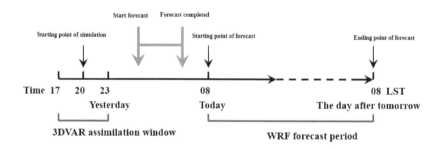

FIGURE 5.62: Diagram of operation time control of the forecast system.

5.4.4.2 Directory structure

The whole real-time forecast system has many complicated files, but its structure is very clear due to the use of modular design concept, as shown in Figure 5.63.

The concrete manifestation of these directories on the server is shown in Figure 5.64 and Figure 5.65, respectively.

5.4.5 Wind Field Forecast Products

5.4.5.1 Grid point data products

Since the WRF model uses map projection, the result is not a regular grid point of latitude and longitude. Therefore, the research group has transformed the results into regular grid points of longitude and latitude coordinates. According to the different needs of ocean current module, ocean wave module and oil spill behavior and fate prediction module, the result data is stored in two formats: ASCII format and NetCDF format. The former is for ocean current module and ocean wave module, and the latter is for oil spill behavior and fate prediction module. Figure 5.66 shows the forecast results of sea surface wind field on April 14, 2010, in which: *.sfc file is ASCII grid point wind field data; GIF is a folder of wind field pictures in GIF format; *.nc is the grid point wind field data in NetCDF format.

5.4.5.2 Graphical products

The graphical products in GIF format are presented by using the GrADS grid analysis and display system developed by the University of Maryland in

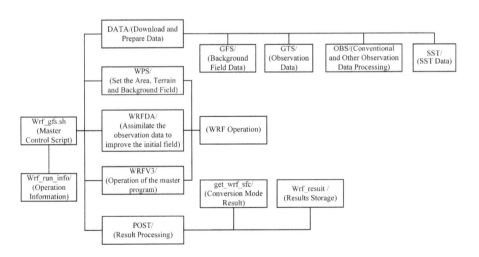

FIGURE 5.63: Directory structure of real-time sea surface wind forecast system.

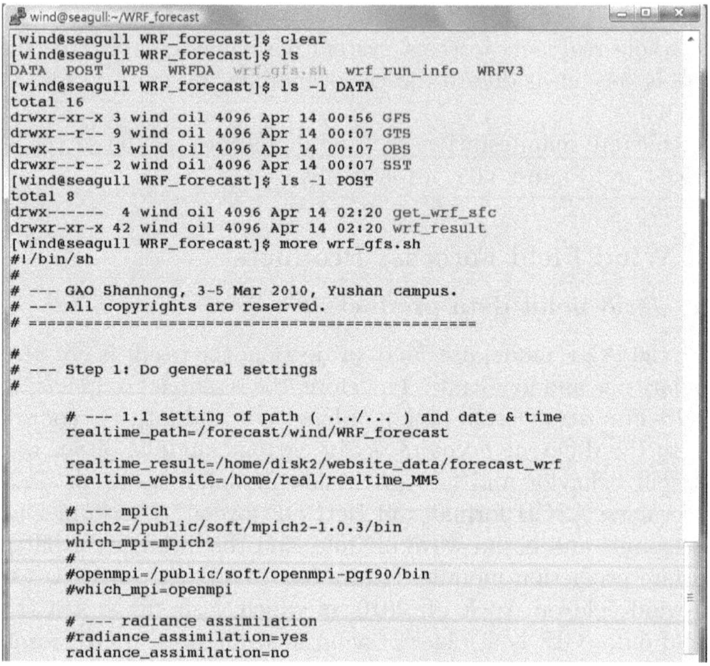

```
[wind@seagull WRF_forecast]$ clear
[wind@seagull WRF_forecast]$ ls
DATA  POST  WPS  WRFDA  wrf_gfs.sh  wrf_run_info  WRFV3
[wind@seagull WRF_forecast]$ ls -l DATA
total 16
drwxr-xr-x 3 wind oil 4096 Apr 14 00:56 GFS
drwxr--r-- 9 wind oil 4096 Apr 14 00:07 GTS
drwx------ 3 wind oil 4096 Apr 14 00:07 OBS
drwxr--r-- 2 wind oil 4096 Apr 14 00:07 SST
[wind@seagull WRF_forecast]$ ls -l POST
total 8
drwx------  4 wind oil 4096 Apr 14 02:20 get_wrf_sfc
drwxr-xr-x 42 wind oil 4096 Apr 14 02:20 wrf_result
[wind@seagull WRF_forecast]$ more wrf_gfs.sh
#!/bin/sh
#
# --- GAO Shanhong, 3-5 Mar 2010, Yushan campus.
# --- All copyrights are reserved.
# ================================================
#
# --- Step 1: Do general settings
#

        # --- 1.1 setting of path ( .../... ) and date & time
        realtime_path=/forecast/wind/WRF_forecast

        realtime_result=/home/disk2/website_data/forecast_wrf
        realtime_website=/home/real/realtime_MM5

        # --- mpich
        mpich2=/public/soft/mpich2-1.0.3/bin
        which_mpi=mpich2
        #
        #openmpi=/public/soft/openmpi-pgf90/bin
        #which_mpi=openmpi

        # --- radiance assimilation
        #radiance_assimilation=yes
        radiance_assimilation=no
```

FIGURE 5.64: Main folders (WRF_forecast) and master control script in the directory structure.

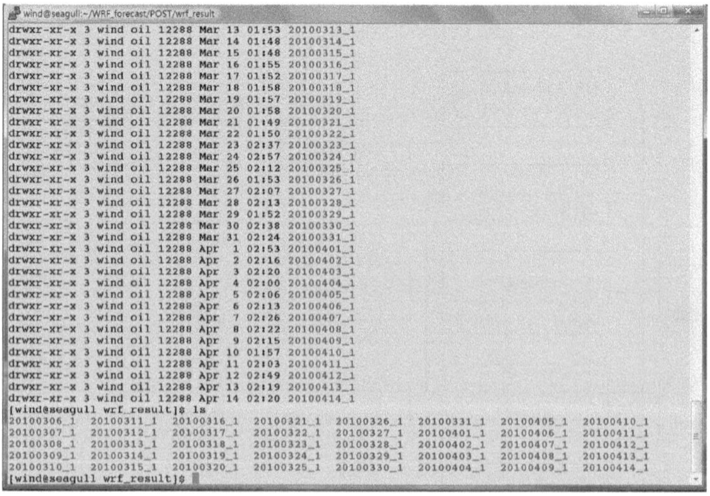

FIGURE 5.65: Forecast results file (wrf_result) and its contents (e.g., 20100306_1).

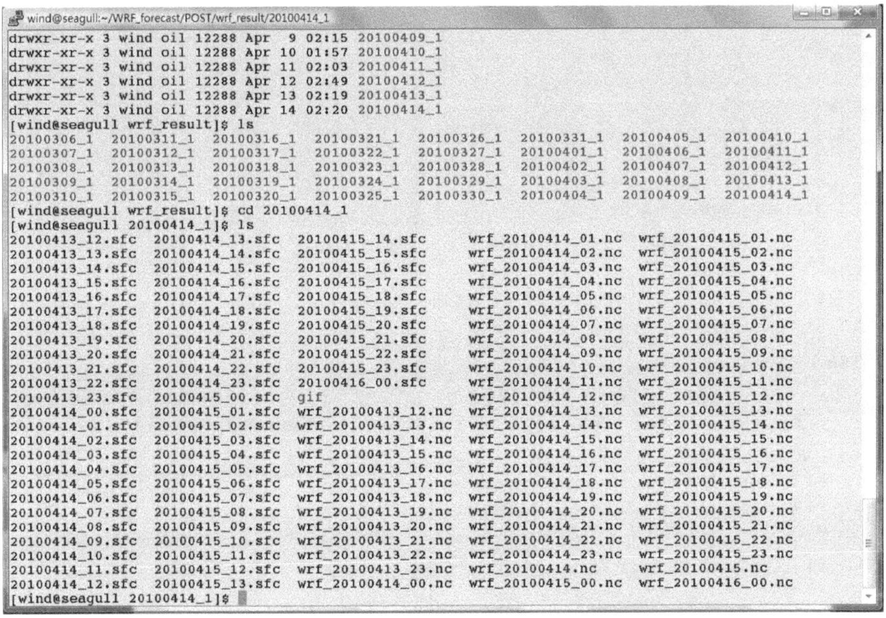

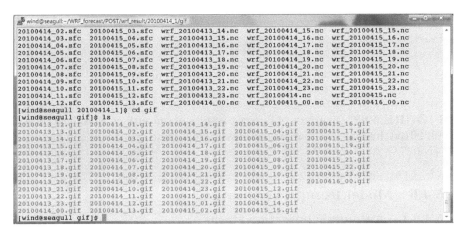

FIGURE 5.66: Forecast results of sea surface wind on April 14, 2010.

FIGURE 5.67: Pictures of forecast results of the sea surface wind field on April 14, 2010.

the United States and the ASCII grid point sea surface wind field data. Figure 5.67 and Figure 5.68 are pictures and graphics of the forecast results of sea surface wind field on April 14, 2010.

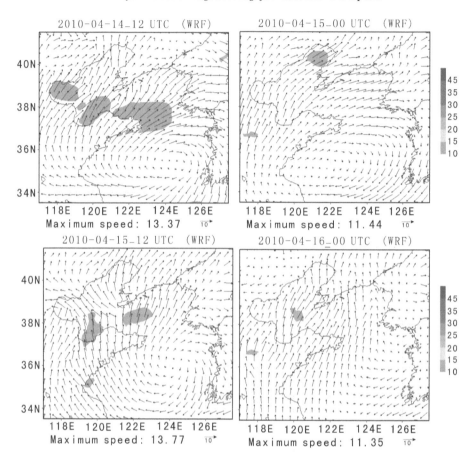

FIGURE 5.68: Diagrams of forecast results of sea surface wind field on April 14, 2010 (the grayscale bar represents different wind speed levels in meters per second).

5.4.6   Model Results Verification

5.4.6.1   Automatic operation status

The forecast system has been running steadily since March 6, 2010. By May 13, the success rate of operation was 100%.

The forecast system runs regularly once a day, with the period of forecast validity as 48 hours. If the actual needs and calculation conditions allow, it can run twice and the period of forecast validity can be extended to 72 hours.

5.4.6.2   Wind field result verification

Based on the wind speed data from the coastal meteorological ground observation stations, offshore ships and islands in the Circum-Bohai Sea Region

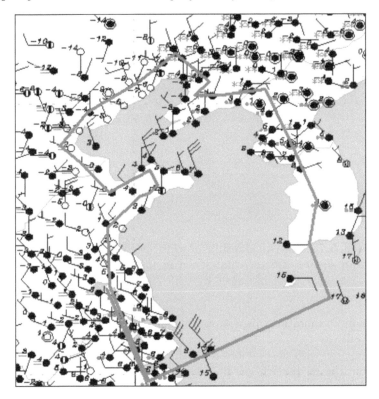

FIGURE 5.69: Distribution of observatories used for statistics. (Observations with the area surrounded by lines in the figure.)

and North Yellow Sea (see Figure 5.69), we had the statistical analysis of the forecast wind field from March 10 to May 10 (two months). The statistical results are shown in Figure 5.70. The results show that the root mean square error of wind speed is between 2.2 and 3.2 m/s, the average error of wind direction is between 24 and 36 degrees, the root mean square error of wind speed within 24 hours is less than 2.5 m/s, and the average error of wind direction is less than 28 degrees. The forecast results are satisfactory.

5.5 Oil Spill Behavior and Fate Prediction Module

Based on the environmental dynamic parameters provided by ocean current module, ocean wave module and atmosphere module, the oil particle model is used to predict the behavior and fate of oil spill. It includes the drift,

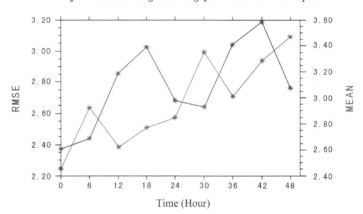

FIGURE 5.70: Statistical results of wind speed and wind direction. (RMSE: root mean square error of wind speed in meters per second; MEAN: Wind direction average error in degree.)

expansion, evaporation, dispersion, emulsification and shoreline adsorption of spilled oil.

Compared with the previous algorithms for solving the convection-diffusion equation, the oil particle method not only solves the deformation and fragmentation process of the oil slick under the environmental dynamic action, but can also accurately predict the expansion process of oil slick and the obvious stretching of the oil slick in the wind direction. It can practically simulate the irregular shape and drift trajectory of oil slick and effectively eliminate the numerical divergence. The drift and weathering of spilled oil on water surface can be simulated by algorithms of surface diffusion, translation, transportation, emulsification and evaporation.

Oil particle method is actually to track the advection and turbulent diffusion of oil particle micro-aggregates along with the surrounding waters. It can be said that this method is a combination of deterministic method and stochastic method since the Lagrange method is used to simulate the advection process and the random walk method is used to simulate the turbulent diffusion process.

5.5.1   Drifting Process

The drifting process of oil slick under the combined action of wind, current and tide is divided into two processes: advection and diffusion. The advection process in the module is simulated by the Lagrange particle tracking equation, while the diffusion process is simulated by the random walk method.

5.5.1.1   Advection process

In accordance with the Lagrange method, the position vector of oil particle at the time point t in such process is expressed as $\vec{X}_t$:

$$\vec{X}_t = \vec{X}_{t-1} + t\vec{U}_{oil} \tag{5.97}$$

Where:

t is the time step in seconds;

$\vec{X}_{t-1}$ is the position of surface oil particle at the time point $t-1$ (m);

$\vec{U}_{oil}$ is the drift velocity of oil slick (m/s);

The drift velocity of oil particle $\vec{U}_{oil}$ is calculated with the following equation:

$$\vec{U}_{oil} = \vec{U}_w + \vec{U}_c \tag{5.98}$$

Where:

$\vec{U}_w$ represents the velocity component produced under wind action (m/s);

$\vec{U}_c$ represents the velocity component under action of current (tidal current and wind-driven current) (m/s);

The velocity component $\vec{U}_c$ can be obtained according to the current field data provided by the ocean current module.

Wind force factor and deflection angle of wind have to be considered in the advection process. By decomposing the drift velocity of oil slick, we can get the east component U_{WC} and north component V_{WC} which can be calculated with Equations 5.99 and 5.100, respectively:

$$U_{WC} = C_1 U_W \tag{5.99}$$

$$V_{WC} = C_1 V_W \tag{5.100}$$

Where:

U_W is the east component of the wind speed (m/s);

V_W is the north component of the wind speed (m/s);

U_W and V_W are provided by the atmospheric module. C_1 is the wind force factor and its value is usually 3% to 3.5% in the conditions of open water areas and medium level of wind force. The default value of the model is 3.5%.

However, in the conditions of closed or semi-closed bay, the smallest C_1 is usually adopted.

If the wind force factor has been considered in calculation of the surface current field, then the wind force factor will be reduced accordingly. This is because the wind-driven drift action has been considered in the surface current field.

The deflection angle of the wind is the angle between the wind direction and the drift direction of oil slick and is positive in the clockwise direction. The east component U_{Wd} and north component V_{Wd} of the wind-driven drift velocity are calculated with Equations 5.101 and 5.102, respectively:

$$\vec{U}_{Wd} = U_{WC}cos\theta + V_{WC}sin\theta \tag{5.101}$$

$$\vec{V}_{Wd} = -U_{WC}sing\theta + V_{WC}cos\theta \tag{5.102}$$

Where:

$\vec{U}_{Wd}$ is the east component of wind speed with the deflection angle of wind taken into consideration (m/s);

$\vec{V}_{Wd}$ is the north component of wind speed with the deflection angle of wind taken into consideration (m/s);

θ is the deflection angle of wind (degree).

Deflection angle of wind is a constant, with default value as 0. Usually, the smaller positive value is adopted in the areas at high latitude.

5.5.1.2   Diffusion process

In order to describe the diffusion process below the resolution scale of the input current field, the random walk method is used in this module. The east component x_{dd} and north component y_{dd} of diffusion distance of the oil slick are calculated with Equations 5.103 and 5.104, respectively:

$$x_{dd} = \gamma\sqrt{6D_x t} \tag{5.103}$$

$$y_{dd} = \gamma\sqrt{6D_y t} \tag{5.104}$$

Where:

D_x is the east-west horizontal diffusion coefficient (m/s);

D_y is the south-north horizontal diffusion coefficient (m/s);

TABLE 5.5: Typical dispersion coefficients.

Environmental conditions	Dispersion coefficient (m^2/s)
A high-energy open environment (e.g. the North Sea of the Atlantic Ocean)	> 10
Medium energy level environment (e.g. Baltic Sea)	$5 \sim 10$
Low energy level environment (Closed bay)	$2 \sim 3$

t is the time step (s);

γ is the random number, with value between -1 and 1.

The horizontal diffusion coefficients D_x and D_y are usually the same. Table 5.5 shows the typical values.

5.5.2 Expansion Process

The expansion of oil slick is the result of joint action of turbulent diffusion, gravity, inertia, viscosity and surface tension.

In the studies on oil slick expansion, Fay's three-stage expansion theory (1971) [170] once was widely applied. Mackay [171, 172] revised Fay's three-stage expansion theory, considered the influence of wind in the equations of the second stage and established the expansion models of thick oil slick and thin oil slick by combining with the measurement results respectively. It is assumed in the model that the thicker oil slick will gradually become thinner and 80% to 90% of the total area of the oil slick is represented by the thin oil slick.

In this module, only the thick oil slick with simulated weight above 90% of the oil slick is considered. The change rate $\tilde{A}_{tk}$ of oil slick area due to expansion of thick oil slick is calculated with Equation 5.105:

$$\tilde{A}_{tk} = \frac{dA_{tk}}{dt} = K_1 A_{tk}^{1/3} \left(\frac{V_m}{A_{tk}}\right)^{4/3} \tag{5.105}$$

Where:

A_{tk} is the surface area of oil slick (m^2);

K_1 is the extension rate constant (s);

V_m is the volume of oil slick (m^3);

t is the time (s).

The sensitivity analysis of Equation 5.105 shows that the number of particles will affect the calculation results. Kolluru [173] proposed the change rate calculation Equation 5.106 of the surface area of single oil particle to correct the impact of changes in the number of particles on the calculation results of Equation 5.105:

$$\tilde{A}_{tk} = \frac{dA_{tk}}{dt} = K_1 A_{tk}^{1/3} \left(\frac{V_m}{A_{tk}}\right)^{4/3} \left(\frac{R_s}{R_e}\right)^{4/3} \tag{5.106}$$

Where:

A_{tk} is the surface area of single oil particle (m^2);

K_1 is the extension rate constant (s);

V_m is the volume of single oil particle (m^3);

R_s is the radius of single oil particle (m);

R_e is the effective radius (m) of oil slick.

The effective radius R_e of oil slick is calculated with Equation 5.107:

$$R_e = \left[\left(\frac{1}{\pi}\sum_{n=1}^{N} A_{tk}\right)\right]^{1/2} \tag{5.107}$$

Where:

A_{tk} is the surface area of single oil particle (m^2);

N is the number of oil particles.

5.5.3 Evaporation Process

For the simulation of oil spill evaporation process, this module uses the parameter equations proposed by Stiver and Mackay [174] to calculate the evaporation rate and evaporation capacity of oil products. The oil evaporation rate F_v can be calculated with the following equation:

$$F_v = \ln\left[1 + B(T_G/T)\theta(A - BT_0/T)\right]\left[T/(BT_G)\right] \tag{5.108}$$

Where:

T_0 is the modified initial boiling point of distillation curve (Kelvin);

T_G is the modified gradient of distillation curve;

T is the ambient temperature (Kelvin);

A and B are the dimensionless constants. For typical crude oil, $A = 6.3$, $B = 10.3$;

θ is the evaporation coefficient.
The evaporation coefficient θ can be calculated with Equation 5.109:

$$\theta = \frac{K_m A t}{V_0} \tag{5.109}$$

Where:

K_m is the mass transfer rate (m/s);

A is the area of oil slick (m^2);

t is the time (s);

V_0 is the volume of oil slick (m^3).

5.5.4 Dispersion Process

Breaking waves play a very important role in the dispersion process of spilled oil and is the main energy source of such process.

For the simulation of spilled oil dispersion process, this model uses the spilled oil dispersion rate calculation equation of Delvigne and Sweeney [175] based on the environmental dynamic parameters provided by the ocean wave module:

$$Q_d = C^* D_d^{0.57} S F d^{0.7} \Delta d \tag{5.110}$$

Where:

C^* is the empirical constant of dispersion rate;

D_d is the energy of surface breaking waves $(Joule/m^2)$;

S is the oil-covered water surface area fraction;

F is the water surface area fraction affected by breaking waves;

d is the diameter of oil particle (m);

Δd is the diameter difference of oil particle (m).

The dispersion rate empirical constant C^* is related to the oil types and weathered state and can be calculated with the equation proposed by Delvigne and Hulsen [176] (Equation 5.111)

$$C^* = \exp\left[a\ln(\mu) + b\right] \tag{5.111}$$

Where:

When $\mu < 132$, a=-0.1023, b=7.572;

When $\mu > 132$, a=-1.8927, b=166.313;

μ is the viscosity of oil (centipoise);

d_{50} is the average diameter of oil particle and can be calculated with Equation 5.112:

$$d_{50} = 1818(E)^{-0.5}\left(\frac{\mu}{\rho_0}\right)^{0.34} \tag{5.112}$$

Where:

E is the wave energy dissipation rate per unit volume (Joule/$m^3 \times s$); for breaking wave, E is 1000 Joule/$m^3 \times s$;

μ is the viscosity of oil (centipoise);

ρ is the oil density (g/cm^3).

The minimum diameter and maximum diameter of oil particle can be calculated with Equations 5.113 and 5.114:

$$d_{\min} = 0.1d_{50} \tag{5.113}$$

$$d_{maz} = 2.0d_{50} \tag{5.114}$$

The dissipative wave energy and the fraction of surface area per unit time affected by breaking wave are the two main parameters of the water carrying process. The dissipative wave energy D_d can be calculated with Equation 5.115:

$$D_d = 3.4 \times 10^{-3}\rho_w g H^2 \tag{5.115}$$

Where:

ρ_w is the water density (kg/m^3);

g is the gravity acceleration (m/s^2);

H is the root mean square of the height of breaking wave.

Fraction of surface area per unit time affected by breaking waves F can be calculated with Equation 5.116:

$$F = 0.032(U_W - U_T)T_W \qquad (5.116)$$

Where:

U_W represents the wind speed at 10 m above the sea level (m/s);

U_T represents the wind speed threshold formed by the breaking wave, and is about 5 m/s;

T_W is the period of typical waves (s).

The total mass of dispersed oil slick M_e can be calculated with Equation 5.117:

$$M_e = A dt \int_{d_{min}}^{d_{max}} Q_d \delta d \qquad (5.117)$$

Where:

A represents the area of oil slick (m^2);

dt is the time step (s);

Q_d represents the water carrying rate $(kg/m^2 \times s)$.
The invasion depth Z_m can be calculated with Equation 5.118:

$$Z_m = 1.5H_b \qquad (5.118)$$

Where:

H_b is the height of breaking wave (m).

The float-up speed W_i of oil particles of different diameters can be calculated with Equation 5.119:

$$Z_m = 1.5H_b \qquad (5.119)$$

Where:

d_i is the diameter of oil particle (m);

g is the gravity acceleration (m/s^2);

ρ_0 is the oil density (kg/m^3);

ρ_w is the water density (kg/m^3);

v_w is the water viscosity (m^2/s).

Stokes law is used in the above relation, and the small value of Reynolds number is taken $(R_e \text{¡} 20)$. The mixing depth Z_i of oil particles of different diameters can be calculated with Equation 5.120:

$$Z_i = \max\left(\frac{D_v}{W_i} Z_m\right) \tag{5.120}$$

Where:

D_v is the vertical dispersion coefficient (m^2/s).

D_v can be calculated with Equation 5.121:

$$D_v = 0.0015W_{10} \tag{5.121}$$

Where:

W_{10} is the wind speed at 10 m above the sea level (m/s).

The mass fraction R_i of oil particles with different diameters floating up to the water surface can be calculated with Equation 5.122:

$$R_i = \frac{W_i dt}{Z_i} \tag{5.122}$$

Where:

dt is the time step (s).

5.5.5 Emulsification Process

The method proposed by Mackay and Zagorski [177] is used to simulate the emulsification process of spilled oil. Mackay and Zagorski [177] put forward the method of calculating the emulsification process by user input parameters (viscosity coefficient and emulsification rate of latex), which can reduce the calculated emulsification rate at the time when the emulsification process will occur. In addition, they also proposed the exponential growth relation of the latex formation process. The rate of water mixing into oil phase can be calculated with Equation 5.123:

$$\tilde{F}_{wc} = \frac{dF_{wc}}{dt} = C_1 U_W^2 \left(1 - \frac{F_{wc}}{C_2}\right) \tag{5.123}$$

Where:

U_W is the wind speed (m/s);

C_1 is the empirical constant: for emulsified oil, the value is 2×10^6; for others, the value is 0;

C_2 is the constant used to control the maximum proportion of moisture: for heavy fuel oil and crude oil, the value is 0.7; for domestic fuel oil, the value is 025;

F_{wc} is the maximum proportion of water in oil phase and is dependent on the oil types.

The viscosity of emulsified oil μ can be calculated with Mooney equation:

$$\mu = \mu_0 \exp\left(\frac{2.5F_{wc}}{1 - C_0 F_{wc}}\right) \tag{5.124}$$

Where:

μ_0 is the initial viscosity of oil (centipoise);

F_{wc} is the maximum proportion of water in oil phase;

C_0 is the emulsification constant and usually is 0.65.

Evaporation will affect the viscosity and the viscosity, can be calculated with Equation 5.125:

$$\mu = \mu_0 \exp(C_4 F_\nu) \tag{5.125}$$

Where:

μ_0 is the initial viscosity of oil (centipoise);

C_4 is a constant: for light oil, it is 1; for heavy oil, it is 10;

F_ν is the evaporation rate of spilled oil.

The exponential growth relation proposed by Mackay and Zagorski [177] has been used in oil spill simulation for many years. It was derived from laboratory data and is suitable for heavy crude oil.

5.5.6 Shoreline Adsorption Process

The oil characteristics, shoreline type and environmental energy status determine the fate state after the oil reaches the shoreline. The weathering process will continue after the oil reaches the shoreline. At this time, the additional shoreline adsorption process (such as, re-floatation, infiltration into the substrate and the retention and migration in the groundwater system) is very important. The oil-infiltrated substrate may become the sediment in the nearshore waters due to erosion. This module can simulate the migration and transformation of the above mentioned oil contaminants in the coastal area.

In this module, each shoreline unit of the water-land grid contains the shoreline type information that characterizes different oil-retention capacity. The grid generation program allows the users to specify the shoreline type for the shoreline grid cell. When the oil particles reach the shoreline, sedimentation occurs. When the oil-retention capacity of the shoreline is saturated, the sedimentation stops. The oil particles that have reached the saturated shoreline cell after that will no longer be retained on the shoreline.

Tables 5.6 and 5.7 show the maximum thickness of oil contaminants on different types of shoreline. The thickness is dependent on three viscosity ranges, average oil infiltration depth, average oil-retention capacity and the time constant of removing the oil contaminants. The constants in the table are derived from the research by Reed, Gundlach and Kana [178]. Each shoreline grid cell has different oil-retention capacity which is mainly decided by the following several parameters: oil type, shoreline type, gradient of bank slope, width of bank slope (see Table 5.8) and length of shoreline grid cell.

The amount of oil contaminants deposited on the shoreline is exponential with the removal time. When the tide is high enough (so high that it can wet the surface of oil contaminants) or there is seaward wind, the removed oil contaminants will return to the water again.

The mass fraction that can be settled on the shoreline F_{sh} can be calculated with Equation 5.126:

$$F_{sh} = \frac{A_{\mathrm{lg}}}{A_s} \tag{5.126}$$

Where:

A_{lg} is the land grid area;

A_s is the area of surface oil particle.

If the total mass of oil contaminants accumulated in a certain shoreline grid cell has not been saturated, the oil particles will settle to the shoreline grid cell. The oil-retention capacity M_{hi} of the shoreline grid cell of i type can be calculated with Equation 5.127:

$$M_{hi} = \rho_0 t_i W_i L_{gi} \tag{5.127}$$

Where:

i is the shoreline type;

ρ_0 is the density of settled oil (kg/m^3);

TABLE 5.6: Relationship between maximum thickness of oil contaminants on different type of shoreline and the oil viscosity.

Shoreline type	Thickness of oil contaminants associated with oil types (mm)[1]		
	Light (<30 centistoke)	Medium (30-2000 centistoke)	Heavy(>2000 centistoke)
1. Exposed rocks	0.5	2	2
2. Wave eroded terrace	0.5	2	2
3. Fine sand beach	4	17	25
4. Coarse sand beach	4	17	25
5. Sand and gravel mixed beach	2	9	15
6. Gravel/pebbles	2	9	15
7. Exposed intertidal zone	3	6	10
8. Sheltered rocky coast	1	5	10
9. Sheltered intertidal zone[2]	6	30	40
10. Sheltered marshland	6	30	40
11. Glacier	2	4	6

[1] Examples of oil types: Light: kerosene, gasoline, diesel; Medium: medium or light crude oil and light fuel oil; Heavy: heavy fuel oil, heavy crude oil or weathered crude oil.

[2] It is assumed that the oil thickness in the sheltered intertidal zone and the sheltered marshland is the same.

t_i is the maximum thickness of oil slick that can be settled on the shore and is decided by the shoreline type and oil viscosity;

W_i is the width of front shoreline polluted by oil;

L_{gi} is the length of the front shoreline polluted by oil.

The mass of oil held-up on the shore M_R can be calculated with Equation 5.128:

$$M_R = M_0 \left[1 - \exp(t/T)\right] \tag{5.128}$$

TABLE 5.7: Time to remove the oil on different type of shoreline.

Shoreline type	Average oil infiltration depth (cm)	Average oil-retention capacity (%)	Time constant of removing the oil contaminants (day)
1. Exposed rocks	NA[1]	NA	1
2. Wave eroded terrace	NA	NA	1
3. Fine sand beach	4.8	9	1.36 to 10 (usually is 5 days)
4. Coarse sand beach	4.8	9	1.36 to 10 (usually is 5 days)
5. Sand and gravel mixed beach	17.8	9	2 to 20 (usually is 10 days)
6. Gravel/pebbles	17.8	9	2 to 20 (usually is 10 days)
7. Exposed intertidal zone	NA	NA	1
8. Sheltered rocky coast	NA	NA	20 to 100 (usually is 50 days)
9. Sheltered intertidal zone	NA	NA	100 to 1000 (usually is 500 days)
10. Sheltered marshland	NA	NA	100 to 1000 (usually is 500 days)
11. Glacier	NA	NA	1

[1] NA means not applicable.

Where:

M_0 is the mass of oil initially settled on the shore (kg);

t is the time (days);

T is the time constant for oil removal related to the shoreline types (days).

In addition to the environmental dynamic element field provided by other modules, this module also requires the oil parameters provided by the oil database.

TABLE 5.8: Gradient and width of bank slope of different type of shoreline.

Shoreline type	Average gradient of bank slope	Average width (m)				
		East coast	Gulf of Mexico	California	Northwest Pacific Ocean	Alaska Bay
1. Exposed rocks	5.7	2	1	2	3	3
2. Wave eroded terrace	5.7	2	1	2	3	3
3. Fine sand beach	0.15	10	5 15	10	15	20
4. Coarse sand beach	0.15	10	5	10	15	20
5. Sand and gravel mixed beach	0.89	5	5	5	7	10
6. Gravel /pebbles	0.89	3	2	3	4	6
7. Exposed intertidal zone	0.15	10	10	10	15	20
8. Sheltered rocky coast	5.7	2	1	2	3	3
9. Sheltered intertidal zone	0.01	140	20	120	210	300
10. Sheltered marshland	0.01	140	50	120	210	300
11. Glacier	5.7	-	-	-	-	-

5.6 Forecast Process of the System

The emergency prediction and warning system of oil spill in the Bohai Sea integrates the ocean current module, ocean wave module and atmosphere module into a forecast system and performs an unattended automatic operation on the computer server (Figure 5.71). It automatically downloads the GFS background field data and GTS observation data from 0:00 every day and processes the data (the processing takes about 25 minutes). It takes about 2 to 3 hours to run the WRF model and the calculated forecast wind field data is sent to FVCOM and SWAN for forecast of current field and ocean wave field in the next 48 hours. This process takes about 2 to 3 hours and will be completed before 7:00 in the morning. That is to say that the staff can get to know the weather, current field and ocean wave field in the next two days after going to work every day. Once there is an oil spill, they can immediately begin the forecast and early warning work. Meanwhile, a six-month simulation result database has been established in the system to enable the trace-back of oil spill.

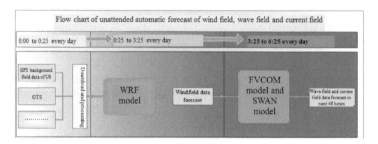

FIGURE 5.71: Automatic forecast process of the emergency prediction and warning system of oil spill in the Bohai Sea.

5.7 Case Verification and Automatic Operation of the System

The system has been applied in marine oil spill accidents many times after it is developed. In case of emergency, more reliable prediction results have been obtained with the system in the shortest time, which provides important reference for handling of the accident.

5.7.1 Performance of Automatic Operation of Such System in "4.19 Accident"

In 2010, according to the satellite images (Figures 5.72 and 5.73), there was a clear anomalous zone between Liaodong Peninsula and North Huangcheng Island, which is identified as likely to be oil spill pollutants. Table 5.9 shows the related satellite monitoring information based on satellite images.

According to the information obtained from the satellite images, the source of pollutants is traced back by using the "emergency prediction and warning system of oil spill in Bohai Sea." Tracing back from 07:00 on April 20, 2010 to 22:00 on April 19, 2010, the results show that the pollutants may come from the north by west of Laotieshan Waterway, as shown in Figure 5.74.

According to the accident report received by Shandong Maritime Safety Administration, at 21:35 on April 19, 2010, "Shengzhou 5" collided with "AAA" ship and sank and the 2t light diesel in the ship overrun. If the collision occurred at point A, the traceability result is about one nautical mile away from the collision location.

Although the traceability result is slightly different from the reality, it is on the whole reasonable and accurate. Analysis suggests that the causes of deviation may be: (1) the specific location where the ship sank is not clear (the traceability is based on the collision location); (2) the accurate time of oil

TABLE 5.9: The remote sensing monitoring information of satellite.

Basic information			
Type of satellite	Radarsat-2	Imaging time	2010-4-20 06:56
Width	$150km \times 150km$	Transmission	2010-4-20 08:31
Forecast wind speed	Northeast wind, scale: 5-6	Production	2010-4-20 09:15
Whether an abnormality was found		Yes	
Number of anomalous zone		1	
Description of anomalous zone			
Anomalous zone 1. The anomalous zone is located in the area under administration of Liaoning Maritime Safety Administration; 2. A suspected oil belt running in the northwest-southeast direction; With point A and point B as the end points, the length of the belt is about 4 nautical miles and the average width is about 500m. he point B is about 12 nautical miles away from the shore and about 10 nautical miles from North Huangcheng Island. The latitude and longitude of each point is as follows: A:120°57′38″, 38°35′47″ B:121°00′57″, 38°32′38″ As shown in Figure 5.74 and Figure 5.75.			

spill is not clear (the traceability is based on 30 minutes after the collision); (3) the wind field and current field are somewhat different from the reality.

In comprehensive consideration of the above factors, it is really commendable for the system to obtain the precision traceability results and find the approximate location of the source of spilled oil with important reference value in a short time.

5.7.2 Application of the System in Traceability of the "4.16" Oil Spill Accident in Kiaochow Bay

At 14:00 on April 16, 2010, marine oil spill occurred at the mouth into Kiaochow Bay. The research group immediately applied the "emergency prediction and warning system of oil spill in Bohai Sea" to trace back the oil pollutants. The results show that the oil spill occurred at Huangdao Oil Terminal (Figure 5.75). Upon investigation, it was determined that it came from the discharge of the water used to wash the oil tank by the cruise at Huangdao Oil Terminal. The successful traceability this time proved again the reliability of the system.

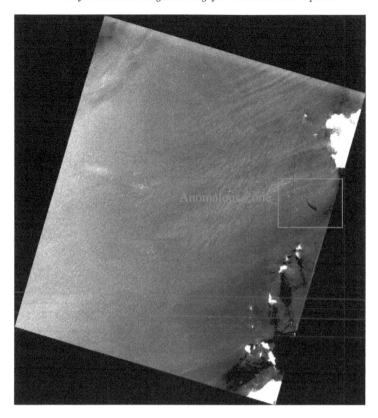

FIGURE 5.72: Distribution of anomalous zone shown on the Radarsat satellite images at 06:56 on April 20, 2010.

5.7.3 Oil Spill Prediction after Collision between "BRIGHT CENTURY" Ship and "SEA SUCCESS" Ship

On May 1, 2010, two ships collided in the Chengshan Cape sea area in Weihai. The DWT bulk carrier "Bright Century" carrying 150,000t iron ore sank after the collision and the head of the other handymax general cargo ship "Sea Success" carrying 30,000t coiled steel was seriously damaged. The oil spill occurred after the collision. To better cooperate with the spilled oil clean-up, the system was applied to predict the drift path. The oil spill occurred at $37°38'30''N/123°09'00''E$, and totally 100t fuel oil was spilled. The forecast period is 07:00 on May 2 to 19:00 on May 3. The wind field used for forecast was WRF real-time forecast wind field (with resolution of 10 km) and the current field was the tidal current and wind current automatically forecast with FVCOM (with maximum resolution less than 50 m).

The prediction results of the system are shown in Figure 5.76. Thirty-six hours after the oil spill, the pollutants drifted to the vicinity of $38°02'57''N$

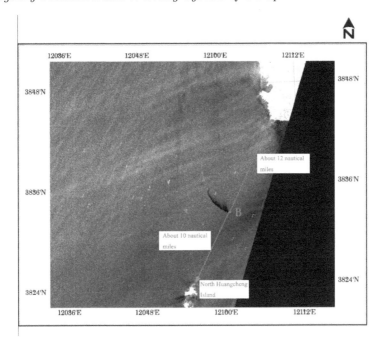

FIGURE 5.73: Details of the anomalous zone.

and $123°23'20''E$ (location in the north by east of and 26 nautical miles away from the collision location). The drift trajectory is represented by the solid lines in the figure: the spilled oil drifted towards northeast first, then it drifted eastwards and then turned to northwest, finally it kept drifting towards northeast in a considerable long distance. Compared with its initial location, the spilled oil drifted to the location northeast of and 26 nautical miles away from the location where the oil spill occurred 36 hours later.

5.7.4 Application of the System in "7.16" Oil Spill Accident in Dalian

Around 18:00 on July 16, 2010, the pipeline in the crude oil tank area of Dalian PetroChina International Warehousing & Transportation Co., Ltd., located near Dalian New Port, Liaoning Province, exploded, causing a fire and some crude oil to leak into the sea. It is preliminarily estimated that the explosion caused 1500 tons of crude oil to flow into the sea. By 13:30 on the 19th, the surveillance results of Chinese ships patrol showed that the polluted sea area was as much as 430 square kilometers, causing heavy damage to local fisheries, aquaculture and tourism. The economic losses were very serious, and the clean-up fee alone was more than one yuan. It can be said that this explosion is the most serious marine oil spill accident in China. After

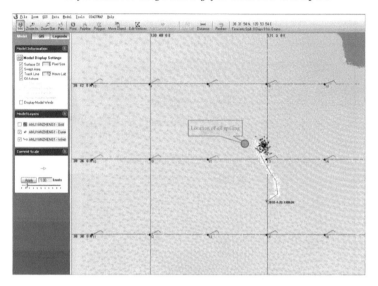

FIGURE 5.74: Traceability results of the system.

investigation, the Ministry of Communications said that the cause was the explosion of 300,000t crude oil vessel "Yuzhoubaoshi" when adding catalyst to land pipelines of crude oil reserve tank during the unloading in crude oil reserve of Dalian PetroChina International Warehousing & Transportation Co., Ltd. which is near Dalian New Port.

When the accident happened, "emergency prediction and warning system of oil spill in Bohai Sea" had been operating for about 6 months. The Maritime Safety Administration of the People's Republic of China used such system to simulate the distribution of spilled oil in Dayao Bay and Dalian Bay, Figure 5.77 (a) at 18:00 on July 19 (about 72 hours after the accident). The simulation results are in good agreement with the satellite image, Figure 5.77 (b) providing forceful technical prediction and warning support for the emergency response of the major oil spill accident.

In the above cases, relatively reasonable and accurate results were obtained effectively in a short time after prediction or traceability with the system. Hence, the system is an environmental prediction system with higher reliability and can provide valuable reference for prevention, control and emergency response of marine oil spill, thus saving the manpower and material input and minimizing the damages of oil spill pollution to the ocean and human health. Currently, the system is being constantly improved. The ultimate goal is to construct prediction and warning system of oil spill pollution in coastal waters of China for major oil terminals and waterways and create the information service platform for early warning decision of oil spill pollution in coastal waters of China, so as to provide information service for minimizing the damages of oil spills and maintaining the sustainable development of ocean economy.

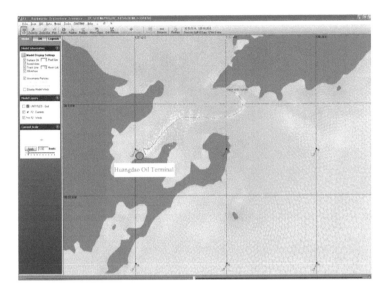

FIGURE 5.75: Traceability on oil spill at mouth into Kiaochow Bay on April 16, 2010.

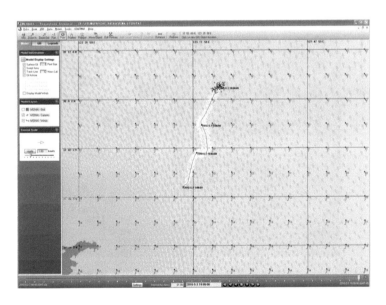

FIGURE 5.76: Location of the spilled oil 36 hours after the collision.

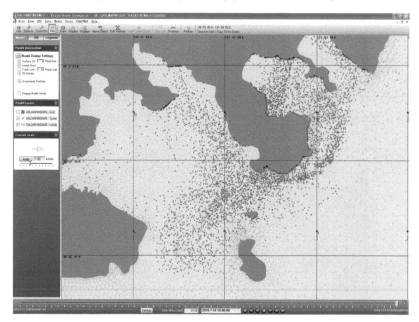

(a) Note: The red dot in the figure represents the oil spill coverage.

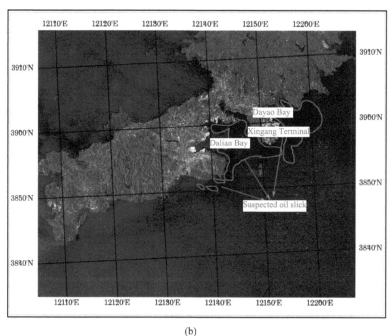

(b)

FIGURE 5.77: The system simulated distribution of spilled oil in Dayao Bay and Dalian Bay 72 hours after the "7.16" major oil spill accident in Dalian (a) and the corresponding satellite image (b).

Chapter 6

Environmentally Sensitive Resources and Emergency Resources of Oil Spills

6.1 Environmentally Sensitive Resources of Oil Spills

Environmentally sensitive resources of oil spills refer to all resources that may be affected by oil spills, including ecological resources, aquatic resources, tourism resources and coastal industrial and mining enterprises. The specific content and scope of environmentally sensitive resources are different under different circumstances. This chapter mainly considers the emergency decision of oil spills, with focus on nature reserves, ecological resources, fishery resources and tourism resources. These resources can be divided into ecological resources, resources of human activities and shoreline resources. There is no strict boundary for this classification, but the angle of classification is different. In some cases, the contents of ecological resources, resources of human activities and shoreline resources are difficult to be clearly distinguished.

6.1.1 Classification of Environmentally Sensitive Resources

6.1.1.1 Biological resources

The biological resources include not only the organism sensitive to oil spills, but also habitats of these organisms and nature reserves at all levels. There are many types of organisms that are susceptible to oil spills, and they are particularly sensitive to oil spills in specific areas at specific times. Their habitats are also wide, including feeding grounds and migratory routes, animal gathering areas such as near-shore activity area of spotted seal, islands where the migrating birds gather, breeding areas such as the spawning ground of seabirds and fishes and the living area of rare and endangered species, as shown in Figure 6.1 and Figure 6.2.

FIGURE 6.1: National first-class protected animal, *Larus saundersi*, in National Nature Reserve at estuary of Yalu River.

6.1.1.2   Resources of human activities

Resources of human activities refer to all resources developed and used by mankind. These resources can be classified into the following categories: (1) highly utilized leisure shoreline and its channels, such as bays, leisure beaches, fishing areas and shallow waters; (2) resources storage and taking sites, such as aquatic sites, living or commercial fisheries, log storage site, leased mining stations and water intakes; (3) and various water or land archaeological, historical and cultural relics, including the cultural relics with special value that are located in the intertidal zone or close to the shore and are easily damaged by the clean-up workers. More specifically, the resources of human activities can include the resources for industry at the ports, residence, aquaculture, fishing, tourism and other purposes (Figures 6.3 and 6.4).

6.1.1.3   Shoreline resources

The shoreline resources are mainly classified according to the shoreline type and include both biological resources and resources of human activities

FIGURE 6.2: Coastal mangrove nature reserve.

(Figures 6.5 and 6.6). The shoreline type is determined according to the shoreline structure, shoreline shape, the extent to which the shoreline is open or sheltered by waves and currents, and the sensitivity of nearby organisms to oil spills. The shoreline structure mainly refers to the external form of the shoreline, and the shoreline is divided into flat shoreline, slopes and cliffs by the shoreline gradient. The extent to which the shoreline is open or sheltered by waves and currents includes: exposed, semi-exposed, semi-sheltered, and sheltered. The shoreline type directly affects the breeding of biotic population and human activities and the shoreline resources contain extensive types of resources.

6.1.2 Sensitivity of Environmentally Sensitive Resources to Oil Spills

The sensitivity of environmentally sensitive resources to oil spills refers to the degree of damages to the resources during the oil spill and emergency response, which is decided by three factors:

- Hazards of oil spills to the resources;

- The possibility that such resources are affected by oil spills;

- The importance of the resources.

FIGURE 6.3: The first beach in Qingdao.

6.1.2.1   Damages of oil spill to environmentally sensitive resources

The oil spill mainly damages the environmentally sensitive resources through physical action and chemical action. The physical action mainly refers to the physical nature of the spilled oil that causes changes in the external form, landscape or composition of environmentally sensitive resources after the spilled oil contacts with the environmentally sensitive resources, thus damaging the normal function of the resources and reducing their original value. The chemical action refers to the chemical components of spilled oil that poison the organism after contacting with the environmentally sensitive resources, thus leading to death of the organism and damaging the original resources. It damages the environmentally sensitive resources of different types and with different functions in different ways. The hazard of oil spills to environmentally sensitive resources directly decides the sensitivity of resources. Generally, the greater the hazards, the higher the sensitivity of such resources to oil spill.

1. Tourism resources and other environmentally sensitive resources related to human activities (non-aquatic activities)

The impacts on this type of resources are mainly exerted through physical action. The color, odor and viscosity of the spilled oil damage the resources, such as:

- Cause pollution to the beach, sand, coastal gardens and other tourist landscape, which prevents them from normal operation.

FIGURE 6.4: Cityscape shoreline.

- Cause pollution to wharfs, ships and docks and other coastal industries, which prevents them from normal operation.

- Threaten the feed inlet of power plants, which affects the normal power supply in the nearby communities.

- Damage the fishing instruments, which affects the normal fishery.

- Cause pollution to the coastal lines and make it lose some of its original functions.

2. Biological resources and aquatic resources
Hazards of oil spills to this type of resources mainly refer to the damages of oil spills to the organism through physical and chemical actions. The physical and chemical actions sometimes exert impact on the organism separately and sometimes jointly. For instance:

- The spilled oil covers on the organism, plankton and aquatic plants in the intertidal zone and thus leads to their death.

- The spilled oil contains many toxic components and the toxicity of different types of oil is different. The toxicity of spilled oil can cause death of plankton and other organisms contacting it.

FIGURE 6.5: Aquaculture shoreline.

- Spilled oil of sublethal level can reduce the immunity of organisms, which may induce gene mutation and cause the organism to have genetic diseases and cancer.

- Long-term contact with the low-concentration spilled oil will cause death of certain sensitive organisms, damages to the structure of biocenosis and disruption of intermediate links of marine food chain, thus damaging the ecological system.

- Spilled oil can kill the cultured organism, and the low-concentration spilled oil can also make seafood contaminated with oil smell or taste funny and thus lose the edible value, causing economic loss.

- The spilled oil has special hazards to the seabirds. It can damage the feather structure of seabirds, cause them to lose the waterproofing ability and the ability to fly and prey, and eventually die.

6.1.2.2   Possibility that oil spill produces damages

This refers to the possibility that the spilled oil makes contact with the environmentally sensitive resources after the accidents. Obviously, no matter how great the hazards of oil spills to environmentally sensitive resources, such hazards will not exist as long as the spilled oil does not make contact with such resources and the sensitivity of such resources to the oil spill can be neglected.

FIGURE 6.6: Reef shoreline.

It includes two types of possibility: temporal and special. For example, oil spills are very harmful to seabirds. Generally speaking, since seabirds live on the sea or look for food mainly on the sea, spatially, there is a high possibility for the seabirds to make contact with oil spills. But for the purpose of evaluating the sensitivity of seabirds to the oil spills, it is also decided by the types of seabird. If the seabirds are migratory birds which live in such region in a certain period of time or season in a year, such seabirds may be negatively affected by the oil spill that occurred at such time, and the so-called sensitivity to seabirds can only be reflected at such time. Take benthic organisms for another example, if they live in the intertidal zone, then the spilled oil may directly cover them and directly poison them or suffocate them to death. If they live in the subtidal zone, then the spilled oil will not directly make contact with them and can only endanger the benthic organism through sedimentation of water-soluble components to the seafloor. Therefore, the sensitivity of these two types of benthic organisms to the oil spill is greatly different. Hence, the temporal and spatial distribution of environmentally sensitive resources directly affects their sensitivity towards the oil spill.

6.1.2.3 Importance of environmentally sensitive resources

Importance of environmentally sensitive resources refers to the value of the resources which includes three aspects: economic value, ecological value and special value.

Economic value refers to the economic loss caused after the resources are polluted by oil spills. It is decided by the economic benefits produced by the resources, the clean-up cost and the cost to restore the resources. The majority of resources related to human activities can be quantitatively assessed with the economic value. For example, the loss due to production suspension of the factory and wharf; the loss due to operation suspension of the beach and the gardens and the loss due to death of aquatic products of the aquafarm.

The ecological value refers to the importance of the resources in the ecological system, the position in the food chain and food cycle and the characteristics of the resources. It mainly qualitatively assesses the biological resources.

The special value mainly assesses the social and political impacts on the environmentally sensitive resources. The resources that can be assessed with this index include the resources with special economic value, special ecological value and important military value.

6.1.3   Determination of Sensitivity Index

6.1.3.1   Principles to determine the sensitivity

The sensitivity of the environmentally sensitive resources, which may be affected by the oil spill, is determined according to the importance of the resources and the degree of hazard that may result from the oil spill is duly considered at the same time. Hence, the sensitivity of environmentally sensitive resources is mainly decided by their value. More specifically, it is dependent on the economic value, ecological value and special value of the resources.

6.1.3.2   Methods to determine the degree of sensitivity and classification of resources

Based on the sensitivity index (SI), the environmentally sensitive resources are classified into four types:

Type A: very important resources which are extremely sensitive to the oil spill.

Type B: important resources which are very sensitive to the oil spill.

Type C: second important resources which are relatively sensitive to the oil spill.

Type D: general resources which are generally sensitive to the oil spill.

The degree of sensitivity is dependent on the value of the resources. When considering the value effect, the economic value and ecological value of the resources are graded into three levels: high, medium and low, and the special values are graded into two levels: with or without. The resources are classified and evaluated according to these grades. Finally, these three types of value of environmentally sensitive resources are comprehensively considered and the resources are classified accordingly. Due to the characteristics of the regional oil spill emergency plan, this study focuses on summary of the natural resources such as ecological and aquatic resources and the ecological value of the resources.

Type A resources mainly refer to the environmentally sensitive resources with high ecological value and special value. Such as, national reserves of rare and endangered species.

Type B resources mainly refer to the environmentally sensitive resources with high ecological value or special value. Such as, national level nature reserve and important ecological resources reserve.

Type C resources mainly refer to the resources with very high economic value and include some important tourist scenic spots, important seafood aquiculture area, provincial and municipal nature and geomorphology reserves.

6.1.4 Comprehensive Sensitivity Index of Shoreline

6.1.4.1 Concept of environmental sensitivity index

The Environmental Sensitivity Index (ESI) is a shoreline sensitivity index developed by the Hazardous Materials Response and Assessment Division of National Oceanic and Atmospheric Administration (NOAA) for the production of sensitive maps for oil spill response in the US coastal regions and Great Lakes. In principle, ESI is graded into ten levels and the sensitivity is gradually increased along with the rise of levels. Meanwhile, a different ESI level represents a different type of shoreline. In addition, there are countermeasures correspondent to different ESI value.

Therefore, ESI is a comprehensive index that includes many aspects such as shoreline sensitivity, shoreline types and shoreline clean-up measures. Since the ESI was used very early, has complete classification system and is applied in many countries, it has greater influence and can be regarded as a standard to some extent.

ESI is variable and changing along with the shoreline types. Since there are only ten ESI levels, it is far from being able to cover all shoreline conditions. Therefore, 1a, 10b, etc. are interpolated. In addition, different countries have different definitions of the specific shoreline type correspondent to the ten levels when applying ESI. The basic classification methods and explanations of NOAA are mainly considered in ESI classification in the sensitive resources database.

6.1.4.2 Relationship between ESI and impacts of oil spill

1. ESI classification method
(1) When ESI = 1, which is correspondent to an exposed rocky cliff and seawall

- Due to repeated actions of the waves, the spilled oil will be washed to the open sea that is far away from the steep rocks.

- Any spilled oil that adheres to the cliffs will be quickly washed away from the rock surface.

- The majority of the retained oil will form an irregular oil band above the high-water line on the shoreline.

- The oil spill usually exerts short-term impacts on the intertidal zone.

- The spilled oil at the junction of the rocky cliffs will infiltrate into the ground.

- The biocenosis in the high-concentration spilled oil zone can be severely damaged.

(2) When ESI = 2, which is correspondent to an exposed rocky terrace

- The spilled oil will generally not adhere to the rocky terrace, but will usually accumulate near the high-water lines.

- The spilled oil will infiltrate into the substrate of the beaches.

- The spilled oil usually retains in the substrates for a short time, except for some spilled oil in the wave-sheltered zones and the large quantity of spilled oil retained.

(3) When ESI = 3, which is correspondent to a fine-sand beach

- The light oil will accumulate along the high-water line in the intertidal zone.

- The heavy oil will cover the whole beach surface.

- The spilled oil will infiltrate about 10 cm into the fine-sand beach.

- Organisms growing in the intertidal zone will be killed.

(4) When ESI = 4, which is correspondent to a coarse-sand beach

- The light oil mainly adheres and accumulates at the high-water line.

- The spilled oil will spread to the whole beach surface.

- The spilled oil will infiltrate up to 25 cm into the coarse-sand beach.

(5) When ESI = 5, which is correspondent to a sand and gravel mixed beach

- In case of small-scale oil spills, the spilled oil will deposit along the high-water line.

- In case of large-scale oil spills, the spilled oil will spread to cover the whole beach.

- The spilled oil can infiltrate up to 50 cm into the beach substrates. If the sands account for more than 40% of the mixed beach, the infiltration of spilled oil in the beach is more similar to that in the sandy beach.

- Landfills in the oil-contaminated beaches must be above the high-water line.

- A shell-like oil layer may be formed on the surface of the more sheltered beaches. Once formed, the oil layer is very stable and can exist for many years.

(6) When ESI = 6, which is correspondent to a rocky beach

- The spilled oil has strong infiltration effect.

- On more exposed beaches, the spilled oil is usually pushed above the high-water line.

- On the more sheltered beaches, a shell-like oil layer is easily to be formed if there is much spilled oil accumulated.

(7) When ESI = 7, which is correspondent to an exposed tidal flat

- The spilled oil will usually not stay on the surface of the tidal flat but adhere near the high-water line.

- If the spilled oil is of high concentration, the spilled oil will deposit and adhere to the surface of the tidal flat at low tide.

(8) When ESI = 8, which is correspondent to a sheltered rocky coast

- The spilled oil is easy to retain on the surface of coarse rocks and forms an obvious oil zone especially at the high-water line.

- Heavy oil and weathered oil will cover the supratidal zone.

- The spilled oil will retain in the fracture zone of the rocks.

- When the rocks on the surface are loose, the spilled oil will infiltrate deeper into the rocky coast and thus cause long-term contamination to the substrates.

- The fresh oil and light refined petroleum products are acutely toxic to the organic matters adhering to the surface.

(9) When ESI = 9, which is correspondent to a sheltered tidal flat

- The spilled oil will usually not retain on the surface of the tidal flat but adhere near the high-water line.

- If there is a large quantity of spilled oil, the spilled oil will stay on the surface of the tidal flat at low tide.

- The oil spills will hardly infiltrate into the substrate.

- In zones with high concentration of suspended solids, the adsorbed oil can cause the contaminated substrate to sink on the tidal flat.

(10) When ESI = 10, which is correspondent to a marshland

- The spilled oil easily adheres in the marshland (the mud has the characteristics of low-energy environment).

- The contaminating oil zone may be relatively wide, which is dependent on the tidal stage and the location of oil slicks.

- The large area of oil slick may cover the range in between the high-water line and low-water line.

- The medium and heavy oil will not easily adhere to or infiltrate into the fine substrates, but are prone to accumulate on the surface of the marshland.

- The light oil will easily infiltrate several centimeters into the substrates and sometimes about 1m into the gap.

2. ESI and the corresponding emergency protection measures and clean-up suggestions

(1) When ESI = 1, which is correspondent to exposed rocky cliffs and artificial coast

- It is often dangerous and difficult to enter a rocky cliff zone, so clean-up is generally not carried out.

- Use high-pressure water to wash the spilled oil in the artificial coastal (artificial structure) zone to prevent the spilled oil from oozing out slowly from the structure.

(2) When ESI = 2, which is correspondent to an exposed rocky terrace

- Clean-up is usually not required.

- If the clean-up staff can enter into the high tidal region, the deposit of heavy oil and oily gravel can be removed.

(3) When ESI = 3, which is correspondent to a fine-sand beach

- This type of beach is the one that is easiest to be cleaned.

- Clean up and transport the spilled oil after all the spilled oil is flushed to the high tidal region.

- When transporting the oily sand from the beach, it should prevent it from scattering around and protect the clean area.

- Regardless of manual clean-up or mechanical transport, care should be taken to minimize the volume of the sand transported (only the contaminated sand on the surface has to be transported), so as to reduce the workload of disposal of oil contaminants.

- Care must also be taken to prevent the oily sand from being stepped into the deeper substrate by machinery and human feet during the clean-up process.

(4) When ESI = 4, which is correspondent to a coarse-sand beach

- Mainly the oil concentrated at the high-water line is cleaned up and transported.

- It should prevent the substrate from scattering around when transporting the substrate and prevent the oily sand from being stepped into a deeper layer.

- Mechanical clean-up and transport may result in an excessive volume of transport (too deep excavation), and manual clean-up may be more effective.

(5) When ESI = 5, which is correspondent to a sand and gravel-mixed beach

- All oily crushed stones must be removed and the removal should be started from the high-water line.

- If possible, minimize the transport of substrate.

- Flushing with low-pressure water can make the oil float from the substrate up to the surface. Then use an oil skimmer or oil absorption material to recover the oil. High-pressure water must be avoided for it may cause the substrate (sand) to be transported to the low-water line in the intertidal zone.

(6) When ESI = 6, which is correspondent to a rocky beach

- All oily crushed stones, especially the oil contaminants deposited at the high-water line must be removed.

- Minimize the transport of substrate.

- Flush with low-pressure water to make the oil float from the substrate up to the surface. Then use an oil skimmer or oil absorption material to recover the oil.

- The heavily polluted gravel can be moved away and replaced with clean gravel.

(7) When ESI = 7, which is correspondent to an exposed tidal flat

- The erosion by tidal current and waves is very effective for the natural dissipation of oil contaminants.

- It is very difficult to take clean-up actions (if possible, it can only be carried out at low tide).

- Heavy machinery is strictly prohibited to prevent oil from mixing with the substrate.

- In sandy beaches, oil will be naturally transported to adjacent beaches and clean-up actions can be considered.

(8) When ESI = 8, which is correspondent to a sheltered rocky coast

- When the spilled oil is not weathered, it is often effective to use low-pressure water or high-pressure water of ambient temperature to flush.

- Special attention should be paid to the areas where organisms grow prosperously in the low tidal region of the intertidal zone and do not flush such areas.

- Do not cut oil-contaminated algae, for the effect of tide will make the oil contaminants float. Consider setting up oil adsorption boom.

(9) When ESI = 9, which is correspondent to a sheltered tidal flat

- High priority should usually be given to the protection of this type of area during oil spills.

- It is often very difficult to take a clean-up action on the tidal flat because the substrate is very soft and many methods are forbidden.

- It may be helpful to manually deploy the oil absorption material with a shallow draft boat.

(10) When ESI = 10, which is correspondent to a sheltered marshland

- It is best to let the oil contaminants on the lightly polluted land dissipate naturally.

- For clean-up in the area with large quantity of oil contaminants deposited, vacuum pump or oil absorption material can be used or flushing with low-pressure water can be implemented. Care should be taken to prevent the oil contaminants from being spread or transported to the sensitive zone of low slope.

- Damages to the vegetation should be avoided during the clean-up process and it is forbidden to mix the oil into the substrate during any clean-up process.

6.1.5 Priority Index of Protection

There are many types of environmentally sensitive resources of which the sensitivity to the oil spill is greatly different. How to determine the sensitive resources to be protected first using a priority index of protection (PI) and how to protect them in the emergency response to an oil spill is an important part of emergency decision making.

Many factors have to be considered and analyzed comprehensively for determining the protection priority of environmentally sensitive resources. Such as sensitivity of the resources to the oil spill, social and political impacts of the oil spill, the feasibility, effectiveness and seasonal factors of the existing emergency measures.

6.1.5.1 Sensitivity of environmentally sensitive resources to oil spills

Sensitivity of the environmentally sensitive resources to an oil spill is an important factor affecting the protection priority. Theoretically, the stronger the sensitivity of the environmentally sensitive resources to an oil spill, the higher the protection priority such resources receive; otherwise, the weaker the sensitivity of the environmentally sensitive resources to oil spills, the lower the protection priority such resources receive. However, it is not always true in reality.

For example, the ESI of the seawater marshland, mangrove forest, freshwater marshland and saltmarsh is 10. That is to say, they are environmentally sensitive resources with stronger sensitivity to an oil spill. In the emergency plan of many countries including China, relatively higher protection priority is given to these resources.

But this is not the case for the tourist terminal. The sensitivity of a tourist terminal to the oil spill is low, with ESI as 1. From the perspective of sensitivity, such resources receive lower protection priority. If an oil spill occurs in the tourist off-season, the protection priority it receives is correspondent to its sensitivity to the oil spill. However, the oil spill occurring in the tourist rush season will directly lead to decrease in tourists and affect the economic benefits. Hence, the tourist terminal which is less sensitive to an oil spill will receive higher protection priority during the tourist rush season. This specific situation must be fully considered when determining the priority order of protection. That is to say, other factors must also be considered when determining the priority order of protecting the oil-spill-sensitive resources.

6.1.5.2 Possible social and political impacts of environmentally sensitive resources

The social and political impacts of environmentally sensitive resources can also be called the special value of the resources. Such value is reflected in the special economic value, special ecological value and main military value of the resources. These resources may be less sensitive to oil spills. However, when they are polluted by oil spills, great social and political impacts will be brought about. Then this factor must be considered when determining the priority order of protecting such resources.

For example, if Yantai's No.1 beach is polluted by an oil spill in the summer, like other beaches used by people, a certain social impact will be produced. Therefore, it receives higher protection priority in the summer and lower protection priority in the winter. However, since it faces Yantai International Expo Center, the value of the beach will be considered when determining the protection priority it should receive in the winter. If the municipal government decides to hold international meetings in the center during the New Year's holiday every year, oil spill pollution to the batching beaches and nearby shoreline at this time will bring about greater political impacts. Therefore, it receives the same protection priority in the winter as that in the summer.

6.1.5.3 Feasibility and effectiveness of existing emergency measures

Specifically speaking, the feasibility of existing emergency measures refers to the possibility of cleaning up the spilled oil and the effectiveness refers to the actual effect of protecting the sensitive areas. For instance, mangrove forest and wetland are environmentally sensitive resources which are strongly sensitive to oil spills. However, it is not feasible to apply the existing emergency technology to clean up spilled oil in the polluted mangrove forest while the application of existing emergency technology would lead to greater damages to the wetland. Hence, in order to avoid oil spill pollution to such resources, the only way is to work out and implement preventive protection measures. Such resources should receive higher protection priority in consideration of their sensitivity index and the feasibility and effectiveness of emergency measures.

6.1.5.4 Seasonal factors

Some environmentally sensitive resources are differently sensitive to the oil spill in different seasons and they therefore have different economic value and ecological value in different seasons. Additionally, there is another type of resources which may also have different sensitivity towards oil spills in different seasons due to the degree of development and utilization by human beings. Therefore, the season-induced impacts must also be analyzed when determining the protection priority that should be given to different resources in the emergency zone, such as migratory bird conservation area, spawning ground of the aquatic species and shoreline tourism resources.

6.1.5.5   Classification of protection priority

The method to determine the protection priority that should be given to different environmentally sensitive resources is as follows: grade the economic value and ecological value of the resources into three levels: high, medium and low; grade the special value into two levels: with or without; classify the resources according to these grades and then evaluate; finally, comprehensively assess the three types of values of the environmentally sensitive resources to determine the protection priority. Many factors affecting the protection priority should be comprehensively considered when evaluating the economic value, ecological value and special value of various resources.

Currently, there is no uniform standard for protection priority which varies from country/region to country/region. The United States grades the protection priority of the coastal sensitive resources into three levels: Class-A, Class-B and Class-C, and the protection priority that the conservation area receives is lowered accordingly. While the regional emergency plans in China's Emergency Plan for Marine Oil Spill also have different provisions on the protection priority, the Emergency Plan for Oil Spill in South China Sea Area classifies the protection priority that should be given to the sensitive resources in the South China Sea area into three classes: top priority, secondary priority and priority. The Emergency Plan for Oil Spill in North China Sea Area classifies the protection priority that should be given to the sensitive resources in the North China Sea area into eleven classes.

6.2   Map of Environmentally Sensitive Resources

Marking the location of different coastal resources, environmentally sensitive areas and suggestions on countermeasures for emergency response to oil spills in the map of environmentally sensitive resources can provide important information for the emergency responders. In the emergency response to oil spills, the emergency decision makers can obtain the information about the sensitive resources in the area of accidents from the map, analyze the areas which may be polluted and the resource types by predicting the drift trajectory of the spilled oil, and quickly make emergency decisions according to the suggestions on countermeasures provided in the map to effectively protect the environmentally sensitive resources which may be damaged and achieve the goal of the emergency plan. Therefore, the production and updating of the map of environmentally sensitive resources is one of the important contents of an oil spill emergency plan. The map of environmentally sensitive resources consists of electronic map and environmentally sensitive resources database and is therefore actually the output after combination of the electronic map and different types of environmentally sensitive resources data.

6.2.1 Basic Information Provided in the Map of Environmentally Sensitive Resources

The map should contain many types of information: the first is the sensitivity related contents, such as the shoreline type based on the ESI, the biological resources reflecting the ecological conditions and the resources of human activities with social and economic characteristics; the second is the protection priority reflecting the importance of environmentally sensitive resources; the third is the suggestions on countermeasures reflecting the oil spill emergency characteristics. The specifics are as follows:

6.2.1.1 Shoreline type

The ten ESI levels of shoreline reflect the extent to which the shoreline is sheltered by the waves, the permeability of oil to the bottom of the beach, the hold-up duration of oil on the shoreline and the propagation of coastal organisms. Hence, ESI is the basic information contained in the map of environmentally sensitive resources. However, since it is a convenient way to summarize information based on ESI, the sensitivity value does not reflect the actual sensitivity of shoreline. For example, ESI 5 does not mean that the resource is five times more sensitive than that of ESI 1. In other words, ESI provides only partial information and does not directly reflect the use of the shoreline by humans and wildlife. For example, the rocky coast exposed on the shoreline of ESI 1 (low sensitivity) may also be the habitat of highly sensitive seabirds, at least during the breeding season of seabirds. Beaches of ESI 3 (relatively low sensitivity) are important for visitors or hatching of turtles at some point during the year. Therefore, specific areas reflecting ecosystem resources should be identified in the map of environmentally sensitive resources.

6.2.1.2 Habitats in intertidal zone

The habitats and the special species in the intertidal zone have proven to be very sensitive to oil spills. Typically, these habitats should be marked in the map of environmentally sensitive resources and usually include coral reefs, seagrass beds, and large seaweed beds. If the oil resides on the surface of coral reefs at low tide, coral organisms will be severely affected, and the chemical action and natural diffusion of oil in shallow sea may affect certain biological species in the deep sea. The seagrass bed may be distributed in the intertidal zone and the shallow coastal waters. Experiments have shown that highly concentrated dispersed oil contaminants are the greatest hazard to seaweed and organisms that grow here. The large seaweed bed (large brown algae) grows in the lower and offshore areas of some rocky coasts. These brown algae are strongly resistant to oil, but other organisms on the algae bed (such as invertebrates, cormorant and sea otters) are very sensitive to oil spills.

6.2.1.3   Wild animals and conservation areas

The map of environmentally sensitive resources should indicate areas where animal populations are highly sensitive to oil spills. For example, predatory feeding areas are important places with concentrated individual animals during a certain period of the year. For endangered species, their main habitat should be highlighted in the map, as severe oil spills in such areas may lead to the extinction of a species. It is necessary to mark these in the map of environmentally sensitive resources, and both the legal and economic factors should be considered when dealing with oil spills in these areas.

Some sensitive biological populations may gather on some coasts with relatively strong resistance against oil pollution. For example, the ESI of the shoreline itself is level-I from the perspective of ESI classification of shoreline, but due to gathering of a large amount of wild animal during the breeding season, it has become a priority protected area. Obviously, in this case, the wildlife is the first thing to be considered in an emergency response to an oil spill.

6.2.1.4   Fish, fisheries, shellfish and aquaculture

Fisheries, aquaculture and trade activities are among the information that should be taken into account in the map of environmentally sensitive resources. The typical information that should be indicated includes: offshore shallow fishing area of species such as baleen whale, crab, lobster and shrimp; seaweed gathering area; shellfish seafloor in tidal belts or shallow coastal waters; fish and crustacean culture areas; beaches for fishing; permanent or semi-permanent fishing nets and fishing platforms; fish, mollusks, shellfish, or seaweed aquaculture facilities, such as tall scaffolds, floating cages and long cables in tidal zones, fish and shellfish ponds, and river estuaries that are important for migratory fish (such as salmon).

From a fishery perspective, sensitive areas include the beaches for fishing, fishing cages and aquaculture facilities such as algae raft and the entrance for migratory fishes.

6.2.1.5   Resources of human activities

Resources of human activities have social-economic characteristics and include ship facilities such as ports, wharves, berthing areas, slipway and ramp, industrial facilities such as water intake and desalination facilities of the power plant, coastal mining facilities and solar salt lakes, entertainment resources like leisure beaches, swimming area, water sports area and recreational fishing area, and onshore or nearshore important historical and cultural scenic spots.

6.2.1.6   Protection priority of environmentally sensitive resources

The resources with protection priority identified by the relevant organization should be marked in the map or the documents attached to the map. Meanwhile, necessary seasonal information must also be indicated in the map for the sensitivity of environmentally sensitive resources varies from season to season.

6.2.1.7   Suggestions on clean-up measures

The suggestions on clean-up measures are the information reflecting the oil spill emergency characteristics. It contains the following information: which areas can the dispersants be used; where can the oil boom or permanent oil boom be set up; which beaches with low sensitivity can the oil be transported to so as to protect the highly sensitive areas if emergency transport of oil is necessary; and the location of passages. To improve the actual emergency capacity, the location of emergency equipment warehouse and the equipment type can also be marked in the map.

6.2.2   Application of GIS Technology in the Map of Environmentally Sensitive Resources

In the early stage, the location of sensitive resources is marked in the paper map of environmentally sensitive resources. With the development of computer technology, especially the application of GIS technology, the map in the oil spill emergency response system is far more valuable than the content of the text itself. The electronic map is based on GIS technology and is an environmentally sensitive resources information management system incorporating all technical data related to oil spill emergency response. Its most prominent feature is that it is easy to be updated and revised.

The use of the map of environmentally sensitive resources in conjunction with the oil spill drift and diffusion models can provide technical support for oil spill emergency response. With the electronic map as the working interface, the map of environmentally sensitive resources is a basic platform providing support to decision making in the oil spill emergency response system.

The map of environmentally sensitive resources is a layered vector diagram and usually an electronic chart. It displays relevant geographic information, resource data and environmental data according to emergency requirements, and provides a basic technical support for emergency decision making. The users can obtain more detailed information from the electronic map produced with GIS technology. The relevant information about a certain location in the map, such as the regulatory details, list of instructions or type of the endangered organisms in the conservation area, the contact information of local experts and the scientists or non-governmental organizations (NGOs), can be displayed. Such other data and statistics as the number of visitors to a certain beach, the nesting season of turtles or the information about tides and currents may also be helpful. All the above information may be displayed in the map in a disordered manner, whereas this kind of problem can be avoided with the well-designed GIS. The GIS technology is more and more open and can now be used in conjunction with many database management systems (DBMS) to realize information and resource sharing.

6.2.3 Database of Environmentally Sensitive Resources

The database of environmentally sensitive resources consists of basic data and technical parameters. The basic data is mainly the raw data of geographic information and resources. The technical parameters are the classification results of the technical indicators formed after processing the raw data according to corresponding technical standards based on the emergency response needs. The basic data reflects the contents of resources, while the technical parameters reflect the response characteristics of resources towards oil spill.

6.2.3.1 Raw data

The environmentally sensitive resources are composed of the following resources: shoreline resources, aquaculture resources, tourist and leisure resources, ecological reserve, port facilities and the residential areas.

1. Shoreline resources

The shoreline resource in the database is expressed in the shoreline type. The database of environmentally sensitive resources is established with the shoreline data as the basis. The composition of shoreline types is shown in Table 6.1.

Database field: shoreline type

Data type: character

Data length: 4

Definition: The shoreline type is a comprehensive description of the geographic information of the shoreline and is defined as the intertidal zone. When the resource data corresponding to the shoreline extend beyond the range of the intertidal zone, the resources are defined as the resources associated with the extension of the intertidal zone of 150 meters perpendicular to the shoreline; that is, the shallow culture resources within 150 meters of the subtidal zone and the pond culture resources within 150 meters from the upper tidal zone to the land all belong to the shoreline associated resources.

Technical description:

- The shoreline type includes the composition of the shoreline material, the shoreline slope and the shape and structure of the shoreline.

- The composition of the shoreline material: the types of substrates forming the shoreline, and they can be divided into: mud, sand and rock.

- Shoreline slope: refers to the steepness degree of shoreline, and based on the shoreline slope, the shoreline can be divided into flat beach, shore-beach, terrace and cliff.

- Shoreline morphology: mainly refers to the external form of the whole shoreline and based on the morphology, the shoreline can be divided into open and sheltered type. The shoreline morphology is not directly reflected in the shoreline type and mainly provides technical basis for relevant technical parameters of shoreline.

- In addition, the artificial shoreline is divided into permeable, semi-permeable and non-permeable shoreline according to its composition and consists of sea dike and seawall according to its slope. The type of shoreline with a specific ecological system or wetland is defined on the basis of its specific wetland characteristics or ecological characteristics, such as saltmarsh and mangrove forest.

Note: the slope of terrace is larger than 20 degrees, that of shore-beach is larger than 5 degrees and that of flat beach is smaller than 5 degrees.

2. Environmentally sensitive resources

It refers to other resources in contrast to the shoreline in the basic data and can be divided into the following according to the emergency plan: aquaculture resources, tourism and leisure resources, port and terminal resources, and ecological reserve resources. In the database, they are expressed in resources, resource extent, resource value, and resource time and so on.

(1) Database field: resource; data type: character; data length: 10

Definition: the content of other environmentally sensitive resources other than shoreline resource and expressed in the resource name, with the resource content as supplement.

According to classification requirements of the emergency plan, the resource contents are explained as follows:

- Aquaculture resources: consist of such two categories as cultured resources and captured resources:
  Captured resources: the captured aquatic specifies.
  Cultured resources: the cultured species, with the breeding methods as supplement.
  Cultured species: fish, shrimp, crab, shellfish, seaweed, etc.
  Breeding methods: pond culture, hanging culture, cage culture, caisson culture and so on.

- Ecological reserve resources: the name of the ecological nature reserves at all levels or protected species and ecosystems.
  The contents include national, provincial and local nature reserves; endangered and protected biological species at the international, national and local levels; protected areas or capture-forbidden area for the rare seafood resources.

TABLE 6.1: The composition of shoreline types.

Composition of shoreline substrate	Shoreline slope	Shoreline type	Remark
Rock	Cliff	1 Rocky cliff	Natural shoreline
	Terrace	2 Rocky terrace	
Mud	Flat beach	3 Mudflat	Natural shoreline
Sand	Flat beach	4 Flat sandy beach	
Stone	Shore-beach	5 Sandy beach	
Sand stone	Flat beach	6 Flat rocky beach	
	Shore-beach	7 Rocky shore-beach	
	Flat beach	8 Flat sandy and stone beach	
	Shore-beach	9 Sandy and stone shore-beach	
Artificial: cement component	Vertical	10 Seawall	Artificial shoreline
Non-permeable	Slope	11 Sea dike	
Stone	Vertical	12 Semi-permeable seawall	
Semi-permeable	Slope	13 Semi-permeable sea dike	
	Vertical	14 Trestle bridge	
Permeable	Slope	15 Breakwater	
Sediment	Wetland, marshland	16 Saltmarsh	Special wetland,
Stone	Mangrove marshland	17 Mangrove forest	ecosystem

- Port and terminal resources: the nature of the port and terminal.
  The contents of the port include: integrated port, passenger port, fishing port, military port.
  The contents of the terminal include: bulk, containers, chemicals, oil, liquefied petroleum gas (LPG), liquefied natural gas (LNG), etc.

- Residential areas: cities, towns and villages.

- Coastal industries: shipyard, repairing yard, warehouse, etc.
  According to the type of storage, the warehouse is divided into the warehouse for crude oil, refined oil, chemicals, LPG, LNG, etc., supplemented by storage methods (ground, underground).

- Other resources: salt fields, water inlet, etc.
  The water inlet consists of industrial water inlet and civil water inlet.

(2) Database field: resource extent; data type: floating-point number; data length: 10

Definition: A numerical description of the corresponding extent of the resource.

- Aquaculture resources: the area of the resources, unit: mu, hectare.
  Captured resources: the areas of captured aquatic specifies or fish.
  Cultured resources: the culture area of the cultured species.

- Tourism and leisure resources: the area of the resources, unit: mu, hectare. Namely, the area of scenic spots and historic sites, parks, resorts, beach swimming pool, and water entertainment facilities and so on.

- Ecological reserve resources: the area of the ecological nature reserves at all levels or area of the place of protected species and ecosystems, unit: hectare.
  The contents include the area of national, provincial and local nature reserves; the activity range of endangered and protected biological species at the international, national and local levels; the area of protected areas or capture-forbidden area for rare seafood resources.

- Port and terminal resources: annual throughput and construction scale of the port and terminal.
  Port scale: annual throughput of port cargo, unit: tons/year.
  Terminal scale: construction scale of terminals of bulks, containers, chemicals, oil, LPG, LNG, etc., unit: tons. For example: 80,000-ton crude oil terminal.

- Scale of residential areas: permanent population in the cities, towns and villages, unit: people.

- Scale of coastal industries: area occupied by the enterprises, unit: mu, hectare. The warehouse scale is expressed in the storage capacity in m^3.

- Other resources:
  Scale of salt field: the area of salt field, unit: mu, hectare.
  Scale of water inlet: daily water intake in tons.

(3) Database field: Time 1, Time 2; data type: date

Definition: time refers to the duration of resource sensitivity. Time 1 is the starting time, and Time 2 is the ending time. It is expressed in months in the background of one year.

Sensitive time is very important for time-sensitive resources, such as aquaculture resources and tourism resources but not so important for the non-time-sensitive resources, such as port and water inlet. It can be regarded that the sensitivity characteristics of the latter exist throughout the year.

- Aquaculture resources: Time 1: the starting month; Time 2: the ending month.
  Captured resources: the time when the aquatic species are captured.
  Cultured resources: the breeding time of the cultured species.

- Tourism and leisure resources: Time 1: the starting month; Time 2: the ending month.
  Scenic spots and historical sites, parks and resorts, etc.: rush season or opening hours.
  Beach swimming pool, water entertainment facilities: opening hours.

- Ecological reserve resources: Time 1: the starting month; Time 2: the ending month.
  National, provincial and local nature reserves: the whole year.
  The endangered and protected biological species at the international, national and local levels: the time when such species stay in the areas.
  Protected areas and capture-prohibited areas of rare seafood resources: time of protection or forbidding the capture specified by the aquatic department.

- Sensitive time of other resources is usually the whole year.

(4) Database field: resource value; data type: floating-point number; data length: 10
Definition: corresponding economic value of the resources. It is expressed in the annual output value of the resources and the unit is RMB ten thousands/year.

- Aquaculture resources: the annual economic value of the resources.
  Captured resources: the annual output value of the captured aquatic species.
  Cultured resources: annual output value of the cultured species.

- Tourism and leisure resources: annual income from such resources. Namely, the annual income from the scenic spots and historic sites, parks, resorts, beach swimming pool, and water entertainment facilities and so on.

- Ecological reserve resources: none.

- Port and terminal resources: the annual output value of the port and terminal.
  Port scale: the annual output value of the port.
  Terminal scale: annual output value of the terminals of bulk and general cargo, container, chemicals, oil, LPG and LNG.

- Residential areas: annual income.

- Scale of coastal industries: annual output value of the enterprises.

- Other resources:
  Salt field: annual output value. Water inlet: annual output value.

3. Other fields

The above are the important contents of the basic data in the database of environmentally sensitive resources: shoreline resources and other environmentally sensitive resources. From the perspective of queries, requirements of emergency response and technical integrity, the database also includes other fields which are briefly described as follows:

(1) Database field: ID
   Data type: figure
   Data length: 4
   Definition: the only numerical serial number corresponding to the shoreline.

The ID value is the only figure corresponding to the shoreline. It can be used for identification and query of different shorelines and can also represent the shoreline section for statistical query and data evaluation.

(2) Database field: location
   Data type: character
   Data length: 150
   Definition: text description of the geographical location of the shoreline.

It refers to the geographical location of the shoreline and the important geographical location is set as the reference point. Like the ID value, it can be used for identification and queries of shoreline.

(3) Database field: shoreline coordinates (longitude LON, latitude LAT)
   Data type: floating-point number
   Data length: 10
   Definition: the specific coordinates of the midpoint of the shoreline, expressed in longitude and latitude.

The coordinates of the midpoint of the shoreline section includes the fields of longitude and latitude and are mainly used for positioning and queries.

(4) Database field: length
   Data type: floating-point number
   Data length: 40
   Definition: the length of the shoreline section in meters.

(5) Database field: contact
   Data type: character
   Data length: 40
   Definition: the name of the person responsible for emergency response of the shoreline section or resources as specified in the emergency plan.

(6) Database field: phone
   Data type: character
   Data length: 40
   Definition: the phone number of the person responsible for emergency response of the shoreline section or resources as specified in the emergency plan.

6.2.3.2 Technical parameters

Technical parameters are an important part of the environmentally sensitive resource database and one of the important contents of the map of environmentally sensitive resources. The technical parameters are the classification values of the technical indicators formed after the basic data are processed according to the technical requirements of the oil spill emergency response and the corresponding standards. They can provide an intuitive reference index for emergency response personnel, and then provide technical basis for emergency response decision making.

Three technical parameters, namely the shoreline environmental sensitivity index (ESI), the resource sensitivity index (SI), and the priority index of protection (PI), are used in the database of environmentally sensitive resources of the Bohai Gulf and the North Yellow Sea.

1. Environmental Sensitivity Index (ESI)
Database field: ESI
Data type: figure
Data length: 2
Definition: ESI of shoreline, levels 1 to 10.
ESI classification method (Table 6.2)

2. Resource Sensitivity Index (SI)
Database field: SI
Data type: character
Data length: 2
Definition: Resource sensitivity index, classified into four types: A, B, C and D.

- Type A: very important resources which are extremely sensitive to the oil spill.

- Type B: important resources which are very sensitive to the oil spill.

- Type C: second important resources which are relatively sensitive to the oil spill.

- Type D: general resources which are generally sensitive to the oil spill.

TABLE 6.2: ESI of the shorelines.

ESI	Shoreline type
1	Open and exposed rocky cliffs, artificial coast
1a	Open rocky cliffs
1b	Open artificial seawall
2	Exposed rocky terrace
3	Fine-sand beach ($slope \geq 5°$)
4	Coarse-sand beach ($slope \geq 5°$)
5	Sand and gravel-mixed beach
6	Rocky beach (large, medium and small stone diameter)
7	Exposed tidal flat
8	Sheltered impermeable and semi-permeable rocky coast and artificial coast
9	Sheltered tidal flat, mud
10	Seawater marshland, mangrove forest, freshwater marshland
10a	Saltmarsh
10b	Mangrove forest

3. Priority Index of Protection (PI)

Database field: PI

Data type: character

Data length: 2

Definition: The priority for protection of the sources, classified into three types: I to III (see Table 6.3).

- Top priority;

- Priority;

- General priority.

6.2.3.3 Development of database

The database should be able to provide information on the location, extent, nature and sensitivity indicators of the resources and the countermeasures, providing important information for emergency decision making and effectively protecting sensitive resources that may be damaged. Therefore, the production and updating of the environmentally sensitive resource databases is an important foundation work in this system. The supplementation and improvement of the database can be continued (except for hardware constraints) and cannot be terminated with the completion of the system development.

The attribute data of sensitive resources is stored in the relational database. The raw data and technical parameters of sensitive resources correspond to spatial data, such as port name, throughput, and port type. The

TABLE 6.3: Classification standard and protection priority of environmentally sensitive resources.

Type	Importance	Sensitivity	Resource value	Protection priority	Examples
I	Very important	Extremely sensitive	Very high ecological value and special value	Top priority or priority	National rare and endangered species conservation areas
II	Important	Very sensitive	Very high economic value	Priority or top priority	Provincial nature reserve, important tourism scenic area
III	General	General	Certain ecological value and economic value	Need protection	Terminal facilities, coastal industrial and mining enterprises

attribute data is stored in the system database in the form of a data sheet, and the system provides a corresponding interface for the user to manage the data.

Sensitive resource information exists in the system in two forms. One is stored as basic ground object information in electronic charts and topographic maps, and the other is stored in a database. The former is mainly static sensitive resources such as historical sites, etc. Sensitive resource information will also be completely stored in the system's database.

The data of the sensitive area and high-value area is collected through graphic editing and database data editing methods according to its different form of representation. The data on the static sensitive resources is collected through graphic editing and the data on the dynamic sensitive resources is collected jointly with graphic editing and database data editing. The graphic editing is used first to position the sensitive resources and the database data editing is then used to control the periodic characteristics of the sensitive resources.

After the completion of sensitive information database, the system provides flexible and diverse query methods, including point selection, rectangle selection, circle selection and polygon selection of the chart and moving targets and string-based fuzzy query of chart database.

6.3   Emergency Resources

The resources for emergency response to an oil spill consist of emergency human resources and emergency material resources. Human resources refer to a well-trained emergency team. For large-scale oil spills, cooperation and support from relevant industries and departments are also needed. Material resources refer to oil spill emergency equipment which are used to control and remove oil spills, mainly including oil booms, oil skimmers, oil absorption material, oil recovery vessels, transport equipment, chemical preparations, etc. These oil spill emergency equipment are distributed in various ports and coastal areas, and their distribution and the equipment list are recorded in the database. The data includes the model, quantity, performance, application occasion and conditions, location and transport mode of the equipment, and communication address of the equipment owner.

6.3.1   Emergency Human Resources

The emergency team is the backbone of emergency human resources and is responsible for dealing with the accident site. It includes a certain scale and quantity of pollutants clean-up equipment and instruments and well-trained operators.

The specific tasks of these emergency teams in the emergency are assigned by the on-site commander. The same emergency team assumes different positions and responsibilities in emergency response at different levels. For example, local emergency teams may independently assume responsibility for the containment and removal in the first-level emergency response to the oil spill accidents, but only partial responsibility for the tasks in the higher level emergency response. Its organizer (or commander) must obey the specific arrangement from the commander of a superior level.

In addition to the oil spill emergency mechanism and salvage system in the northern sea area invested and constructed by the Ministry of Transport, China's marine pollution emergency resources are mainly from the port, terminal and marine petroleum exploitation enterprises and the professional decontamination companies. The use of corporate decontamination equipment is basically limited to the corporate and the corporate's system. With the gradual implementation of the national marine pollution emergency plan and the further improvement of the emergency mechanism, marine pollution emergency equipment configuration will be further enriched and emergency agencies and emergency professional companies will play a greater role.

6.3.2 Emergency Equipment

The emergency equipment includes recovery vessels, working ships, oil booms, oil skimmers, oil absorption materials and chemical preparations. This will be detailed in Chapter 7.

6.4 Oil Spill Emergency Response System

6.4.1 System Introduction

6.4.1.1 System overview

The Oil Spill Emergency Response System 2.0 (OSERS2.0) was developed by China Waterborne Transport Research Institute. Based on the GIS platform, this system integrates the information database of environmentally sensitive resources, the information database of emergency equipment and team, and the oil spill drift models to provide daily management of oil-spill-sensitive resources and emergency resources for the oil spill emergency response, realizes superposition and coupling of the oil spill drift prediction results and the map of sensitive resource, and achieves the rapid warning of environmentally sensitive resources. Figure 6.7 shows the full picture of OSERS2.0.

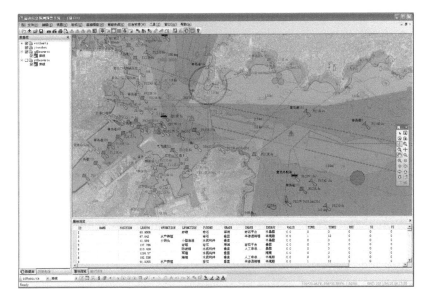

FIGURE 6.7: Full picture of OSERS2.0.

6.4.1.2 System positioning

OSERS2.0 provides basic geographic information, the spatial distribution, sensitivity grading and suggested protection priority order of the environmentally sensitive resources for the emergency personnel and provides the information about the emergency facilities, equipment and teams to provide basic environment for daily emergency training and practices. The system integrates the drift model of leaked oil on the water, which provides technical support for the emergency personnel to accurately grasp the oil spill trend when an accident happens and work out an emergency plan based on the information about the environmentally sensitive resources and the emergency teams and equipment.

6.4.1.3 System details

1. The system's function

OSERS2.0 is a multi-functional service system which integrates the comprehensive information about the sensitive resources in the Bohai Gulf and North Yellow Sea, the oil spill drift model and emergency resource information management, and the oil spill pollution prediction and warning based on NAVIGIS platform. OSERS2.0 has the following main functions:

- The system uses the advanced NAVIGIS platform to realize synchronous layered display of S57 charts, conventional charts, conventional land electronic maps and satellite remote sensing images.

- Based on NAVIGIS, the system sets up a multi-media information management system of oil-spill-sensitive resources which includes the raw technical and image data, the sensitivity, protection priority and other technical indicators of the environmentally sensitive resources.

- The system sets up an oil spill emergency resource information management system which includes information about the distribution of emergency equipment and teams.

- It integrates the oil spill drift model.

- It realizes the synchronous display of all kinds of basic geospatial attribute data, various resource data, technical indicators and dynamic prediction results and multi-indicator superposition analysis to provide comprehensive technical support for oil spill pollution prediction and warning.

2. Key technical issues addressed

Key technical issues addressed for establishing such system include:

- Study to establish a management system of oil-spill-sensitive resources in the Yellow Sea and the Bohai Sea: (1) determine the classification method, standard and database structure of the sensitive resources; (2) data survey and on-site acquisition of shoreline information and data; (3) establish GIS system based oil-spill-sensitive resources database which includes the basic information about shoreline resources and resources related to human activities, NOAA ESI shoreline index and the protection priority and other technical parameters.

- Set up GIS system based oil spill emergency resource management system which includes information about the types, distribution and management of the emergency resources.

- Develop the technology of superposing and coupling the oil spill drift prediction results and the map of environmentally sensitive resources to achieve quick warning of environmentally sensitive resources.

6.4.1.4 Technological innovation

The system has the following innovations:

- The embedded integration of GIS platform and OILMAP oil spill drift model is realized.

- A multi-parameter oil-spill-sensitive resources database based on GIS system is established. Such system includes the basic information about the shoreline resources and human activity related resources and so on and technical parameters such as ESI and PI.

- The superposition and coupling of the simulation results of oil spill drift model and the sensitive resource map is realized to make the system have rapid oil spill pollution prediction and warning function specific to sensitive resources.

- The ESI was applied to China's Bohai Sea and North Yellow Sea for the first time.

- An ESI-based multi-media sensitive resource database is established for the first time. Such database includes the photos, videos, vector chart and high-resolution satellite remote sensing images of the shores.

- The system integrates the S57 chart, Arcinfo vector chart, Mapinfo vector chart and high-resolution satellite images to solve the compatibility problem.

- The oil spill drift module of the system contains current field files in many formats.

6.4.2   Composition of OSERS2.0

The system is based on NAVIGIS to establish a spatial location-based environmentally sensitive resource information system and emergency resource information system; it integrates the oil spill drift tracking model based on the wind and current field and oil spill fate simulation in Bohai Gulf; it can provide the prediction on drift direction and impact range, the information on the environmentally sensitive information and the sensitive resources with protection priority in such range, the distribution information of the emergency equipment bases, and emergency equipment and teams for the emergency personnel, thus achieving the goal of providing sensitive resources and emergency resources to the emergency personnel and support for decision-making about prediction and warning of oil spill pollution.

The system composition and basic functions are shown in Figure 6.8 and Figure 6.9.

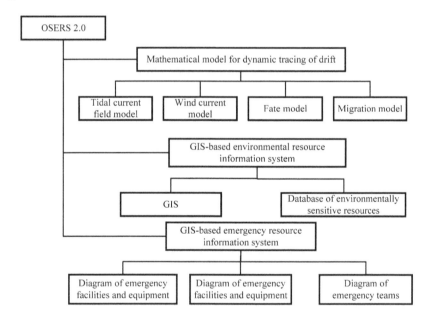

FIGURE 6.8: Composition of OSERS2.0.

6.4.2.1   Oil spill drift simulation system

Oil spill drift simulation system consists of hydrodynamic model and oil spill drift model, as shown in Figure 6.10. The OSERS2.0 only provides the data interface of oil spill drift model and tidal current field model but does not have the tidal current field calculation function. The environmental fate model of oil spill includes the oil weathering model, shoreline adsorption model and water dissolution model which mainly simulate the physical transformation

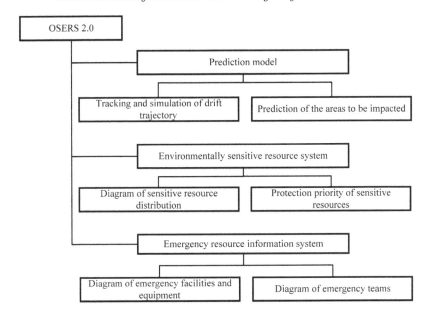

FIGURE 6.9: Functions of OSERS2.0.

processes such as volatilization, dissolution and adsorption after oil spills. The oil spill drift model locates and tracks the oil spill by equally dividing the spilled oil and tracking the temporal and spatial distribution of the equally divided oil particles under hydrodynamic conditions and in its own transformation process.

By tracking the oil particles, the oil spill drift model can predict the location of an oil spill, dynamically display the prediction results of oil slick drift, conduct early warning on the impact range of oil spills and assess the emergency measures.

6.4.2.2   Environmentally sensitive resource system

The system contains data with four aspects:

- Raw data: photos, videos and satellite images of the protected areas, beaches, mudflats, marshland, breeding areas, bathing places, salt field, water intake and prohibited areas.

- Professional technical data: NOAA ESI, EI and PI.

- Diagram of sensitive resource distribution.

- Protection requirements of sensitive resources, such as protection time and specific technical requirements.

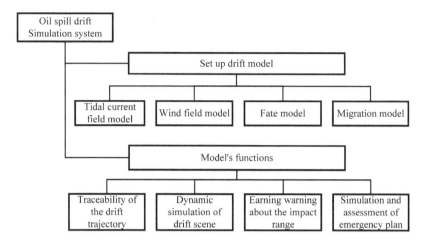

FIGURE 6.10: Composition of oil spill drift simulation system.

The composition of an environmentally sensitive resource system is shown in Figure 6.11.

6.4.2.3   Emergency resource system

The emergency resource system mainly includes the type, distribution and management of emergency resources. Combined with the GIS system, it is saved or displayed in the OSERS2.0 system in both space and attribute. The composition of emergency resource system is shown in Figure 6.12.

6.4.3   Setting of Oil Spill Simulation Scene and Model Operation

6.4.3.1   Build new oil spill simulation scene

When simulating the oil spill drift, it is first necessary to determine the location of the oil spill accident. In the OILMAP model, the location of the accident can be determined by clicking the map with the mouse, or by manually inputting the latitude and longitude coordinates. After clicking on the "New Scene" button and on the location of the oil spill accident with the mouse, the dialog box of Figure 6.13 appears.

The following functions can be achieved in the "New Scene" dialog box:

- Specify the name of the simulation scene.

- Display the latitude and longitude coordinates of the location of oil spill accident and manually modify them if necessary.

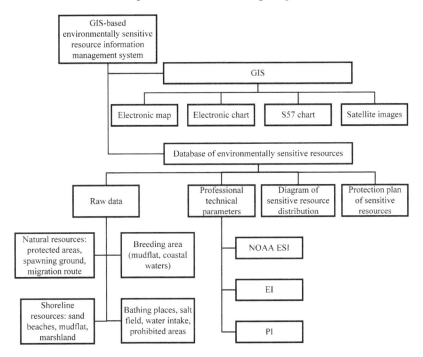

FIGURE 6.11: Diagram of composition of the environmentally sensitive resource system.

- Specify the water and land grid used for the simulation (need to be defined in the area concerned before simulation), or directly choose the GIS layer to define the water and land boundary.

6.4.3.2 Specify the spillage parameter–"Spillage" dialog box

After determining the location of the oil spill, the spillage related parameter has to be specified, as shown in Figure 6.14.

The following functions can be achieved in the "Spillage" dialog box:

- Specify the starting time of spillage and the time zone to which the simulation scene belongs as well as the simulation duration.

- Select the simulation mode of oil spill model which consists of normal mode and reverse deduction mode. In the reverse deduction mode, the simulation duration means the "arrival time," namely the time when the spilled oil arrives in the location of the designated latitude and longitude, while OILMAP will reversely calculate the drift trajectory of spilled oil.

- Specify the water temperature of the simulation scene which can be expressed in centigrade or Fahrenheit.

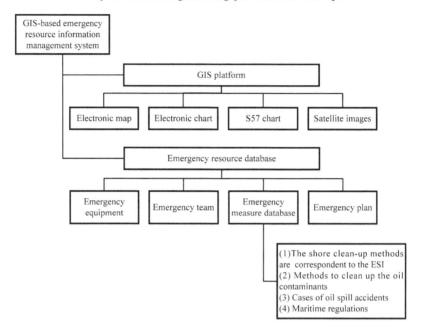

FIGURE 6.12: The emergency resource system.

- Select the emergency measures to be used. In the OILMAP model, the emergency measures can be used are:

  - Oil dispersant
  - Oil boom
  - Manual clean-up

The "Monitor Zone" option under the "Emergency Options" is the function provided by the OILMAP model to correct the simulated oil spill drift trajectory in real time with the measured data.

6.4.3.3   Oil database–"Oil" dialog box

The OILMAP model provides a database of different types of oils that can be selected for oil spill simulations, see Figure 6.15.

The following functions can be achieved in the "Oil" dialog box:

- Specify the leakage amount of an oil spill accident and provide a variety of measurement units for ease of use.

- Specify the spillage duration in an oil spill scene.

- Specify the types of oil spilled. The users can choose from the oil variety provided in the OILMAP oil database or Adios database. The Adios

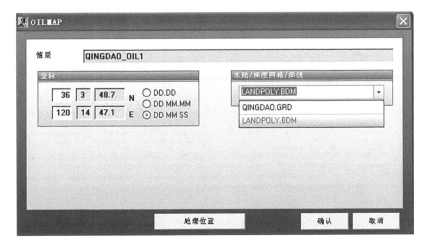

FIGURE 6.13: "New Scene" dialog box.

database provides more oil varieties. However, some parameters have to be filled according to the reality. The users can also define the oil database by themselves.

At present, the oil types in the system that can be directly simulated include: gasoline, diesel, light crude oil, heavy crude oil, Prudhoe Bay crude oil, heavy crude oil, 6# marine fuel oil and 1# fuel oil (JP-4). In addition, the users can define the oil database and specify various parameters or add new oil types in the Adios database.

6.4.3.4 Environmental wind field–"Wind Field" dialog box

The "Wind Field" dialog box of the OILMAP model is shown in Figure 6.16. In the OILMAP model, the wind field of the oil spill scene can be specified in the following ways:

- Manual generation of wind field files with "Edit the Cell" (see Figure 6.17) or "Use the Wind Field Tool" (see Figure 6.18).

- Opening the existing wind field file. That is, using the wind field files of the previous simulation.

- Environment data of the server. That is, obtaining the real-time wind field data through EDS.

- Choosing not to use the wind field data in the simulation, namely, ignoring the wind effect.

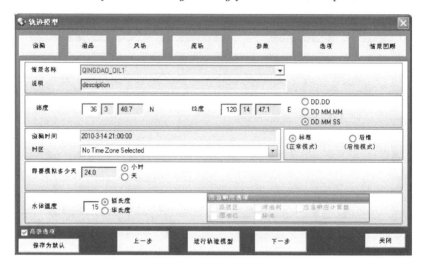

FIGURE 6.14: "Spillage" dialog box.

6.4.3.5 Tidal current field–"Tidal Current Field" dialog box

OILMAP provides three ways to get current field data, namely:

- Historical current field files, namely the current field calculated with hydrodynamic model, as shown in Figure 6.19.

- Environment data of the server, namely, obtaining the real-time current field data through EDS (Environment Data Service of ASA), see Figure 6.20.

- Specifying the constant current field, see Figure 6.21.

6.4.3.6 Calculation parameters–"Parameters" dialog box

Specify the following calculation parameters in the "Parameters" dialog box (Figure 6.22):

- Model calculation step in minutes.

- Model output time interval in minutes.

- Number of the oil particles simulated, unit: pcs.

- Wind force factor, unit: %.

- Deflection angle of wind, unit: degree.

- Horizontal dispersion coefficient, unit: $m^{3/8}$.

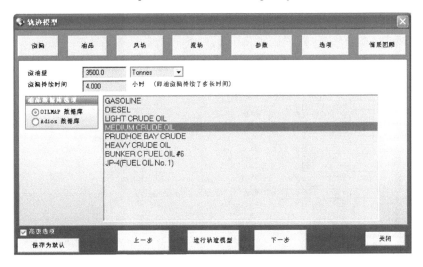

FIGURE 6.15: "Oil" database dialog box.

Model options are mainly for additional options for model calculations and output options for simulation results files, as shown in Figure 6.23. In the "Options" dialog box, the items available for choice include:

- Attribute information of the particles tracked by the model;

- Additional simulation functions can be selected when there are reeds or sea ice in the simulated waters;

- Smooth shoreline can be selected for simulation to complement the shoreline algorithm;

- The output file option is for the users to decide whether to output the synchronized current field files and whether to create contour data of the oil slick thickness;

- Uncertainty analysis of wind direction, wind speed, current direction and current velocity and so on can also be chosen.

The "Scene Review" dialog box is shown in Figure 6.24.

The purpose of the "Scene Review" dialog box is to confirm the consistency of the simulation duration, the current field time and the wind field time on the timeline so as to ensure that no loss of current or wind field data occurs in the specified simulation duration. In addition, the dialog box is also used to specify the default shoreline types which mainly consist of four types: wetland or mangrove forest, sandy beach, intertidal zone and rocks.

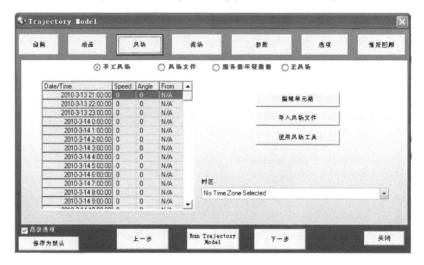

FIGURE 6.16: "Wind Field" dialog box.

6.4.3.7   Input and output of GIS files

OILMAP provides the interfaces to input and output GIS files in a variety of formats and can convert files into the format acceptable to the conventional geographic information systems like ArcGIS, Mapinfo and C-map, or even directly read the files output by these systems. The files that can be served as the base maps of OILMAP include:

- ASA Map file, *.bdm;

- ASA GIS layer file, *.gdw;

- ESRI SHP file, *.shp;

- Mapinfo Mif file, *.mif;

- GIS server data files: WMS, ArcIMS.

The current field data files that OILMAP can directly read include:

- Real-time monitored current field data;

- Hydromap current field file;

- NetCDF current field file;

- LLU current field file;

- MDB current field file;

- Current field files in Canadian format.

OILMAP's simulation result files can be directly output into SHP files, making them well compatible with most geographic information systems.

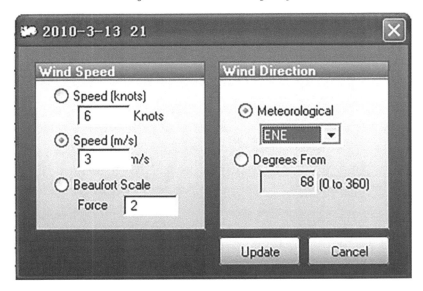

FIGURE 6.17: "Edit the Cell" under "Wind Field" dialog box.

6.4.4   Contents of Sensitive Resources

6.4.4.1   Sea area division

The sea area of the sensitive resources involved in the project of "prediction and warning of oil spill pollution in Bohai Sea" refers to the Bohai Sea and North Yellow Sea and includes the sea areas under administration of the Liaoning, Shandong, Hebei and Tianjin Marine Safety Administrations. In accordance with Official Reply of the State Council on Nationwide Marine Functional Zoning (2002), the key environmentally sensitive areas related to this project are as follows:

1. Eastern waters of Liaodong Peninsula

It includes the adjacent sea area from Yalu River estuary in Dandong City, Liaoning Province to Laotieshan Cape in Dalian. The key functional zones are Dalian, Dandong and Zhuanghe port areas and the related waterways, Jinshitan Tourist Resort, the tourist areas in southern Dalian, lvshun South Road, and Dalu Island in Dandong, the aquaculture areas in south of Dagushan Peninsula and west of Lingshui River Estuary, and Yalu River Estuary Wetland Nature Reserve. For this area, the emphasis should be on guaranteeing the demands of container terminal at Dalian Port and construction of large-scale professional terminals for the seawaters, active development of coastal tourism, construction of rare seafood breeding base and protection of coastal wetland ecosystem.

2. Western waters of Liaodong Peninsula

It includes the adjacent sea area from Liaotieshan Cape to Daqing River estuary in Yingkou City. The key functional zones are Yingkou, lvshun and

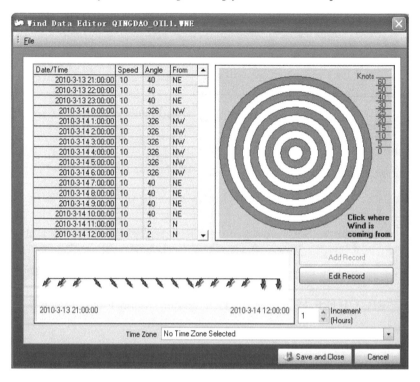

FIGURE 6.18: "Use the Wind Field Tool" under the "Wind Field" dialog box.

Bachagou port areas and the related waterways, Fuzhou Bay and Jinzhou Salt Fields, Gaizhou and Changxing Island Aquaculture Zones, Xianyu Bay and Changxing Island Tourist Areas, spotted seal, Shedao-Laotieshan and Yingkou marine erosion landscapes reserves in Dalian and sand dike nature reserve in Fudu River Estuary. For this area, the emphasis should be on development of port and marine transportation, utilization and conservation of fishery resources, and protection and conservation of sandy shores and island ecosystems.

3. Sea area adjacent to Liaohe River Estuary

It includes the adjacent sea area from Daqing River Estuary in Yingkou City, Liaoning Province to Housanjiao Mountain in Jinzhou City. The key functional zones are Bijialing and Taiyangdao oil and gas fields, Yingkou and Jinzhou salt fields, Gaizhoutan and Erjiegou aquaculture zones, and the nature reserves in Shuangtaizi River Estuary and Daling River Estuary. For this area, the emphasis should be laid on strengthened exploration and development of oil and gas resources in shallow waters, reasonable utilization, propagation and recovery of fishery resources, protection of wetland ecosystem, strengthened potentiality exploitation and technical transformation of

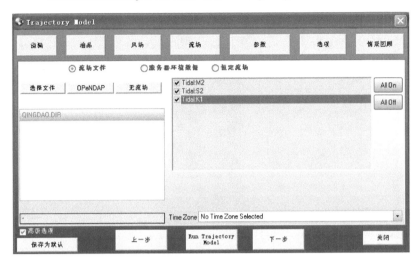

FIGURE 6.19: Historical current field files.

the salt fields, strengthened comprehensive environment governess of the old port areas, Liaodong Bay and the sea areas adjacent to the estuaries.

4. The sea area from West of Liaoning Province to East of Hebei Province

It includes the adjacent sea area from Housanjiao Mountain in Jinzhou City, Liaoning Province to Jianhe River Estuary in Tangshan City, Hebei Province. The key functional zones are Qinhuangdao, Jingtang and Jinzhou port areas and the related waterways, Beidaihe, Nandaihe, Shanhaiguan, Xingcheng coastal areas, Jinzhou Bijia Mountain tourist areas, Changli and Juhuadao sea areas, Luanhe River Estuary aquaculture zone, Changli and Beidaihe nature reserves, Suizhong, Jinzhou and Jidong oil and gas fields, and Luannan and Daqing River salt fields. For this area, the emphasis should be on guaranteeing the demands of Qinhuangdao and Jinzhou port and terminals for seawaters, guaranteeing the demands of the oil and gas resources prospecting and development and the utilization of fishery resources for seawaters, development of coastal tourism and protection and conservation of coastal ecosystem.

5. The sea area from Tianjin to Huanghua

It includes the adjacent sea area from Jianhe River Estuary in Hebei Province to the border between Hebei Province and Shandong Province. The key functional zones are Tianjin and Huanghua port areas and the related waterways, Changlu, Hangu and Cangzhou salt fields, Xingang and Madong oil and gas fields, the Shanggulin and Qingtuozi shell dike core areas in Tianjin ancient coastal and wetland nature reserve, Tanggu and Hangu propagation and breeding zones, and Hangu, Dagang and Beitang River Estuary special conservation areas. For this area, the emphasis should be on guaranteeing

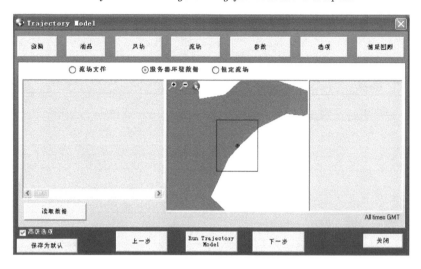

FIGURE 6.20: Read real-time current field data through EDS.

the demands of the construction of professional terminals like Tianjin Port and Huanghua Port, the oil and gas resource development and utilization of fishery resources in shallow waters for the seawaters, protection of the water quality and environment for the salt industry, protection of the ecological environment in the utilization zone of fishery resources, establishment of Hangu shallow water ecosystem, ecosystem in intertidal zones of Luju River, Dagang ancient lagoon wetland, Dagang coastal wetland and Huanghua shell dike nature reserves and vigorous development of integrated seawater use.

6. The sea area from Laizhou Bay to Yellow River Estuary

It includes the sea area from the border between Hebei Province and Shandong Province to Longkou of Yantai. The key functional zones are Yellow River Estuary and Hutouya aquaculture zones, oil and gas fields in west of Yellow River estuary, Penglai 19-3 oil and gas fields, Zimaihe-Hutouya salt field, Wudi shell dike and wetland, Yellow River Estuary wetland nature reserve and Longkou port area. For this area, the emphasis should be on guaranteeing the demands of the oil and gas exploration and development and the aquaculture for seawaters and protection of wetland ecosystem.

7. Miaodao archipelago waters

It includes the adjacent sea areas from Changdao County in Yantai to Penglai, Shandong Province. The key functional zones are Nanwu Island and Beisi Island aquaculture zones, Penglai and Changdao tourist areas, marine ecosystem and rare seafood nature reserves around the archipelago, and Penglai port area. For the area, the emphasis should be on construction of Changdao aquaculture base, development of characteristic island tourism, strengthened protection of ecological environment and improvement of the land-island transportation.

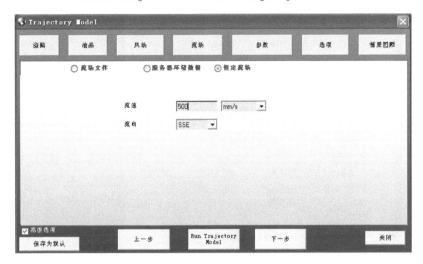

FIGURE 6.21: Define the constant current field.

8. Changshan archipelago waters

It includes the adjacent sea area from Dalian to Changhai County. The key functional zones are Zhangzi Island and Xiaochangshan Island aquaculture zones, Sikuaishi and Miaodi port areas, and Dachangshan Island and Wangjia Island tourist areas. For the area, the emphasis should be on vigorous development of breeding, propagation and release, construction of marine farming and ranching base, accelerated construction of land-island transportation infrastructures, active development of island tourism, comprehensive utilization of seawater and strengthened protection of island ecological environment and marine organism diversity.

9. The sea area from Yantai to Weihai

It includes the sea area from Yantai urban areas to Haiyang, Shandong Province. The key functional zones are Yantai, Mouping and Weihai port areas and the related waterway, Jinsha Beach, Zhifu Island, Tian'e Lake and Liugong Island tourist areas, and Taozi Bay, Sishili Bay and Weihai Bay aquaculture zones. For the area, the emphasis should be on guaranteeing the demands of the port construction and the utilization of fishery resources for seawaters, vigorous development of coastal tourism and aquaculture, active development of marine pharmaceuticals and comprehensive utilization of seawater.

6.4.4.2   Distribution of nature reserves

There are 27 nature reserves of municipal level and above, with a total area of 2,015,062 hectares, of which 10 are national level nature reserves and 17 are provincial or municipal nature reserves (see Table 6.4).

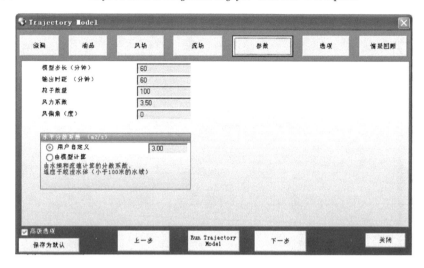

FIGURE 6.22: "Model Calculation Parameters" dialog box.

6.4.4.3   Distribution of fishery resources

1. Sea area near Shandong Province

Such area has abundant fishery resources. The main marine fishery resources distributed in or migrating in Shandong coastal waters are: fish, prawns, crabs, cephalopoda, shellfish and echinoderms. Specifically, there are 527 swimming species and nearly 109 relatively important economic fish species and invertebrates; more than 90 shellfish species distributed in the tidal flat and shallow seawaters, of which more than 20 species have higher economic value; up to 602 types of fishery resources that can be cultured, with more than 70 species that have been cultured. The available aquaculture area in shallow seawaters is nearly $2,000\,km^2$, and that in tidal flat is nearly $1,500\,km^2$. Shandong coastal waters are one of the important rare seafood production areas of China.

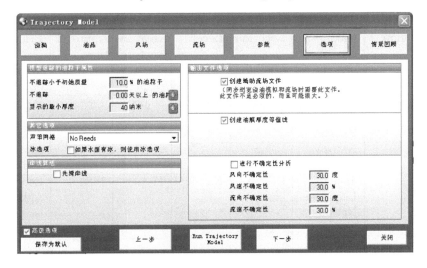

FIGURE 6.23: "Options" dialog box.

TABLE 6.4: Nature reserves in Bohai Sea and North Yellow Sea areas.

Region	Name of the nature reserve	Location	Level	Area (hectare)	Approval time	Primary protection object
Liaoning Province	Dandong Yalu River Estuary Wetland National Nature Reserve	Dandong	National	101,000	1997	Rare migratory birds and wetland ecosystems
	Dalian Spotted Seal National Nature Reserve	Dalian	National	909,000	1997	Spotted seal
						Continued on next page

TABLE 6.4 – continued from previous page

Region	Name of the nature reserve	Location	Level	Area (hectare)	Approval time	Primary protection object
	Liaoning Shuangtai Estuary Nature Reserve	Panjin	National	80,000	1988	Wetland ecosystems and spotted seal
	Shedao-Laotieshan National Nature Reserve	Lvshun-kou	National	17,800	1980	Ecosystem, Agkistrodon halys, and migratory birds in Shedao and Laotieshan
	Dalian Cheng-shan Cape Coastal Landform National Nature Reserve	Dalian	National	1350	2001	Coastal karst landforms and birds
						Continued on next page

TABLE 6.4 – continued from previous page

Region	Name of the nature reserve	Location	Level	Area (hectare)	Approval time	Primary protection object
	Rare Marine Species Nature Reserve	Chang-hai County	Provincial	220	1998	Breeding areas of sea cucumber, Haliotis discus hannai Ino and Chlamys farreri and migration site of shrimp
	Dalian Laopian-dao–Yuhuang-ding Marine Ecology Nature Reserve	Dalian	Municipal	1580	2000	Marine life and marine ecosys-tems, karst and marine erosion landscape
	Dalian Hai-wangjiu Island Marine Land-scape Nature Reserve	Dalian	Municipal	2143	2000	Waterfront landform, Chinese egret
						Continued on next page

TABLE 6.4 – continued from previous page

Region	Name of the nature reserve	Location	Level	Area (hectare)	Approval time	Primary protection object
	Liaodong Bay Wetland Marine Nature Reserve	Panjin	Provincial	80,000	Unknown	Rare animals such as birds in wetland ecosystem and spotted seals
	Zhujiatun Marine Erosion Zone Nature Reserve	Dalian, Liaoning	County	1350	1989	Marine erosion landscape
	Jinshitan Geological Rare Relics Nature Reserve	Dalian	Municipal	2200	1987	Typical geological structure, paleontological fossils, peculiar coastal landforms
	Daling Estuary Coastal Wetland Nature Reserve	Jinzhou	Provincial	83,556	2005	Wetland ecosystems, migratory birds, spotted seals in Western Pacific
						Continued on next page

TABLE 6.4 – continued from previous page

Region	Name of the nature reserve	Location	Level	Area (hectare)	Approval time	Primary protection object
	Tuanshan Marine Erosion Landscape Nature Reserve	Yingkou	Municipal	1495	2005	Marine erosion landscape
	Dalian Changshan Archipelago Rare Marine Organism Nature Reserve	Changhai County, Dalian	Municipal	413	2003	Special rare species in North Yellow Sea of China, such as Haliotis discus hannai Ino, Stichopus japonicus, Strongylocentrotus nudus, Chlamys farreri, Ostrea plicatula Gmelin and bastard halibut

Continued on next page

TABLE 6.4 – continued from previous page

Region	Name of the nature reserve	Location	Level	Area (hectare)	Approval time	Primary protection object
	Suizhong Primary Sandy Coast and Bio-diversity Nature Reserve	Suizhong	County	207,700	1996	Sandy coast and marine ecosystem
Shandong Province	Yellow River Delta Nature Reserve	Dongying	National	153,000	1992	Primary wetland ecosystem and rare poultry
	Binzhou Shell Dike Island and Wetland National Nature Reserve	Wudi County	National	80,480	2002	Shell dike island and wetland ecosystem
	Miaodao Archipelago Marine Nature Reserve	Chang-dao County	Provincial	5250	1991	Island ecosystem in warm temperature zone
	Miaodao Archipelago Seal Nature Reserve	Chang-dao County	Provincial	173100	2001	Spotted seal
						Continued on next page

TABLE 6.4 – continued from previous page

Region	Name of the nature reserve	Location	Level	Area (hectare)	Approval time	Primary pro-tection object
	Shandong Changdao National Nature Reserve	Chang-dao County	National	5300	1988	Habits of predatory birds, like hawks and falcons, and migratory birds
	Weifang Laizhou Bay Ostrea rivularis Breeder Seed Nature Reserve	Weifang	Munici-pal	813	2005	Ostrea rivularis and its living environ-ment
Hebei Province	Huanghua Ancient Shell Dike Provincial Level Marine Nature Reserve	Huang-hua	Provin-cial	117	1998	Ancient shell dike, shell sand and vegeta-tion in the reserve
	Changli Gold Coast National Nature Reserve	Changli	National	30,000	1992	Coast dune and am-phioxus
						Continued on next page

TABLE 6.4 – continued from previous page

Region	Name of the nature reserve	Location	Level	Area (hectare)	Approval time	Primary protection object
	Laoting Shijiutuo Archipelago Provincial Nature Reserve	Yueting	Provincial	3775	2002	Animal and plant resources and birds
	Nandagang Coastal Wetland and Birds Provincial Nature Reserve	Cangzhou	Provincial	8000	2002	Wetlands and birds
Tianjin	Tianjin Ancient Coast and Wetlands National Nature Reserve Sea area near	Tianjin	National	21,180	1992	Shell dike, oyster bank, ancient coastal historical sites, coastal wetland ecosystems
	Dagang Ancient Lagoon Nature Reserve	Tianjin	Municipal	44,240	2002	Coastal wetland ecosystems

(1) 48 fishing ports and fishery infrastructure bases. The key ports are: Yangkou Fishing Port, Longkou Fishing Port, Yantai Fishing Ship Port Area, Weihai Fishing Port, Shidao Fishing Port, Haiyang Dabuquan Fishing Port, Jimo Shanzikou Fishing Port and Shijiu Fishing Port.

(2) 147 aquaculture zones which mainly include:

- Tidal flat aquaculture. The coastal waters of Binzhou, Dongying and Weifang have tidal flats with average width of $3.64\,km$. The area of the

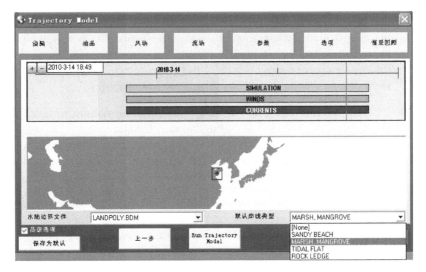

FIGURE 6.24: "Scene Review" dialog box.

tidal flat is $2,215 \, km^2$, accounting for 68.7% of the tidal flat of Shandong Province. These three coastal waters have abundant clam, scapharca subcrenata and mactra veneriformis, and are the key sea areas of tidal flat aquaculture in Shandong Province. The tidal flats in Yellow Sea section are mostly natural bays and have average width of $0.4 \, km$. The area is $1,009 \, km^2$, accounting for 31.3% of the mudflat of Shandong Province. They have good conditions for culture and are therefore also the key aquaculture zone.

- Shallow sea aquaculture. The Bohai shallow water is fertile and of little transparency and is therefore the key shellfish aquaculture zone. The shallow water of the Yellow Sea has stable salinity and has rich nutrient salt and prosperously growing planktons. It has large transparency, densely distributed rocks in the seafloor, and large number of algae. Hence, it is the key aquaculture zone of the seafood like Stichopus japonicus, Haliotis discus hannai Ino and scallop and the economic algae like kelp and Undaria pinnatifida.

- 29 propagation zones. The shallow water and tidal flat in the Bohai Gulf and Laizhou Bay are vast and have fertile water and rich plankton species. They therefore are the ideal place for feeding, spawning and breeding of multiple types of migratory fishes and have become the important propagation zone of fishery resources like prawns in China. The Miaodao Archipelago waters and coastal waters of the Yellow Sea have bedrock embayment shores, rich nutrient salt, and they have large transparency and rich and large number of plankton species. The rocks and algae are densely distributed in the seafloor. Therefore, such waters are key seafood aquaculture zones. The southern area of the Yellow Sea is the key artificial releasing and propagation zone of prawns, Sepiella maindroni, Portunus trituberculatus, Pagrosomus major and Sparus macrocephalus, key tidal flat propagation zone of shellfish, and key bottom sowing propagation zone of rare seafood.

- 5 fishing zones. The main fishing zones are the fishing grounds in the Bohai Sea, Yanwei, Shidao, Qingdao and Haizhou Bay. The areas other than aquaculture zone, port navigation zone, marine engineering zone, dump zone, nature reserves, tourist zones and areas for special purpose are all classified as the fishing zones.

- 4 protected areas of important fishery varieties. It mainly refers to the fishery-prohibited areas set for protecting the migratory marine swimming organism with important economic value, namely: the prohibited trawl fishing zone in the Bohai Sea and Yellow Sea and the hot season fishing-off zone in the Bohai Sea and Yellow Sea.

2. Sea area near Tianjin

There are about more than 80 identified species of fishery resources in Tianjin coastal waters and more than 30 species captured, of which the demersal fish species are bass, Sphyraenus and Collichthys lucidus; the pelagic fish species are Konosirus punctatus, Harengula zunasi Bleeker and Setipinna tenuifilis; the invertebrates are prawns, Acetes chinensis and Exopalaemon carinicauda; the demersal shellfish species are Scapharca subcrenata, oysters and Rapana bezona Linnaeus. However, in recent years, with the massive increase of pollutants into the sea, blocking by the river dams and overfishing, the fishery resources in coastal waters of Tianjin have obviously declined.

(1) 9 fishing ports and fishery infrastructure bases: Beitang Fishing Port, Donggu Fishing Port, Xindonggu Fishing Port, Dashentang Fishing Port, Caijiabao Fishing Port, Tangjiahe Fishing Port, New Mapengkou Fishing Port, Old Mapengkou Fishing Port and Tianjin Central Fishing Port.

(2) 4 aquaculture zones: Sajintuo aquaculture zone, new Mapengkou aquaculture zone, old Mapengkou aquaculture zone and aquaculture zone of Tianjin Bohai Resource Propagation Station.

(3) 3 propagation zones: Dagang Shellfish Resource Recovery and Propagation Zone, Hangu Shallow Water Shellfish Resources Recovery and Propagation Zone, Hangu Dashentang Clam and Polychaetes Resources Recovery and Propagation Zone.

(4) 3 fishing zones: Sajintuo Fixed Fishing Instrument Operation Zone, Mapengkou Fixed Fishing Instrument Operation Zone and Mobile Fishing Instrument Operation Zone.

3. Sea area near Liaoning Province

This sea area has 251 fishery resources utilization and conservation zones.

(1) The aquaculture zones include Changhai, Donggang, Zhuanghe, Dawa, Linghai and Xingcheng marine fishery seedling bases; Donggang, Zhuanghe,

Linghai and Dawa tidal flat shellfish aquaculture bases; Dalian, Dandong and Yingkou distant fishery bases; fish processing bases for export in six coastal cities; Changhai, Jinzhou, Lvshunkou, Zhuanghe, Pulandian, Donggang, Laobian, Dawa and Panshan rare seafood and algae culture and propagation bases. The seawater aquaculture zones include the coastal water culture in south end of Dagushan Peninsula, west of Lingshui River Estuary, Changshan Archipelago, Juhua Island in Xingcheng and Guizhou in Yingkou.

(2) Propagation zones include sea areas in Liaodong Bay, along Liaodong Peninsula and near some islands.

6.4.4.4 Distribution of tourism resources

1. Sea area near Shandong Province

The beautiful scenery, pleasant climate and profound Qilu cultural heritage along the coasts of Shandong and on its islands created many tourism resources. Three tourist zones have been formed along the coast, namely Yantai, Weihai, Qingdao and Rizhao Costal Tourist Area, Weifang Folk Customs Tourist Zone and Yellow River Estuary Tourist Zone. The main tourism resources are: port and bay beaches and bathing beach (such as Jinsha Beach, Yinsha Beach, Wanmi Beach, etc.) composed of medium and fine sand; Miaodao Archipelago which is as beautiful as pearl; scenic spots and historical sites like Laoshan, Qingdao which is the holy place of Taoist, Penglai Pavilion which is a wonderland on earth and Liugong island in Weihai city which was the Qing Northern Naval Base; the magnificent Yellow River Estuary; and the fishing village with unique characteristics.

(1) 36 scenic tourist areas. The key tourist sites are: Yellow River Estuary wetland ecological tourist zone, Changdao National Forest Park, Changdao Jiuzhangya and Banyuewan Scenic Spots, Penglai Pavilion Scenic Spot, Yantai Coastal Tourist Sightseeing Area, Weihai Coastal Park, Liugong Island Tourist Area, Chengshan Cape Coastal Scenic Spot, Qingdao Qianhai Scenic Spot and Laoshan Scenic Spot.

(2) 20 tourist resorts. The key tourist resorts are: Dakou River-Wangzi Island Coastal Tourist Resort, Penglai Haishi Tourist Resort, Yantai Jinsha Beach Tourist Resort, Yangma Island Tourist Resort, Weihai Northern Urban Coastal Resort, Huancui Tourist Resort, Tian'e Lake Tourist Resort, Shilaoren Tourist Resort, Liuqing River Tourist Resort, Yangkou Tourist Resort, Tianheng Island Tourist Resort, and Shanhaitian Tourist Resort.

2. Sea area near Tianjin

Tourist areas refer to the sea areas delineated for the development and utilization of coastal and marine tourism resources and for satisfying the demands of tourism development and include coastal and marine scenic spots and tourist resorts. Tianjin has great potential of coastal tourism resources, vast sea areas, rivers and lakes which are good for water sport activities. The coastal zones have low terrain, many depressions and rivers. Unique natural ecosystem has been formed in some depression and meanders, making these areas become spots with beautiful scenery. Tianjin also well maintains the gifted tourism resources such as ancient coast shell dike. It has the humanistic tourism resources like Taku Forts. These tourism resources provide better resource conditions for tourism development.

(1) 7 scenic tourist areas. The key tourist sites are: Taku Forts Tourist Area, Beitang Tourist Area, Chaoyin Temple Folk Customs Tourist Area, Salt Field Scenic Tourist Area, Shanggulin Shell Dike Tourist Area and Guanxiang Forest Park Tourist Area.

(2) 2 tourist resorts are: Bagua Beach Coastal Tourist Area and Dongjiang East Coastal Tourist Area.

3. Sea area near Liaoning

Tourist Area: The national and provincial key scenic tourist areas in Dalian coastal zone, Lvshunkou, Jinshitan, Xingcheng coastal zone, Dandong Yalu River Estuary, Jinzhou Big and Small Bijia Mountain, Wafangdian Xianyu Bay, Changhai Haiwangjiu Island and Donggang.

6.4.5   Case Studies of Typical Shoreline

Qingdao shoreline is taken as an example to introduce the work content and method of establishing a sensitive resource database. The following sections will introduce according to the sequence from the data collection and field survey at the initial stage to the data processing in the later period and finally to the formation of the sensitive resource database of Qingdao shoreline.

6.4.5.1   Site survey and data acquisition

The project research group investigated the resources of Qingdao shoreline from October 25 to 27, 2009.

- Investigation scope: The shoreline from Dingjiazui ($N35°54'36.67''$, $E120°8'0.63''$) to Shilaoren ($N36°5'37.29''$, $E120°28'38.68''$), with a total length of about $268,689.2\,m$. The survey objects are the shoreline and ancillary facilities within a range of approximately 100 meters above intertidal zone.

- Survey content: Mainly the contents of raw data of sensitive resources, including natural conditions like shoreline type and shoreline slope, and shoreline utilization, type of shoreline resource and other information.

- Survey method: Record the information of shoreline resource by means of on-site recording forms, handheld GPS positioning, photography and video recording. The vehicles were used in the survey of the shoreline that can be landed with land vehicles, and ships were used in the survey of the shoreline which is far from the land.

- In addition to field survey, the research group also collected the data on the sensitive resources in the Qingdao sea areas, such as the marine functional zoning.

6.4.5.2 Data analysis and organization

- Organize the data collected through field survey, locate the shoreline by the location information established on GIS and then summarize the environmentally sensitive data of shoreline to obtain the sensitive information database of Qingdao shoreline, as shown in Figure 6.25 and 6.26.

- This study uses a satellite image with a resolution of 1 m (covering Qingdao shoreline) as the base map, and marks the sensitive information of such shoreline in the GIS software. Meanwhile, the data and information obtained in the field survey is input so as to facilitate the data updating and inquiry.

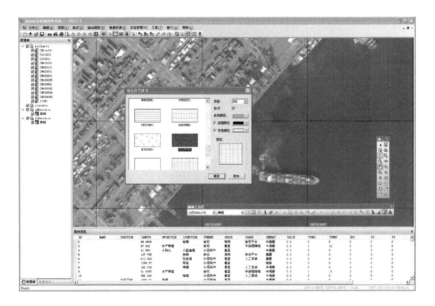

FIGURE 6.25: The marked shoreline data and information.

6.4.6 The System's Function

System data was collected in two ways: First, the system provided a corresponding interface, and the user input the specified data information according to the interface prompt; second, the system provided an interface to connect with other systems, so that users can input a large quantity of sensitive resource information. It can be set to record multimedia information such as pictures and videos.

FIGURE 6.26: Shoreline data.

6.4.6.1 Manual data input

For temporary and small amounts of data entry or modification of part of the data, the system provides a human-computer interaction interface for collecting sensitive resource data, and the users can input the specified information according to the prompt of the interface. Once the information is submitted to the system, the system will enter the data into the database.

The process of manual data input is illustrated in the following with an example (as shown in Figures 6.27 and 6.28).

If you plan to enter the relevant data of Qingdao No. 1 Bathing Beach, select the drop-down menu "Emergency Management" on the menu bar first and choose "Data Maintenance" from it. The dialog box "Emergency Information System Data Maintenance" will appear at this time, click the left mouse button to select "Tourist Area" in "Sensitive Resources," and then click "New" button to input the sensitive resource data.

6.4.6.2 Bulk import

It would be lowly efficient to manually input and maintain the large quantity of sensitive resource data. The system can input the data into the database through importing the data in the standard system data format (data in Excel format for this system). The process of bulk import of data is illustrated in the following with an example (as shown in Figures 6.29 and 6.30):

- First of all, we should ensure the unification of data format and system format. The system accepts data in Excel format and the data to be

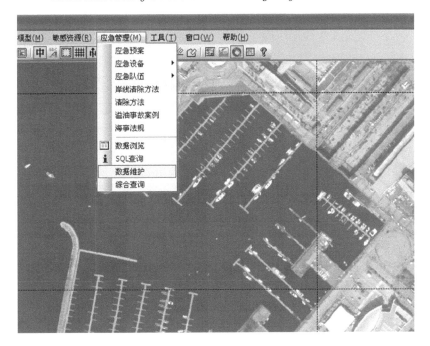

FIGURE 6.27: Open data maintenance.

input must be classified into an Excel sheet and the data order of each sheet should be consistent with the data order in the system.

- Select the drop-down menu "Emergency Management" on the menu bar and select "Data Maintenance". The dialog box "Emergency Information System Data Maintenance" will appear at this time, click the left mouse button to select "Tourist Area" in "Sensitive Resources" and then click "Import Excel Sheet" to import the sensitive resource data.

6.4.6.3 Output

The site information can be regarded as the data and stored in the system. What's more, it can also be displayed and printed out in the form of a table to facilitate use. The sensitive resource data can be exported or directly printed in the form of an Excel sheet.

6.4.6.4 Query

In the oil spill emergency process, the site information will be reviewed as needed, and the site information query can reproduce the existing data in the system. Meanwhile, for visual display, the data can also be directly displaced

FIGURE 6.28: Manual data input.

on the electronic chart in addition to display in text, making the sensitive resource information more intuitive.

The system provides flexible and diverse query methods, including point selection, rectangle selection, circle selection and polygon selection of the chart and moving targets and string-based fuzzy query of chart database.

1. Graphic query

Graphical query is more intuitive and convenient, and can quickly query sensitive resource data information within the area of concern. The users can select the area they want to query on the chart and satellite images by clicking on the graphic icon (rectangle selection, circle selection, polygon, etc.) at the top left of the system. The sensitive resource information of the selected area is displayed on the left side of the system, as shown in Figure 6.31.

2. Character query

Character query can be used when the resource information is well understood, but the location information is undefined, or when searching for resource information with the same attributes. There are two types of character queries: SQL query and integrated query.

(1) SQL query

Apply the query ability of the SQL database in the background of the system to simply select and query the information (see Figures 6.32 and 6.33). For example, if the user wants to search the shoreline with high sensitivity after May, the user can choose the dataset "shoreline" and choose "sensitive period", then select the operator ">", fill in the number 5 on the left side and finally make a selection.

(2) Integrated query

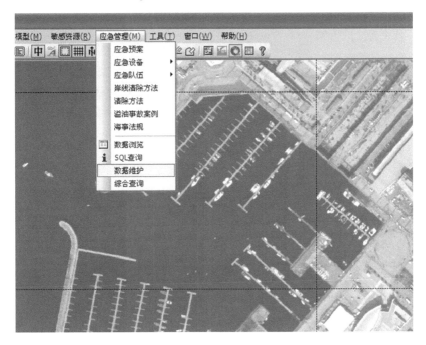

FIGURE 6.29: Open data maintenance.

The system supports setting of a range of conditions to find the data you need. This kind of query method is characterized by quick speed, accuracy and complete information, see Figures 6.34 and 6.35.

6.4.7 Emergency Resource System

6.4.7.1 Overview of emergency resources

1. Classification of emergency measures

- Emergency unloading
- Oil absorption material
- Control with oil boom
- Oil dispersant
- Oil recovery equipment

2. Distribution of emergency resources in the sea area studied

According to investigation, the current distribution of the emergency equipment for oil spills in the Bohai Sea is shown in Table 6.5.

FIGURE 6.30: Import Excel sheet.

TABLE 6.5: The summary of emergency equipment for oil spill in the Bohai Sea (to end of 2018).

Equipment	Administered areas				
	Liaoning	Hebei	Tianjin	Shandong	Total
Ships	31	16	5	23	75
Oil skimmer (set)	21	13	33	28	95
Oil boom (m)	31590	9700	7150	22000	70440
Emergency transfer pump (set)	1	1	-	-	2
Sprinkling device (set)	14	8	4	46	72
Oil dispersant (tons)	37.16	11	7.8	128	183.96

6.4.7.2 Emergency resource database

The emergency resource database mainly includes emergency team database, emergency equipment database, emergency plan database, shoreline clean-up method database, oil spill clean-up method database, oil spill accident cases and maritime regulations database, see Figures 6.36 and 6.37.

6.4.7.3 Emergency equipment database

The structure of emergency equipment database is as follows:

- Recovery ship and working ship. The specifics are: Patrol surveillance ship, recovery ship, oil boom deployment vessel, oil boom towing vessel, oil contaminants collection and treatment vessel, oil barge, oil spill clean-up ship, rescue ship, and floating recovery ship.

FIGURE 6.31: Graphical query.

- Collection equipment. The specifics are: light oil sac and floating oil sac.

- Oil skimmer. The specifics are: oil skimmer, new type oil skimmer and oil trawler.

- Oil boom. The specifics are: oil boom and inflatable oil boom.

- Sewage treatment facilities. The specifics are: Ballast water treatment site, oily sewage treatment tank, oily sewage tank, refined oil treatment tank.

- Oil adsorption materials and oil spill dispersants. The specifics are: demulsifier, oil absorption boom, oil absorption material, oil dispersant spraying device, dispersing agent, curing agent, sedimentation agent.

- Other equipment.

The OSERS 2.0 system also provides image and photos of emergency equipment, as shown in Figure 6.38.

6.4.7.4 Emergency team database

The fields of emergency team database include the entity's name, all departments, the name and type of emergency equipment, contact person, contact phone number, quantity of the equipment, main motor power, total tonnage, speed, recovery capacity, tank capacity, oil recovery method, oil viscosity and scope of use.

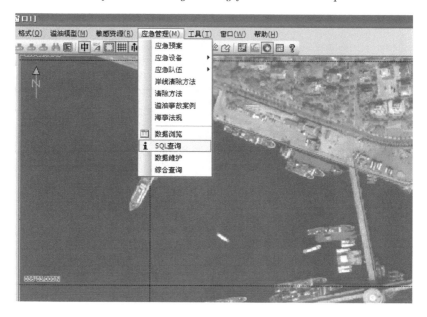

FIGURE 6.32: Open an SQL query.

The name of the emergency equipment is matched with the equipment field in the emergency equipment database. That is, the information on the manager of each set of emergency equipment and the emergency capacity of each emergency team can be queried.

6.4.7.5   Functions of the emergency system

The main functions of the emergency system are query, statistics and update of the emergency equipment and the emergency team, which are similar to the functions of an environmentally sensitive resource system. All data of the emergency system is stored with access data and the users can input, edit and modify the data, as shown in Figure 6.39.

6.4.8   System Maintenance

6.4.8.1   Update of base map data

The base map data mainly includes basic data such as chart, land map and satellite image, and the OSER2.0 system provides a method for the user to define the map and its attribute data. The users can create the following types of databases:

- Plotting database;

- S57 chart database;

FIGURE 6.33: SQL settings.

- Map database;

- Image database.

There are three ways to generate data from the above databases:

- Loading the ECIVMS system chart;

- Importing electronic maps in a common format, including tab files in Mapinfo format and .shp files from ESRI;

- The users can define the plot layers, the layer name and attribute fields of the plotting database and use the point, line and plane editing function of OSER2.0, as shown in Figure 6.40.

6.4.8.2   Update of sensitive resources and emergency resources

Sensitive resources and emergency resources are data that users need to update frequently according to actual conditions. OSER2.0 provides a variety of update methods.

6.4.8.3   Update of wind field and current field data

The update of wind field and current field data refers to the update of the wind field and current field data required for calculation of the oil spill module and there are two update ways:

1. Static update

Since the OILMAP model can read current field data in a variety of formats, including the current field files in HYDROMAP, NetCDF, LLU, MDB and Canadian formats, researchers are required to calculate with a hydrodynamic model.

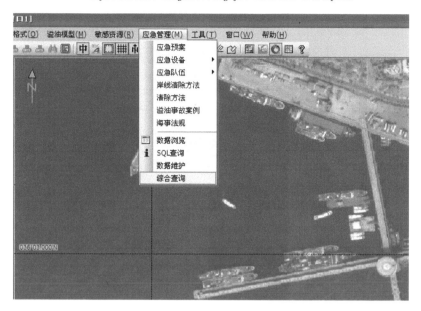

FIGURE 6.34: Open an integrated query.

2. Dynamic update

The OILMAP model provides tools for reading real-time current field, that is EDS (ASA's Environment Data Service) or access to real-time monitored current field data.

6.4.9 System Application

6.4.9.1 Oil spill accident scenario analysis

The simulation results of OILMAP are presented here in the form of scenario simulations. The oil spill scenario analysis mainly analyzes the location of oil spill, quantity and type of spilled oil and the meteorological marine conditions at the time of the accident. The analysis selects north of Kiaochow Bay mouth as the location of the oil spill accident for simulation and the 500 tons as the quantity of spilled oil (lower limit of major pollution accident by ship) (as specified in Administration Regulations on Preventing and Controlling Marine Environment Pollution from Ships (No.561 Order of the State Council of the People's Republic of China)).

OILMAP is used to simulate the above accident scenario. The wind coefficient of the model is 3%, and the accident scenario analysis is shown in Table 6.6.

FIGURE 6.35: Integrated query.

6.4.9.2 Scenario simulation results of oil spill accident at Kiaochow Bay mouth

1. Forward mode

Under the forward mode, the oil spill drift trajectory and thickness of oil slick after 3 hours, 6 hours, 12 hours and 24 hours of the oil spill accident are shown in Figures 6.41-6.44.

It can be seen from the simulation results that the oil slick will drift towards the inside of Kiaochow Bay under the action of tidal current and wind when an oil spill accident happens at Kiaochow Bay mouth and finally the majority of spilled oil will be adsorbed on the east and south shoreline of Tuandao Island and some other spilled oil will drift into Kiaochow Bay.

According to the superposition and coupling of the prediction results of oil spill drift by OSERS2.0, it can be concluded that the oil spill will cause severe pollution to the ecological reserve, fishing area and tourist shoreline in Kiaochow Bay.

Distribution of spilled oil after 24 hours is shown in Figure 6.45.

2. Reverse deduction mode

This mode is to reversely deduce the possible oil spill location and historical oil spill drift trajectory on the basis of the current field and wind field (E is the wind direction, 5 m/s) in the past 24 hours when an oil spill is found in Kiaochow Bay, thus obtaining the information on sensitive resources that may be polluted. The 24-hour simulation results are shown in Figure 6.46 to Figure 6.48.

When using this mode, the tidal current field and wind field before the time of the oil spill accident must be input. In this simulation, the constant wind field is adopted. It can be seen from the results that the mode can quickly deduce the historical oil spill drift trajectory and thus provides basis for de-

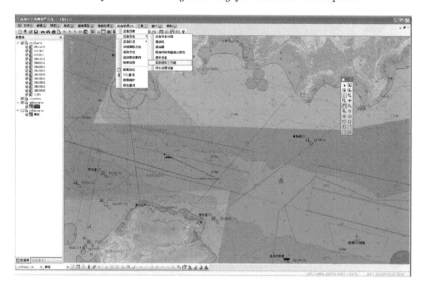

FIGURE 6.36: Contents of Emergency Resource Database of the OSERS2.0.

termining the oil spill drift trajectory and quickly judge the environmentally sensitive resources that may have been affected along the route.

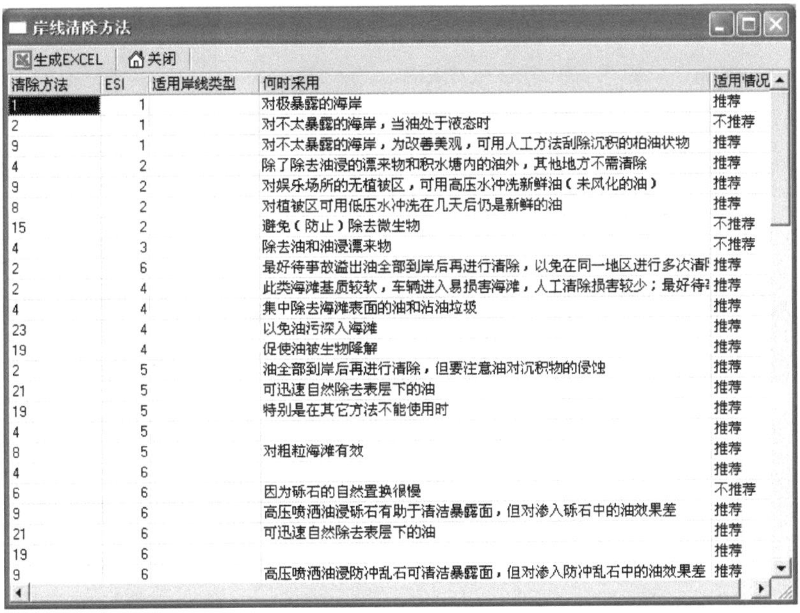

清除方法	ESI	适用岸线类型	何时采用	适用情况
1	1		对极暴露的海岸	推荐
2	1		对不太暴露的海岸，当油处于液态时	不推荐
9	1		对不太暴露的海滩，为改善美观，可用人工方法刮除沉积的柏油状物	推荐
4	2		除了除去油浸的漂来物和积水塘内的油外，其他地方不需清除	推荐
9	2		对娱乐场所的无植被区，可用高压水冲洗新鲜油（未风化的油）	推荐
8	2		对植被区可用低压水冲洗在几天后仍是新鲜的油	推荐
15	2		避免（防止）除去微生物	不推荐
4	3		除去油和油浸漂来物	推荐
2	6		最好待事故溢出油全部到岸后再进行清除，以免在同一地区进行多次清除	推荐
2	4		此类海滩基质较软，车辆进入易损害海滩，人工清除损害较少；最好待事	推荐
2	4		集中除去海滩表面的油和沾油垃圾	推荐
23	4		以免油污深入海滩	推荐
19	4		促使油被生物降解	推荐
2	5		油全都到岸后再进行清除，但要注意油对沉积物的侵蚀	推荐
21	5		可迅速自然除去表层下的油	推荐
19	5		特别是在其它方法不能使用时	推荐
4	5			推荐
8	5		对粗粒海滩有效	推荐
4	6			推荐
6	6		因为砾石的自然置换很慢	不推荐
9	6		高压喷洒油浸砾石有助于清洁暴露面，但对渗入砾石中的油效果差	推荐
21	6		可迅速自然除去表层下的油	推荐
19	6			推荐
9	6		高压喷洒油浸防冲乱石可清洁暴露面，但对渗入防冲乱石中的油效果差	推荐

FIGURE 6.37: Database of shoreline clean-up methods.

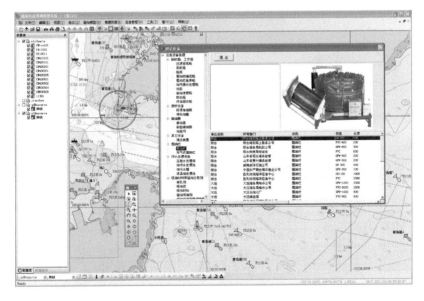

FIGURE 6.38: Emergency equipment database of OSERS2.0.

FIGURE 6.39: Database maintenance interface of the emergency information system.

TABLE 6.6: The composition of shoreline types.

Accident parameter		Accident at mouth of Kiaochow Bay
Accident scenario	Spillage duration	6 hours
	Oil type	Medium crude oil
	Quantity	500 tons
	Oil spill location	Kiaochow Bay mouth
Model function		Forward mode/reverse deduction mode
Meteorological marine conditions	Wind speed	5 m/s
	Temperature	13.3 degrees (average temperature in 2009)
	Tidal current field	Historical tidal current field is adopted

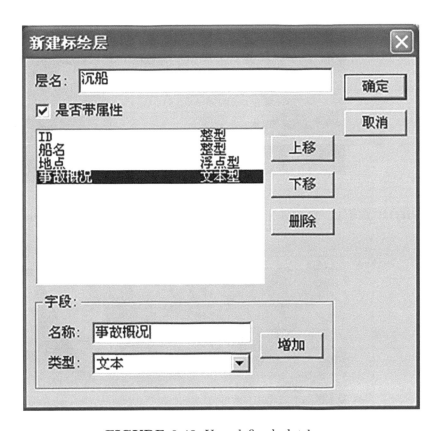

FIGURE 6.40: User-defined plot layer.

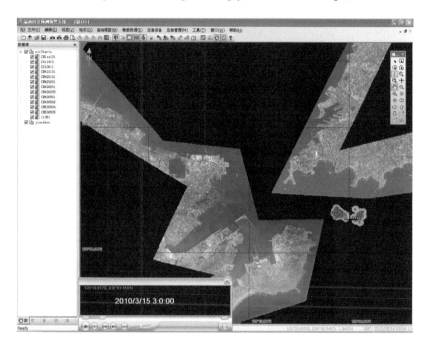

FIGURE 6.41: Thickness of oil slick 3 hours after the oil spill accident.

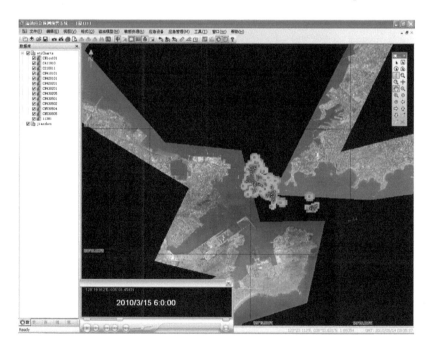

FIGURE 6.42: Thickness of oil slick 6 hours after the oil spill accident.

FIGURE 6.43: Thickness of oil slick 12 hours after the oil spill accident.

FIGURE 6.44: Thickness of oil slick and shoreline pollution 24 hours after the oil spill accident.

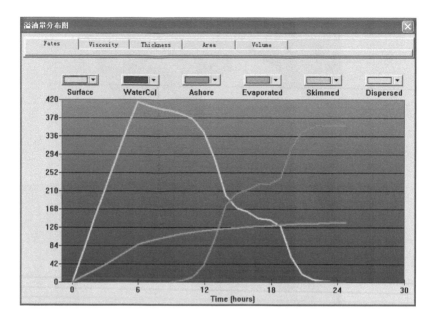

FIGURE 6.45: Distribution of spilled oil after 24 hours.

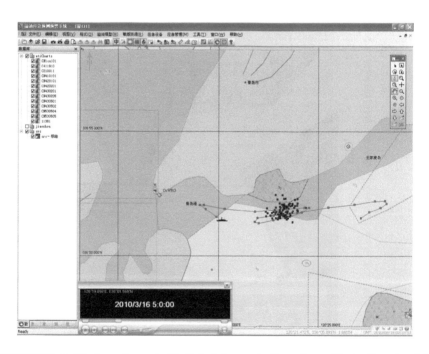

FIGURE 6.46: Location of oil slick 6 hours before the spilled oil arrived in the location where it is found.

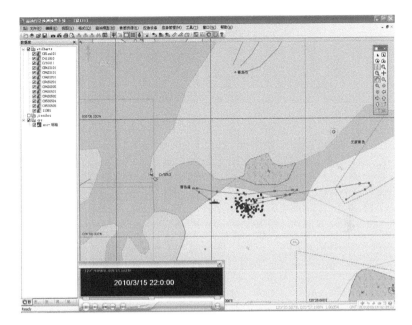

FIGURE 6.47: Location of oil slick 12 hours before the spilled oil arrived in the location where it is found.

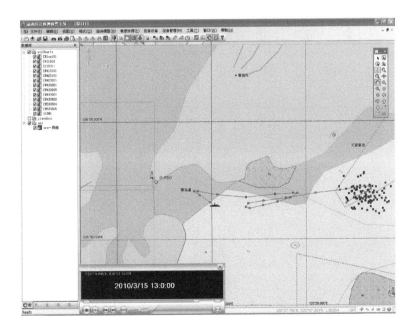

FIGURE 6.48: Location of oil slick 24 hours before the spilled oil arrived in the location where it is found.

Chapter 7

Emergency Treatment of Marine Oil Spill

After the marine oil spill accident, the spilled oil will quickly drift and spread around from the location of accident to form a large area of oil slick and oil band on the sea under action of the surface wind, current, tides and waves. It is much more difficult to treat the oil spill when it reaches the shore. Therefore, generally, the emergency treatment of oil spill on the sea must be carried out first after an oil spill accident happens in order to timely and effectively minimize the pollution range of oil spill.

The conventional emergency treatment methods are containment, recovery, on-site combustion, treatment with chemical preparations, and biodegradation of spilled oil. Equipment and materials used are oil boom, oil skimmer, oil delivery pump, oil storage equipment, oil recovery vessel, oil absorption material and chemical preparations, etc. The oil boom is mainly used to control spilled oil to a small range to prevent it from further diffusion and drift. The oil skimmer, oil delivery pump, oil storage equipment, oil recovery vessel and oil absorption materials are used to recover the spilled oil with certain thickness on the sea surface. When performing oil recovery, the oil recovery level needs to be based on the weather conditions of the accident location, the range of oil spill, the oil type, the capacity of recovery equipment and the storage capacity of the oil storage equipment. Each factor is important. For a very thin oil slick, the recovery effect of above equipment and materials is negligible and chemical preparations or biodegradation technology must be used. The chemical preparations used to treat the spilled oil include oil spill dispersants, oil coagulant and oil-collecting agent and the like. Chemical preparations change the existing form of spilled oil in the marine environment by thickening or solidifying the spilled oil, thereby reducing the contamination degree of oil spills in the seawater. However, the use of chemical preparations also has some disadvantages, as evidenced by the fact that the chemical preparations themselves may also cause pollution to the marine environment. It should be noted that the staff will generally choose on-site combustion to treat the large marine oil spill accidents in addition to the use of chemical preparations. Biodegradation is generally used to treat the partially unrecoverable spilled oil or the spilled oil of which the existing form is changed by chemical preparations so as to eliminate it.

In conclusion, the emergency treatment methods of oil spill can be roughly classified into three categories, namely physical, chemical and biological treatment methods. When treating the oil spill, the scheme should be optimized according to the specific conditions of oil contaminants and attention should be paid to protection of the marine environment and minimizing the pollution to the surrounding environment at the same time.

7.1 Physical Treatment Method

After a large-scale oil spill, the spilled oil must be contained and recovered first by physical means. The usual method is to use the oil boom to contain and intercept the spilled oil to prevent it from spreading and drifting around and concentrate it to make it thicker for recovery, or guide the oil slick to the relatively clam water which has less impacts on the environmentally sensitive areas to avoid pollution or cause less pollution to the environmentally sensitive areas. Then, use the machinery equipment like oil skimmer and oil recovery vessel and the oil absorption material to recover the spilled oil on the sea and store the recovered oil in the oil storage equipment. The contained oil must be timely removed to prevent emulsification or sedimentation and damages to the surrounding ecological environment. In the process of containment and recovery, it is necessary to select the equipment according to the factors such as weather, oil spill range, oil types, etc. and select the appropriate operation mode according to the model of selected equipment.

7.1.1 Oil Boom

Oil boom is a device used to block and control the spread of spilled oil in a large area and maintain the thickness of oil slick to facilitate recovery. Oil spill control with oil boom has the advantages that the equipment required is simple, the investment required is small and it is convenient to operate.

It mainly has the following three functions: (1) Oil spill containment. Control the spilled ill in diffusion in a small range with the oil boom to prevent it from further spreading and increase and maintain the thickness of oil slick so as to recover and remove it. (2) Oil spill diversion. Set the oil boom according to the design route to prevent the spilled oil from flowing to the designated location or facilitate the recovery. In the offshore rushing water, diversion with oil boom can effectively prevent the spilled oil from entering into the environmentally sensitive areas. (3) Potential oil spill prevention. Set the oil boom in the water where an oil spill accident may happen according to the local water conditions ahead of time to prevent and control the oil spill. In this way, it will play the role of preventing oil spill diffusion and creating conditions to timely recover the spilled oil in containment once an oil spill accident happens [179].

7.1.1.1 Structure and type of oil boom

1. Structure of oil boom

The oil boom is generally a ribbon-like substance made of PVC materials. It can be made of other materials such as rubber, foamed plastics, straw, timber and metals. It generally consists of floatation, skirt, tension belt and ballast. The floatation is to provide buoyancy force for the oil boom by use of air or buoyance materials so as to fix the oil boom on the seawater. The skirt is an underwater barrier below the floatation and is used to prevent or reduce the oil contaminants from flowing away through the part below the oil boom. The tension belt refers to some long-strip members (such as chains, belt, and so on) that can withstand the horizontal pull exerted on the oil boom and is mainly used to withstand the wind, waves and tidal current, and the towing-generated tension. Ballast refers to the one that can make the oil boom droop and improve the performance of the oil boom. It generally hangs under the skirt and is used to maintain the vertical balance of the oil boom. The oil boom must have certain wind, current and wave resistance, and the skirt must also have a certain tensile strength. In general, the performance of the oil boom needs to be determined by testing. There are two main test methods: (1) Directly placing the oil boom in the actual sea and testing the maximum standard of wind speed, wave height and current velocity enabling it to block the spilled oil. (2) Carrying out wind-resistance, wave-resistance and current-resistance test of the oil boom in the laboratory and calculating its performance according to the experimental results [180].

2. Type of oil boom

The existing oil booms are divided into two types, namely curtain booms and shore seal booms.

(1) Curtain boom

The curtain boom is mainly used in the case of calm sea surface and good coastal conditions and is classified into inflatable boom and solid floatation boom.

The floatation of inflatable boom is inflatable and is generally made of PVC materials. Based on the inflation method, it can be further divided into pressurized inflation type and self-inflation type. Based on the structure of the gas chamber, the inflatable boom can also be divided into single-chamber type and multi-chamber type (the length of a gas chamber is two to four meters). The application facts show that multi-chamber oil boom has stronger floatability. In case that one of the gas chambers is broken, the overall oil boom will not sink and this type of oil boom is thereby more widely applied. The deployment of the inflatable boom is slow. After deflation, it is small in volume, flat in surface and easy to clean, of which the long-chamber inflatable boom is easy to inflate quickly but is sensitive to puncture and scratch and has poor flow ability [179].

The floatation shell of solid floatation boom is generally made of PVC material or rubber, steel material or the like (as shown in Figure 7.1), and filled with columnar or granular foam. Compared with the inflatable boom, the laying of this type of oil boom is faster, but it requires large space and large workload to take it back.

The floatation of solid floatation boom is made of heat-resistant steel material and is also called fireproof oil boom. The underwater skirt of this oil boom is generally formed by two layers of high-strength fabric coated with oil-resistant and aging-resistant flame-retardant rubber. The stainless steel wire rope at the top, the high-strength reinforcing belt at the waist and the tension ballast chain at the bottom can constitute a strong tensile system. The oil boom is generally used in combination with incineration technology. Due to its fireproof properties, the oil boom is particularly suitable for oil ports, oil terminals, petroleum drilling platform and other sensitive areas with high fire-rating and for being used to tow the spilled oil to a suitable location for combustion processing.

FIGURE 7.1: A rubber solid floatation boom.

(2) The shore seal boom

The shore seal boom has the advantage of being difficult to turn over in shallow waters, and is particularly suitable for being placed near shore beaches. Most of the shore seal booms are made of polyurethane (PU) which consists of three separate tubes. One tube is located at the upper part and the other two are located at the lower part. The upper tube is inflated and used as floatation. The lower tubes are filled with water and used as a skirt to provide sufficient weight to keep the oil boom tightly sealed with the ground or beach. The oil boom is uniquely constructed, and the freeboard (the minimum vertical height above the oil boom waterline) is the height of the upper tube; the draft is the vertical height after the lower tube is filled with water and accounts for about half of the total height of oil boom; the tension belt is its own structural material; and the water in the bottom two tubes is served as ballast.

Perhaps it is because of the unique structure of this oil boom that it is suitable for being placed in intertidal zone or the land and water junction.

When placing this oil boom, it is generally necessary to select the location first, and then inject the lower tube and inflate the upper tube, respectively. The volume of water to be injected must be well controlled, and too much water will affect its sealing with the ground.

This oil boom has the following characteristics:

- Its application scope is small and it is suitable for being placed in intertidal zone and the water and land junction.

- There are limitations on its deployment and it is usually placed on a flat terrain to ensure the sealing effect.

- It can be connected to other oil booms for use.

- Due to its unique structure, its appearance is fragile and it is easy to be stabbed and scratched.

In general, quality oil booms should have the following characteristics:

- Easy to unfold and take back;

- High float-sink ratio;

- Good static and dynamic stability;

- Good hydrodynamic properties;

- Dense and not blocked;

- Easy to wash;

- The joints can resist ultraviolet ray and hydrolysis;

- Anti-wear, puncture resistant and oil resistant. It is required that the oil boom should be able to prevent the diffusion of spilled oil in the horizontal direction and prevent the crude oil from condensing into tar ball and its diffusion in the vertical direction.

7.1.1.2   Research progress of oil boom at home and abroad

The research on oil boom at home and abroad mainly focuses on the experiment of oil boom performance and the design improvement of oil boom. S. Badesha evaluated the structural properties of a given oil boom by establishing gas/hydrodynamic model and finite element model and concluded the maximum load, pressure and deformation of three types of oil booms [181]. His study results can be used as a reference for the strength requirements of individual components of an oil boom. Zhan Dexin used circulating water tank and wind-wave tank to carry out experiments on anti-current, anti-wind and anti-wave of floatation PVC plastic oil boom and used universal material testing machine to carry out tensile experiment on the skirt of the oil boom

[180]. The study results provided important parameters for the performance improvement of the oil boom. Cheng-Yi Zhang et al. carried out a model study on the effective depth of the solid floatation boom under the action of water flow and the oil containment performance under the combined action of water flow and waves and found out that the effective depth of the model is 0.204 m [183]. In addition, the plastic granules floating on the water surface move up and down with the waves under action of the waves and the solid floatation boom model also showed similar motion. Namely, the up and down movements of the two are basically synchronous, indicating that the oil containment performance is good. Gui-Feng Yu et al. based on the N-S equation and volume-of-fluid fraction (VOF) method of incompressible viscous fluids, used the CFD commercial software Fluent to conduct numerical experiments on the containment performance of oil boom of typical structures and carried out comparative analysis of the maximum containment capacity of monomer boom and double boom [184].

Zhang Yindong et al. improved the inflation valve of existing oil boom on the basis of "plane" recovery concept so that it has a constant pressure exhaust function [185]. The "plane" recovery concept refers to using the oil boom and treating the entire oil spill plane as the oil recovery object, narrowing the containment area and increasing the thickness of oil layer after combined use of the ships and machinery equipment so as to give play of the best performance of oil skimmer while maintaining the thickness of the oil layers by continuous reduction of the contained area to ensure that oil skimmer continuously works in an optimal state, thus achieving the goal of quickly recovering the spilled oil. The "plane" recovery can avoid the drawbacks of low efficient oil recovery due to limited thickness and gradually thinning of the oil slick during the "point" or "line" operation within the oil spill diffusion lane.

Yin-Dong Zhang et al. transformed the original inflation valve of oil boom [185]. The oil boom with transformed inflation valve can automatically recover the oil on the water surface, ensuring that the efficient "plane" recovery can be applied in the practices.

7.1.1.3   Deployment of oil boom

In order to best prevent oil spill diffusion, it is also very important to place oil boom according to the oil spill nature, the hydrological and meteorological conditions of the sea area and the surrounding environment. There are five ways to deploy oil booms:

- Containment method

  This method is used to contain the oil spill source at the initial stage of oil spill or when there is not so much spillage per unit time and the wind and tidal current exert little impact. Considering the effect of wind and tidal current, the spilled oil may drift outside the oil boom, so two oil booms can be placed. In addition, based on the needs of oil recovery, the entrance for operation ships and oil recovery vessel should be provided.

- Waiting method

  This method is used when there is much spillage, insufficient oil booms or large impacts exerted by wind and tidal current and difficulty in containing the spilled oil. The oil boom can be placed at a certain distance from the oil spill source according to the wind direction and current to wait for the oil. Two or three oil booms can also be placed according to the specific conditions.

- Blocking method

  This method is used when oil spill occurs in a narrow waterway and canals in the port area. Deployment while keeping a central opening or deployment of two or three oil booms can be adopted when the water has large current rate, there is difficulty to block the water area or complete blocking will affect the traffic.

- Induction method

  When there is large spillage and large impact of wind and tidal current and it is impossible to contain the oil with oil booms on the oil spill site, or in order to protect the coastal and aquatic resource, an oil boom can be used to induce the spilled oil to the sea surface where the recovery can be carried out or has less pollution effect and multiple oil booms can be deployed according to the realities of the site.

- Moving method

  In the case of deep sea or large wind and tidal current and it is impossible to use the anchors or the oil has spread to a large area, this method is usually used. This method requires two boats to cooperate with each other [186].

The above five methods are the basic oil boom deployment methods. In the actual emergency treatment, the operation teams are required to timely and quickly take appropriate number of oil booms from the storage site and effectively place them into the sea. The oil booms can be deployed flexibly according to the actual situation and multiple methods can be used at the same time. However, in consideration of the changes in natural conditions, it is necessary to take corresponding measures in a planned manner.

7.1.1.4　Conclusion

Oil spill control with oil boom requires simple equipment and low investment and is easy to operate. However, the oil boom can work effectively only when the current velocity is relatively small and the sea surface is relatively calm. When the oil spill scale or current rate exceeds the critical value, the oil boom will fail. Oil boom is only a preliminary measure of oil spill treatment.

Recovery or combustion is required thereafter to deal with the oil spill. Which method to use should be determined according to the actual situation and the impacts exerted.

7.1.2   Oil Skimmer

An oil skimmer is the main mechanical device to recover an oil slick on the sea surface. It has wide application range and good oil collection effect. It is suitable for centralized oil recovery of medium scale or above [187]. It is mainly composed of three parts: skimmer head, transmission system and power system. The skimmer head is used to separate the oil from water. The transmission system consists of pump or vacuum device, hose and connecting pieces and is used to transmit power and pump out the recovered liquid. The power system is used to provide power to the skimmer head and pump.

The work performance of an oil skimmer is mainly described by recovery rate and recovery efficiency. The recovery rate refers to the quantity of pure oil recovered by the oil skimmer in unit time (m3/h) when the oil layer is thick, there is no wind and waves and the viscosity is ideal. Recovery efficiency refers to the percentage of pure oil recovered by the oil skimmer in a unit time to the total amount of oil-water mixture. Due to the performance of oil skimmer and restriction by surrounding environment, the recovered oil contains a lot of water (up to 80%), which often makes high recovery rate but low recovery efficiency. Due to the working environment, oil spill type, thickness of oil slick, marine conditions or marine wastes and other factors, it is difficult to achieve ideal recovery rate and recovery efficiency when the oil skimmer is working. The oil skimmer even fails in the case of severe marine conditions or too thin oil slick [179].

The oil skimmer has many working principles. Which one to follow is mainly decided by the oil viscosity. The working principles of oil skimmers of different structures may be different or may be the same. Regardless of their structure, there is huge difference in oil viscosity that they are suitable for. Their application conditions and recovery efficiency are also different.

The following sections introduce the breakdown of oil skimmers based on the working principle.

7.1.2.1   Lipophilic/adsorption principle

This principle refers to using the lipophilic materials to adsorb oil slick to a moving surface so as to make them escape from the water surface, and then transferring the adsorbed oil into the oil storage device by scraping or squeezing [187]. The oil skimmers following this working principle are of various structures and can be disc type, brush type, belt type and rope type.

When the disc skimmer is working, the disc made of lipophilic material rotates in the oil-water mixture. When the disc is rotated out, the adsorbed oil is scraped into the oil collector and pumped into the oil storage device.

This type of oil skimmer is mainly composed of disc, scraper, oil collector, delivery hose, power system and pump. Its advantages are that it is suitable for recovery of light oil, is less affected by the wastes, has high recovery efficiency and is easy to maintain. Its disadvantages are that it has low recovery rate (generally 10 to 60 cubic meters per hour), it is not suitable for recovery of oil with high viscosity and has poor adaptability to seaweed and waves. This type of oil skimmer is mainly used in ports and near shore waters.

When the brush skimmer works, the spilled oil adheres to the rotating brush and is scraped into the collector by the scraper, and then the spilled oil is introduced into the oil storage device through the pump. This type of oil skimmer is mainly composed of several groups or rows of brushes, scraper and oil collector. The advantages are that it has high recovery efficiency and recovery rate and good flow ability (can be applied to 1 m high waves), and is easy to operate and maintain. The disadvantages are that it has large volume, requires high cost, is not easy to clean and is only suitable for recovery of oil with high viscosity.

The belt skimmer referred to here is mainly the adsorption belt skimmer. When the belt skimmer works, the spilled oil on the water surface adheres to the operating lipophilic adsorption belt, then scraped into the oil collector by the scraper and finally introduced into the oil storage device through the pump. This type of oil skimmer is mainly composed of adsorption belt, scraper (or roller), transmission device and oil collector. The advantages are that it has high recovery rate and recovery efficiency, can be widely applied and has good flow ability. The disadvantages are that it has a complicated structure and large volume, and requires high cost and lifting equipment to work together.

When the rope skimmer works, the spilled oil is adsorbed by a certain length of rope made of lipophilic material and then squeezed into the oil collector by the roller extrusion device. This type of oil skimmer is mainly composed of rope mop, squeeze roller, oil collector, power station and hydraulic motor. The advantages are that it has good flow ability and high recovery efficiency, is less affected by wastes, covers a large area, is easy to maintain, requires low cost and the rope mop can be repeatedly used. The disadvantages are that it has poor adaptability to seaweed, has low recovery rate and can be applied to recovery of oil with limited range of viscosity and it is difficult to deploy the rope mop. This type of oil skimmer is used more frequently in shallow sea areas and more polluted rivers and port areas. When collecting oil in narrow waters, such oil skimmer is usually fixed on land.

7.1.2.2   Principle of mechanical transmission

Any mechanical operation, such as shoveling, brushing, starching and screwing, to transfer the oil slick follows the principle of mechanical transmission, all propelling the oil and water into a recovery container along a horizontal axis. This principle is generally used in combination with other principles, which can greatly improve the ability of an oil skimmer to recover oil with high viscosity and scum and improve the recovery efficiency to some extent. The use of the principle of mechanical transmission can well solve the difficulty of recovery in severe weather and recovery of water-containing oil. The bucket conveyor is the best tool of oil skimmer working by the mechanical transmission principle [187].

The conveying belt skimmer is of simple structure. There is toothed structure on the conveying belt, which is basically the same as that of belt skimmer. But their working principles are different. The conveying belt skimmer uses the rotation of the conveying belt placed into the oil-water mixture and the gear on the conveying belt to lift the spilled oil. This type of oil skimmer is generally placed on the ship. The advantages are that it can be applied in large areas and less affected by the waves and wastes on the water surface, is suitable for recovery of oil with high viscosity, emulsified oil and waste-containing oil during the sailing, has best operation effect in calm waters and has high recovery efficiency. The disadvantages are that it has low startup speed and requires other equipment to work together.

7.1.2.3   Principle of pumping

The pumping includes air transmission and underwater pumping.

The oil skimmer working by the air transmission principle is also commonly referred to as vacuum skimmer. It usually has a vacuum oil storage tank which is connected to the skimmer head via the sucker. In addition, it also has vacuum pump and power system. The oil recovery principle of this oil skimmer is that it sucks in air while sucking in oil, the high-speed flow of air in the suction inlet and the pipe take the oil away from the water surface and transfer it into the oil storage tank. Due to the friction loss in the suction arm, when the maximum suction pressure is 80 to 90kPa, such oil skimmer is almost ineffective for the oil with high viscosity. If the small oil blocks can enter the suction inlet, this type of oil skimmer can also be used to recover solid oil [187].

The vacuum skimmer generally works well in calm water and has high recovery efficiency. But if there are waves on the sea surface, the wave will increase the water content of the oil and reduce its recovery efficiency. Therefore, this oil skimmer is more suitable for use in calm water. The advantages are that its operation device is small, it does not require high operation skills, it is not sensitive to wastes, is easy to maintain and low in cost. The disadvantages are that it is sensitive to waves and only effective for very light oils.

The underwater pumping type skimmer is also called the underwater suction or weir type skimmer. Weir skimmer is generally composed of floatation, oil collector, edge height adjustment device, drain pump, oil suction pump and power system. It appears more like a funnel. The bottom is a drain pump, and the upper part is an oil suction pump. The suction inlet is deep into the oil slick layer, and the funnel is fixed on floatation for buoyancy force adjustment. The drain pump continuously absorbs water from the funnel during operation and drains it out from the bottom of the funnel to lower the water level in the funnel. The seawater and oil slick continuously enter from the upper part so that the oil layer in the funnel thickens and the oil suction pump sucks the oil and transfers it to an oil collector.

Most of the weir edge of the existing weir skimmer can be vertically adjusted under action of the water or adjusted up and down with the air ballasting. But usually most of the weir edges are of the self-regulating type and can be adjusted high or low with the speed of the pump. The weir skimmer can be fixed on the floatation or is hand-held or fixed on the lifting arm. It can also be used in conjunction with other oil skimmers.

A weir skimmer has the advantage of being able to be refitted and reorganized, for example: if installing the weir, funnel and oil delivery pump on the top of the oil boom, the weir skimmer is then transformed into a weir boom. In the actual application, this oil boom can be equipped with two or more weirs. There are also some oil skimmers which can be combined with some different discs, bristle brushes and other transmission devices to form a multi-functional oil skimmer.

7.1.2.4 Conclusion

During oil spill control with any type of oil booms, the oil skimmer is required to recover the oil slick in the boom and the finally recovered oil-water mixture needs to be further separated. Setting an oil boom around a vessel that has spilled volatile substances may increase the risk of fire or explosion, so only timely recovery of the oil slick contained by the oil boom can ensure the safe and reliable use of the oil boom. Containment without recovery will inevitably lead to the failure of the oil boom. So, it is undeniable that an oil skimmer is the core part of an oil recovery system. However, the vast majority of oil skimmers can only be used normally when the wind speed is less than 10 m/s and the wave height is less than 0.5 seconds. It is impossible to carry out effective oil recovery when there is large wind and wave.

7.1.3   Oil Recovery Vessel

Oil recovery vessel is the vessel specially designed to recover spilled oil and oil wastes on the water surface. The oil recovery vessel mainly consists of oil recovery device, recovered oil storage tank, lightering device, mechanical power system and waste collection device.

The requirements for the function oil recovery vessel depend on the sea area where it is to be applied and the sea environment. The oil recovery device, lightering device and storage capacity should be matched. The power device should have quick response capacity, can sail at a low speed during the oil recovery and should have a certain towing capacity.

According to the shipbuilding specifications, the oil recovery vessel can be divided into vessels suitable for offshore waters, vessels suitable for port areas and vessels suitable for sheltered waters.

The performance and characteristics of an oil recovery vessel are determined by the performance and characteristics of the oil recovery device. Therefore, based on the performance and characteristics of the oil recovery device, the oil recovery vessel can be classified into four types: pumping type, adsorbing type, hydrodynamic type and weir type.

7.1.3.1 Pumping type oil recovery vessel

The pumping type oil recovery vessel is a kind of ship that uses pump to pump the oil slick on the water surface and has two types: floatation pumping type and vacuum pumping type. The working principle of oil recovery device of the pumping type oil recovery vessel is similar to that of vacuum skimmer. When the vessel advances, the oil guiding arm in front of the vessel is used to guide the spilled oil on the water surface to the oil catcher and then the suction head is used to pump the spilled oil into the storage tank. Therefore, it is suitable for being used in calm water and the recovery efficiency is significantly decreased when there is a large wave, and is particularly suitable for recovering low-viscosity oil with relatively thick oil layer. This oil recovery vessel is simple in structure and low in cost, and is suitable for being used in port waters.

7.1.3.2 Adsorbing oil recovery vessel

Adsorbing oil recovery vessel is a kind of ship that uses the rope, belt and barrel made of lipophilic material to adsorb the spilled oil and thus complete the oil recovery. Its working principle is the same as that of a rope skimmer, belt skimmer, chain-brush skimmer and barrel skimmer as set forth above. This oil recovery vessel is simple to manufacture and is small. It is suitable for recovering various types of oil in the waters of port areas and can also work with the mother ship in the offshore waters.

7.1.3.3 Hydrodynamic oil recovery vessel

Hydrodynamic oil recovery vessel is a kind of ship that uses the hydrodynamic force produced by the rotating conveying belt to guide the spilled oil into the oil catcher, takes advantage of oil-water specific gravity difference to make the spilled oil float on the water surface again and pumps it into the storage tank after it accumulates to a certain thickness. Based on the deployment of rotating conveying belt, it can be divided into immersion type and floating type. Its working principle is similar to that of dynamic inclined plane (DIP) skimmer and the non-adsorption belt skimmer. This kind of ship can be built into a large offshore oil recovery vessel and can be added the emergency command functions such as deploying and towing oil boom, spraying oil spill dispersants and fire and rescue. It can also carry other types of oil skimmers to work.

7.1.3.4   Weir type oil recovery vessel

The weir type oil recovery vessel is a kind of ship that uses the weir or oil-sweeping arm to introduce the oil-water mixture and then separates the oil from water. When the vessel advances, the spilled oil on the water surface enters into the oil-water separation tank through the adjustable weir or oil-sweeping arm and is naturally separated from the oil under action of the oil-water specific gravity difference. The separated water is then drained from the tank bottom, thus completing the oil recovery.

7.1.3.5   Conclusion

The above oil recovery vessels can also work together with the oil booms or work independently. If the users want to fully play the role of oil recovery vessel, attention should also be paid to the following: (1) The weather and marine conditions on the oil spill site. The recovery efficiency and recovery rate of oil recovery vessel are related to the weather and marine conditions. The sailing capacity of the oil recovery vessel is also affected by the weather and marine conditions. And the navigation area and wind-resistance level of the oil recovery vessel should be considered. (2) The type of spilled oil that the oil recovery vessel is suitable for varies along with the type of oil recovery device. Hence, the oil recovery vessel with different oil recovery devices should be selected according to the type and scale of the spilled oil. For example, the belt oil recovery vessel should be chosen for recovery of the large amount of spilled oil with high viscosity; and the hydrodynamic and weir type oil recovery vessel should be chosen for recovery of the spilled oil with medium and low viscosity. (3) The emergency response capacity of oil recovery vessel should also be considered during oil recovery. Whether the oil recovery vessel can play its role mainly depends on whether it can reach the scene before large-area diffusion of spilled oil and track the spilled oil. The storage capacity of oil recovery vessel should also be considered during the oil recovery. During offshore oil recovery, it is also necessary to consider its cooperation with the mother ship.

7.1.4   Oil Absorption Material

Using oil absorption materials to adsorb the spilled oil is a simple and effective method for treating marine oil spills and is suitable for shallow seawaters, shore and relatively calm waters [188]. Adsorption refers to the attachment of oil to the materials' surface while absorption refers to absorbing the oil into the materials. As an alternative, oil absorption material is mainly used in areas where the use of oil skimmers is difficult or restricted, such as shallow seawaters or the sea area where the oil skimmer is unable to get in.

The oil absorption material is generally lipophilic and hydrophobic, and has good absorption properties. The oil absorption capacity is usually measured by the maximum oil absorption capacity per gram of oil absorption material. That is, the ratio of the weight of the oil that can be absorbed to its own weight. The higher the value, the lighter the material and the better the oil absorption performance. Generally speaking, the capacity of oil absorption materials to absorb the high-viscosity oil is greater than that of low-viscosity oil. The oil absorption materials must be salvaged to the shore for treatment after oil absorption. If it sinks to the bottom after oil absorption, it is likely that oil contaminants and the material itself will exert serious impact on benthic organism of the waters.

7.1.4.1 Classification of oil absorption material

Based on different classification standards, the oil absorption material can be classified into different types. According to the type of material, oil absorption material can be divided into: natural organic oil absorption material, natural inorganic oil absorption material and synthetic oil absorption material. Based on the oil absorption mechanism, the oil absorption material can be divided into occlusion type, gel type and occlusion-gel composite type.

1. Classified by material type

(1) Natural organic oil absorption material

Such materials include straw, wheat straw, wood chips, grass ash, reeds, peat, crushed corn cobs, feathers and other charcoal matrix products, which are characterized by being cheap and readily available. These materials can generally adsorb oil of 3 to 15 times their own weight. But there will be a large amount of thin oil re-flowing out during oil recovery and some materials absorb water while absorbing oil, so they are easy to sink. There are also many other organic oil absorption materials (such as sawdust, etc.) which are bulky particles that are difficult to recover after being sprinkled on the water surface. Therefore, some measures must be taken to overcome their sinking for recovery. Another disadvantage of this type of material is that it is not easy to store.

(2) Natural inorganic oil absorption material

Such materials include clay, perlite, vermiculite, glass fiber, sand or volcanic ash, etc., which can generally adsorb oil of 4 to 10 times their own weight, and the adsorption capacity is weak. Similar to the natural organic oil absorption material, it is cheap and readily available.

(3) Synthetic oil absorption material

Such materials include polyurethanes, polyethylene, nylon fibers, etc., most of which can absorb oil 70 times their own weight and some of which can be washed and reused. Although such materials have a high oil absorption rate and large absorption capacity, they also bring about some negative effects, such as polluting the environment, which is characterized by various harmful gases and pollution to the atmosphere during incineration. Since they are generally difficult to degrade or are non-degradable materials, they will contaminate soil and groundwater resources during landfill [189].

It can be concluded from the above that the natural oil absorption material has the advantages of rich raw materials and low price, and being biodegraded, and the disadvantages of small oil absorption capacity, poor oil-water separation performance and having the possibility to absorb water and sink and further render difficulties to oil recovery. The advantages of synthetic oil absorption materials are that they have large oil absorption capacity, good oil-water separation performance and good oil retention performance, while the disadvantages are that they are high in cost and most of them cannot be biodegraded and will therefore cause pollution. In the actual oil recovery, the suitable oil absorption materials will usually be selected according to the range and extent of oil spill, oil viscosity and the surrounding environment.

2. Classified by oil absorption mechanism

The occlusion-type oil absorption material is often a substance having a loose porous structure and absorbs oil through a capillary phenomenon. The inorganic oil absorption material is mainly of occlusion type based on the oil absorption mechanism, such as, cotton, peat bog, kapok, pulp, activated carbon and the like. The gel-type oil absorption material is mostly low-cross-linking lipophilic polymer such as a high oil-absorbing resin. The oil absorbed is stored in the network space inside the resin. The lower the cross-linking degree of the polymer, the larger its networked space and the greater the oil absorption and storage capacity. However, the lowering of the cross-linking degree will lead to increase of solubility of polymer in oil. The mechanism of occlusion-gel composite oil absorption mechanism is a combination of the above two mechanisms [190].

7.1.4.2   Development of oil absorption material

The oil absorption material has undergone an evolution process from the traditional type to the high performance type. At first, people used porous materials such as straw, sponge and clay to absorb oil. The result was not satisfactory. This kind of materials has small oil absorption rate and does not absorb a large total amount of oil. And it often absorbs water while absorbing oil and it is impossible to separate oil from water. The key point is that these materials have poor oil-retention performance after oil absorption and are prone to have oil leakage. These shortcomings limit their application in the actual oil spill treatment.

An important development of oil absorption material is synthetic high oil-absorbing resin. The oil absorption mechanism of this resin is similar to the water absorption mechanism of high water-absorbing resin. That is, the hydrophilic group in the high water-absorbing resin is replaced by lipophilic group, making the original high water absorbability converted into high oil absorbability. High-absorbing resin uses lipophilic monomer to form a networked structure through moderate cross-linking, and utilizes the interaction force of lipophilic group with oil molecules as an oil-absorbing driving force. Its advantages are that it has high oil absorption rate, good oil-water selectivity, and greatly improved oil retention performance, and it is not easy to leak oil.

The synthesis of high oil-absorbing resin is basically divided into three categories: (1) The polymer prepared by cross-linking suspension polymerization of alkyl methacrylate or alkyl styrene monomer. (2) The copolymers prepared by cross-linking copolymerization of monomer with a vinyl group, such as vinyl acetate-vinyl chloride copolymer and alkyl styrene-alkyl acrylate copolymer. (3) Graft copolymers prepared by graft copolymerization of the monomers such as alkyl methacrylate and styrene with natural cellulose. This kind of high oil-absorbing resin has the characteristics of natural polymer and synthetic polymer, and has a good development prospect [191].

Foreign countries, especially the United States and Japan, have an early start of research on high oil-absorbing resin and have achieved excellent results. For example, the U.S.'s Dow Chemical Company prepared a non-polar high oil-absorbing resin by divinylbenzene cross-link with alkyl ethylene as monomer in 1966; Japan's Mitsui Petrochemical prepared a polar high oil-absorbing resin by divinylbenzene, ethylene glycol diacrylate and glycol dimethacrylate cross-linking and suspension polymerization with alkyl methacrylate and alkyl styrene as the basic monomers in 1973; the vinyl acetate-vinyl chloride copolymer prepared by Japan's Murakami Corporation in 1989 through cross-linking with triisopropylphenyl peroxide is also a polar high oil-absorbing resin. NIPPON SHOKUBAI Co., Ltd. used acrylic monomers as raw materials and prepared a medium polar high oil-absorbing resin with a long-chain alkyl group on its side chain in 1990 [191]. In China, research on high oil-absorbing resin started later but developed rapidly in recent years. For example, Guo Sanwei et al. synthesized high oil-absorbing resin by emulsion polymerization, and studied the effects of types of emulsifier, initiator dosage and cross-linking agent dosage on oil absorption performance of the product and the particle size and surface topography of oil-absorbing resin [192]; Wang Shulei et al. synthesized a high oil-absorbing resin by suspension polymerization with n-Lauryl acrylate and styrene as monomers, ethylene glycol acrylate as cross-linking agent and ethyl acetate as pore-foaming agent and investigated the effects of comonomer composition, cross-linking agent dosage, initiator dosage and dosage of pore-foaming agent on the oil absorption performance of oil-absorbing resin [193]. It can be seen from Table 7.1 that the initiator dosage has an important influence on the oil absorption performance of the resin when synthesizing the high oil-absorbing resin.

In addition to high oil-absorbing resin, the research on oil absorption material at home and abroad still mainly focus on synthetic materials such as polyethylene, polyurethane foam, and polystyrene fiber and some new natural oil absorption materials. A research institute in Canada tested the absorption effect of polyurethane oil absorption material on crude oil with different viscosity and found that it takes a short time for half of the polyurethane oil absorption material to reach saturation when it absorbs the crude oil with higher viscosity and the absorption capacity is dozens of times the capacity of absorbing low-viscosity crude oil. When the viscosity is smaller than 3000 centipoise, the oil absorption rate can reach 47 g/g. In 1999, Japan discovered that cedar bark has a strong oil absorption capacity. The scientific research carried out experiments and found out that the oil absorption capacity of cedar bark increases with the increase of bark fiber tissue size. When the fiber size is 600 g, the oil absorption rate can reach 10.3 g/g, and the dry bark absorbs twice as much as the wet bark. For the low- and high-viscosity crude oil, the oil absorption rate is 10 to 15 g/g and 15 to 20 g/g, respectively. Hence, Japan has put it into production in 2001 [188].

In recent years, developed countries have made in-depth research on new biodegradable oil absorption material and developed corresponding products. The biodegradable oil absorption material can be biodegraded after oil absorption and the absorbed oil is processed, without polluting the environment. Spill-Sorb is a non-toxic and natural oil absorption product produced by a Canadian company. It is biodegradable under natural conditions and is composed of 100% organic matters. The weight ratio of oil absorption is 1/7.955 to 1/8.091 and the oil absorption rate varies with the uniformity, composition and specific gravity of the oil, and is also affected by the external temperature and oil absorption duration. Aminoplast Capillary adsorbent produced by another Canadian company, Landmark, has a unique nitrogen-containing structure that can absorb oil up to 60 times its own weight and can be slowly decomposed into the nutrients needed by plants under the action of light and air in the natural environment. Prozorb is a highly efficient and degradable oil absorption material produced by a U.S. company. It can absorb oil up to 60 times its own weight. Since ProZorb contains many nutrients required by microorganisms, it can accelerate the natural degradation of microorganisms. Hence, it will not cause secondary pollution to the environment. Sea Sweep, another degradable oil absorption material invested and developed by a company in Denver, Colorado, USA, is prepared by a thermal decomposition with the saw residue as raw material. Such material can always float on the water surface and is easy to recover from the water surface. In addition, the studies found that the hydroxy butyrate-valeric copolymer prepared by microbial fermentation of polysaccharides and extraction from the cell metabolites with organic solvent not only has the basic characteristics of high-molecular compound, but more importantly, has biodegradability and biocompatibility, and will be new type of oil absorption material with broad prospects.

7.1.4.3 Conclusion

The oil absorption material is easy to carry and operate and can be used to well handle the very thin oil layers. At present, how to enhance its use frequency, its oil absorption effect and its capacity to absorb high-viscosity oil is the hot issue in the research of oil absorption material. In addition, with increase of people's consciousness about environmental protection, the development of biodegradable oil absorption material has received increasing attention. As people continue to deepen their research and development, various low-cost, high-performance biodegradable oil absorption materials will continue to be introduced into the market and become the dominant oil absorption material.

As far as the current situation is concerned, once oil absorption material is used to handle the oil spill on the sea surface, it must be cleaned up and disposed of properly or reused to avoid new environmental pollution. It is believed that in the near future, the wide application of biodegradable oil absorption material will play a positive role in avoiding the occurrence of secondary pollution and protecting the environment.

7.2 Chemical Treatment Method

At present, the physical methods such as oil containment and recovery with oil boom, oil skimmer and oil absorption material are mainly used in large-scale oil spill accidents. However, the physical methods are suitable for recovering spilled oil with a certain thickness. The light oils like gasoline, kerosene and diesel oil have low density, small viscosity and diffuse fast and the formed oil slick is very thin, so it is often difficult to recover it with physical methods. At this time, the relevant emergency department will consider incinerating these difficult-to-handle spilled oils, or changing the existing form of the spilled oil in marine environment with some chemical preparations, thus reducing its contamination level in seawater. Therefore, the physical methods are usually used in combination with on-site combustion and treatment with chemical preparations to minimize the environmental damage of oil spill.

7.2.1 On-Site Combustion

7.2.1.1 Concept of on-site combustion

On-site combustion refers to the burning of oil floating on the sea surface of an oil spill site. This technology is mainly used to deal with large-scale marine oil spill accidents. Meanwhile, the emergency response personnel gradually realize that this method can quickly, safely and effectively handle oil spill accidents due to offshore oil exploration and production, and accidents of marine oil pipelines and tankers and can also be applied in oil spill accidents happening in a specific river environment. In recent years, although the equipment and technology used for on-site combustion have been greatly developed, the combustion of spilled oil cannot replace the oil containment and recovery and natural disappearance of spilled oil. Although the emergency department will use the traditional oil boom and oil skimmer to contain and mechanically recover the spilled oil after occurrence of oil spill accidents and can safely and effectively give play of the role of oil boom and oil skimmer, on-site combustion (or the use of oil spill dispersants) in some cases may be the only way to fast and safely eliminate a large quantity of spilled oil [194].

1. Development of combustion technology

The earliest case of using on-site combustion technology is the "Torrey Canyon" accident in Cornwall, England in 1967. At that time, in order to quickly burn the spilled oil, the emergency personnel used many combustion methods including various bombs, rockets and other burners. Due to the rapid volatilization rate, severe marine conditions, drastic ignition mode and lack of containment measures, the combustion that time was not successful. In the following years, no significant progress on the combustion technology to handle oil spills was made. In the oil spill accidents that occurred in the 1970s and early 1980s, limited success was achieved only when burning an absolutely thick oil layer. By the mid-1980s, combustion technology gradually matured due to proper containment measures and ignition technology. The United States and Canada have carried out experiments on aero-ignition combustion technology in 1978, 1979, 1983, and 1986, respectively, and experiments on fire suppression containment and combustion of spilled oil in 1981, 1986, and 1988, respectively. In 1988, 3M Company of the United States and SINTEF of Norway conducted a combustion experiment on containment of 2t crude oil with 90m 3M fireproof oil boom in open water. The fireproof oil was towed in a U-shape and ignited by a helicopter. 95% of the contained spilled oil was burned in about only half an hour. On March 25, 1989, the second night after the Exxon Valdez tanker was stranded on Prince William Strait, Alaska. About 60 to 120 tons of crude oil contained by 3M fireproof oil was burned and the combustion efficiency of crude oil was about 98%.

Texas, Alaska, Norway, and Canada have carried out a large number of experiments on on-site combustion. This technology is still in the experimental phase in other countries and regions, such as the Canadian Environmental Protection Agency's near-shore experiments in Newfoundland; experiments planned by the United States and Canada on the Gulf of Mexico; experiments jointly carried out by the United States Coast Guard and the Russia Administration of Maritime Anti-pollution, Search and Rescue in Russia. China also conducted the diesel oil combustion test during the oil spill emergency drill in Shanghai, and achieved limited success.

2. Relevant regulations

In foreign countries, combustion of spilled oil needs to be approved. However, the Marine Environment Protection Law of the People's Republic of China has no explicit stipulation of such aspect. On-site combustion is only a commonly chosen measure in the oil spill emergency decision making and no fixed theory has been formed. However, such measures have been widely accepted by foreign environmental protection agencies. In particular, when dealing with oil spill accidents occurring in the waters of the contiguous zone, the administrative decision-making department prioritizes the emergency plan of using such measures. On the basis of considering the ability of the emergency team in such region, the on-site commander approves the use of on-site combustion technology. In the United States, the requirements for approving the use of on-site combustion technology in the territorial waters vary from state to state. In several states with long coastlines, guidelines for the use of on-site combustion technology and operational guidelines for emergency teams have been developed to avoid the loss of approval of and the best geographical location and timing for use of on-site combustion technology due to administrative decision making.

7.2.1.2 Basic theory of on-site combustion

1. Requirements of ignition and combustion for oil and water

If the oil layer is thick enough, the ignition area is large enough and the ignition temperature is sufficient to make the oil evaporate, then most of the crude oil and refined oil on the sea surface can be burned. In order to prevent the heat loss during the heat transmission from the oil layer to the water below the oil layer, the minimum thickness of the oil layer should be less than 2 to 3 mm. All the results of fresh crude oil combustion test showed that the combustion would stop quickly when the thickness of the oil layer was reduced to about 1-2 mm. If the oil layer is too thin, the water is cooled quickly and the heat loss is fast so that the oil temperature is lowered below the oil temperature of evaporation, then the conditions of oil combustion are not satisfied. If evaporating the diesel oil of which the content is less, the weathered and emulsified crude oil and C-grade fuel oil, the minimum combustion thickness may reach 8 to 10 mm. The key to maintaining combustion is to maintain this thickness using fireproof oil boom or using natural conditions such as ice or shoreline to contain the spilled oil. Tests have shown that the most effective oil spill ignition device is the one that transfers as much heat as possible to the oil layer and produces very little fluctuations. Using aircraft to deploy gel fuel is one of the most efficient, fast, safe, and least expensive methods of ignition.

The most suitable environmental conditions for on-site combustion are that the wave height is less than 1 m and wind speed is less than 7 m/s. The time range suitable for combustion can last from a few hours to a few days. If the spilled oil is rapidly diffused due to wind and marine conditions, then the opportunity for on-site combustion will be lost. Even if it is possible to control a part of the spilled oil, then the spilled oil may have been weathered and cannot be ignited and burned. Studies have shown that spilled oil with water content of 50% to 70% can also be ignited and burned, but it is extremely difficult to ignite and burn the spilled oil with water content of only 10% to 20%.

2. Combustion rate and efficiency

Combustion rate of spilled oil on the water surface refers to the reduction rate of oil layer thickness or the reduction rate of spilled oil volume within unit area. Combustion efficiency refers to the percentage of of oil that disappears from the water surface due to combustion. The combustion rate is mainly determined by the oil layer thickness at the beginning of combustion and after the combustion. The combustion experiments show that the combustion efficiency is generally higher than 90%, and some combustion experiments also showed that the combustion efficiency can be as high as 98% to 99%.

7.2.1.3   Possible harsh conditions

In many cases, the use of combustion technology is a very effective way to eliminate a large number of spilled oil due to offshore oil exploration, loading and unloading and leakage of offshore oil pipelines and tankers. On-site combustion undergoes three phases: (1) Containment of the spilled oil to be burned. (2) Combustion of contained or naturally contained spilled oil. (3) Control and suppression of accidental fires.

When using fireproof oil boom and the aero-ignition system to control the collected oil for combustion, care must be given to some key elements: (1) Maintain a safe distance between the spilled oil in combustion and the oil spill source. (2) Control and burn the spilled oil at downstream of the oil spill source. (3) Immediately contain and burn the spilled oil at the downwind of the blowout or damaged tanker. (4) Collect, accumulate and ignite the already diffused oil. (5) Control and burn the spilled oil at the shoreline intercept.

The spilled oil has a relative thickness in the following cases, for example the wind blows the spilled oil to a barrier similar to a shoreline, floe-ice, etc.; at the downwind of large ships, islands, etc.; a lot of oil is instantly leaked on the relatively calm water surface, forming a temporary balanced oil layer several millimeters thick. Once the spilled oil is ignited, the thermal force generated by the combustion causes the spilled oil on the edge to flow towards the fire region, thereby increasing the thickness of the oil layer and promoting combustion.

On the oil spill site, in order to reduce the danger caused by fire and explosion, the flammable substances in oil are sometimes naturally volatilized. On the other hand, if the immediate oil containment may lead to fire and damages to the ship, facility or equipment nearest to the oil spill site, second ignition of oil spill source and more damages may be caused. In general, using a fireproof oil boom can prevent the diffusion of spilled oil or intercept the oil slick to be burned so as to isolate the burning oil from the nearby ships, docks and facilities. The fireproof oil boom can also be used to control the burning oil so as to facilitate the fire-fighting with foam.

7.2.1.4   On-site combustion system

Successfully burning the spilled oil on the water surface requires two prerequisites: first, the means to increase the thickness of spilled oil layer; second, a safe ignition method. The spilled oil can be naturally contained by shoreline, floe-ice or other objects. Fireproof can also be used to control the oil, increase the thickness of spilled oil layer and achieve on-site combustion or combustion far away from the oil spill resource.

1. Fireproof oil boom

As mentioned in the first section, the floatation of fireproof oil boom is generally made of heat-resistant steel material and has extremely high fire resistance and tensile properties. As early as in the 1970s, Dome Petroleum of Canada developed a fireproof oil boom made of stainless steel, which is expensive and cumbersome. In the early 1980s, Shell Petroleum developed a fireproof oil boom with oil spill emergency response capability under icebreaking conditions. After that, the latest generation of fireproof oil boom developed by 3M Company of the United States is firmer and has longer fire resistance. The oil boom consists of ceramic fibers and heat-resistant floating core. Its structural design and operation are very similar to those of the conventional oil boom and passed the 48-hour combustion test. The Oilstop Company of the United States has recently developed a smaller inflatable fireproof oil boom. In the 1990s, China also began to develop fireproof oil booms and has achieved some success.

2. Ignition system

Once fireproof oil boom is used to control the spilled oil, the controlled oil must be burned and the ignited device should be selected according to the type of the spilled oil to be burned. It is sufficient to use one ignition device for large-scale oil spills, layers with continuously oil spillage or the spilled oil already controlled by the fireproof oil boom. If spilled oil has been weathered or agitated by wind waves, multiple ignition devices may be required to ignite successfully. However, in order to cause the ignition device to transfer sufficient heat to the oil slick and evaporate and burn the spilled oil, it is very important to ensure the minimum disturbance caused by the ignition device to the oil layer in any cases.

In some cases, the ignition source may be a rag impregnated with oil. Of course, there are some special ignition devices, such as the helicopter-thrown igniting torch device produced by Simplex of the United States, which sprays a large amount of solidified fuel drips to the oil layer surface for combustion; two manual ignition devices developed in Canada, Pyroid and Dome ignition devices; a standard ignition device mounted on a helicopter can spray a lot of polystyrene pellets and the chemical heat released by spraying these polystyrene pellets can ignite the oil. In addition, some countries use commercial lasers as ignition devices which are effective in some calm water conditions.

7.2.1.5 Safety notes

Personal safety is of the utmost importance in all oil spill emergency response. A safety plan must be prioritized when developing any combustion program, on-site combustion or special combustion plan. Commanders must consider the potential explosion hazard of on-site combustion in a comprehensive manner. The hazards upon assessment include at least the following factors:

the combustion scale and duration predicted according to the estimated quantity of spilled oil to be burned, combustion rate and the combustion control measures; changes or possible changes in the combustion location that should be considered due to changes in tidal current, equipment failure, or towing of oil boom; it is possible to lose control of the movement and duration of oil combustion from the combustion control zone due to accidental or emergency response; it should be considered that some fixed facilities or locations such as anchored ships, docks, bridges, etc. will be directly exposed to the flame of oil combustion or concentrated areas of combustion products.

Anyone involved in on-site combustion should be professionally trained and capable of dealing with urgent situations; they should be entitled to the right to be protected by the provisions of the national occupational safety and health management, including the provisions on handling of toxic and hazardous waste and its emergency response. The relevant national regulations are also applicable and the relevant safety procedures should be formulated, including the use of waterproof oil boom with sufficient length to tow the oil boom, the use of aircraft for reconnaissance, that the ship should be located at the upwind, monitoring of the air quality in the people's living areas, and provision of adequate personnel and safety equipment.

7.2.1.6 Environmental precautions

The most relevant concern about on-site combustion is the smoke produced by combustion. The main products of oil burning are: soot granular, flue gas carbon monoxide, carbon dioxide and sulfur dioxide, unburned hydrocarbons and the remaining residue of combustion. Although soot only accounts for a small proportion of combustion products, people are more concerned about it since it is clearly visible. Since the combustion is confined to the area controlled by fireproof oil boom, the typical flame diameter is not more than 30 meters, and the soot formed by combustion is about 10 nautical miles from the burning site. At the beginning, the soot is concentrated and then gradually dispersed and turned into gray smoke cloud. In the absence of wind, the diffusion rate will gradually decrease. If it is slightly windy, the smoke generated by burning the spilled oil of the same quantity will be diluted, and it is difficult to perform standard test using the same testing equipment.

The researchers compared the combustion of oil controlled by the oil boom and the combustion in the oil pool and found that the content of carbon monoxide and nitrogen dioxide in the smoke generated by on-site combustion is very low. The cycloparaffin about which people are concerned is also burned to a low level, reaching the national atmospheric standard [179].

For a specific oil spill accident, comparison of the air impact on the oil combustion region and the potential impact on the non-combustion region must be made. If no combustion measures are taken for the oil spill accident, the volatile components in the spilled oil will be evaporated into the atmosphere. Unburned oil may diffuse into shallow waters and environmentally sensitive areas. In addition, spilled oil will also affect the shoreline or deposit onto the seafloor. In some cases, the negative impact of combustion is small, and the combustion products (temperature, air pollution) have a short impact time and little damage.

Regardless of the type of combustion, the number of residues produced by oil combustion is much less than that of original spilled oil and generally only accounts for 2% to 5% of the spilled oil. Due to cooling of the water, the residue produced by combustion solidifies, has high viscosity and accumulates under the action of wind and current, making it easy to contain with an oil boom. In any case, in order to ensure safe on-site combustion and reduce environmental impact, the scale of containment and combustion should be evaluated and the on-site combustion should be carefully planned. Only in this way can the combustion be prevented from adversely affecting people's concentrated areas and natural resources, and from doing harm to the staff, ships, equipment and shoreline of the site.

7.2.1.7   Conclusion

In short, on-site combustion can quickly deal with the large oil spill accidents and reduce many subsequent disposal measures. Therefore, it is one of the technical measures to deal with large oil spill accidents. Therefore, the promulgation of relevant work guidelines and policies on the use of combustion technology is necessary. These policies should clarify the sea area where combustion technology can be used to increase the success rate of using on-site combustion technology.

7.2.2   Chemical Preparations

The chemical preparations can change the physical and chemical properties of oil and make the oil easy to be removed. Hence, they have the advantages of being easy to operate and having good treatment effect and low residual oil content. The commonly used chemical preparations for oil spill treatment at present are dispersant, oil coagulant and oil-collecting agent.

7.2.2.1   Oil spill dispersant

1. Composition and mechanism of dispersant

The dispersant consists of surfactant and solvent. Surfactant is the main component of the dispersant. It uses lipophilic group and hydrophilic group to change the action of oil-water interface, thus greatly reducing the surface tension of an oil slick. Except for a very small number of anionic surfactant, the surfactant used in the commonly used dispersant is mostly non-ionic. The solvent is generally alcohol and hydrocarbon, and sometimes water can also be used to dilute the oil and lower the freezing point of oil and can reduce the viscosity of surfactant and oil to facilitate emulsification [195]. The compatibility of the solvent with the oil directly affects the dispersing ability of the dispersant and, therefore, the dispersant has good oil solubility.

In addition to dispersing the oil into small oil droplets, the dispersant can also prevent the oil droplets from re-polymerization. This is because surfactant stays at the oil-water interface for a long time, just like a barrier between oil droplets, which prevents irregular collisions between oil droplets and thus prevents re-polymerization of oil droplets.

2. Advantages and disadvantages of dispersants

The dispersant has the advantages of being non-flammable, stable in chemical properties, non-corrosive to metals, and safe to transport. The emulsion after use of the dispersant is more favorable for biodegradation. Therefore, it accelerates the elimination of spilled oil from the ocean while increasing the surface area of spilled oil, reducing the evaporation of its volatile components, preventing the formation of tar balls and reducing the risk of acute toxicity of spilled oil and potential danger of fire, providing time guarantees for emergency response. Meanwhile, the dispersants can reduce the oil viscosity, thereby reducing the chance that oil adheres to the deposits, organisms and shorelines. Compared with other emergency measures, the dispersants have a quick effect. In harsh climates, the mechanical measures are limited while strong winds and rapid current can effectively improve the effectiveness of the dispersant so that the use of dispersant can process a large area of spilled oil in a short time. Moreover, the dispersant can be quickly sprayed onto the sea surface by aircraft. Hence, spraying the dispersant is the most suitable choice for dealing with an oil spill in remote areas and areas difficult to access.

Disadvantages of the dispersant are that it is only effective for the oil with medium and low viscosity; various countries have special regulatory restrictions on its use since there are no sufficient studies on the bio-toxicity of a dispersant, especially in terms of its direct and indirect potential harm to the bio-chain. In addition, its dispersion efficiency to spilled oil is affected by many factors, including viscosity of oil, weathering degree of oil, salinity and temperature of seawater, oil slick thickness and properties of dispersant itself, etc. Therefore, the use ratio of the dispersant is also different in different cases.

3. Types of dispersants

In accordance with Oil Spill Dispersant Technical Regulations (GB 18188.1-2000), the oil spill dispersants can be classified into ordinary dispersant and concentrated dispersant according to the proportion of surfactant and solvent.

In general, the surfactant content of the ordinary dispersant is only 10% to 20%, while the solvent content is as high as 80% to 90%. The ordinary dispersant has a strong ability of oil dissolution and has better effect of treating high-viscosity oil and weathered oil. The surfactant of the concentrated dispersant is mostly fatty acid extracted from natural oils and fats, and the sorbitol extracted from sugar, corn and sugar beet and is basically non-toxic. Its surfactant content is high and is usually 40% to 50% so that it can rapidly disperse the spilled oil while its solvent content is low, accounting for 50% to

60%. The concentrated dispersant has high oil dispersion efficiency but poor effect in treating high-viscosity oil [196]. In order to improve and enhance the oil dispersion effect of the concentrated dispersant, a small amount of wetting agent and oxidizing are sometimes added thereto [195].

In addition to ordinary dispersant and concentrated dispersant, there is also another type of dispersant: diluted dispersant. Such dispersant only contains surfactant and uses seawater as the solvent when being used. This dispersant should be diluted with seawater in a ratio of 1:10 before use. It can be used directly on beaches and to rub the shore walls and reefs.

The dispersants used in China are mainly domestically produced ordinary dispersant and imported concentrated dispersant, and the diluted dispersant is rarely used.

4. National standards for dispersants and the regulations on their use

The national standard Oil Spill Dispersant–Technical Regulations (GB18188.1-2000) and Oil Spill Dispersant–Application Criteria (GB18188.2-2000) were implemented as of October 1, 2001.

Oil Spill Dispersant–Application Criteria specifies that the oil spill dispersants can be used in the following cases:

(1) The oil slick on water surface or the oil spilled in an accident may move to coastal, aquaculture and other waters sensitive to oil spills, threatening the commercial, environmental or comfort benefits and can neither be dispersed by natural evaporation or by the action of wind, waves and current nor can it be contained or recovered with physical methods before arriving in the above sensitive areas.

(2) For the spilled oil which is difficult to handle by physical and mechanical methods, the total damages caused by oil dispersion into the water body with use of the oil spill dispersant is less than the damages caused by leaving the oil on the water surface and not treating it.

(3) When the oil spill occurs in non-portal waters with a water depth greater than 20 meters, the oil spill dispersants can be used first and then reported to the competent authority.

(4) The type of oil slick on water surface or oil spilled in an accident and the water temperature are suitable for chemical dispersion, and the environmental conditions such as meteorology and marine conditions are suitable for dispersion with oil spill dispersants.

(5) In the case that an event of force majeure such as fire, explosions, etc. that endanger human life or facility safety has happened or may happen.

Oil Spill Dispersant–Application Criteria also specifies that the use of oil spill dispersants is prohibited in the following cases:

(1) When the spilled oil is volatile light oil such as gasoline and kerosene, or thin oil slick exhibiting rainbow characteristics.

(2) The spilled oil is the type of oil with high wax content and high pour point and that is difficult to disperse chemically.

(3) The spilled oil does not show a flow state at ambient water temperature or is formed into a thick piece of water-in-oil emulsion after several days of wind erosion.

(4) The oil spill occurs in the closed shallow water area or in calm waters.

(5) The oil spill occurs in areas of freshwater sources or in the areas that have significant impacts on aquatic resources.

5. Research progress on dispersants in China

The development of dispersants can be dated back to the 1960s and has gone through three generations. The first generation is ether dispersant which uses highly toxic anionic surfactants, such as alkylphenol polyoxyethylene. This dispersant is easy to penetrate and dissolved and will hence cause adverse impacts to the marine ecological environment. The second generation is ester dispersant which is non-ionic. The raw materials it used are mostly natural animal and vegetable oils which are less toxic and have good emulsification effect, such as polyethylene glycol oleate and polyoxyethylene sorbitan oleate. The third generation is concentrated dispersant. Its raw materials are sorbitol and fatty acids which are derived from agricultural and sideline products. Its solvent is synthetic polyethylene glycol which is less toxic, only 1 of that of the first generation [197]. Meanwhile, since its active ingredients include emulsifier, wetting agents and oxidants, the activity has been greatly improved, so its efficiency is the highest [188]. In addition, there is a type of dispersant of which the surfactant is first co-dissolved with oil and then co-dissolved with water. Since the particle size of the oil droplets produced is very small, oil slick will rapidly diffuse with the natural movement of water body. So the dispersant does not have to be mechanically mixed during use, avoiding the danger of operation when the ship sails in harsh marine conditions. It can be directly sprayed by the aircraft. Some of the oil spill dispersants currently developed are extracted from vegetable oil, sugar, sugar beet and other raw materials. The solvent is a certain synthetic agent, and has low toxicity. Zongting Wang et al. developed a concentrated dispersant with low toxicity and low freezing point. Such dispersant has good emulsification performance and the required amount is small during use. It is suitable for low temperature environment of minus 20 degrees [198]. Hao Ling developed a series of high-efficiency marine oil spill dispersants and determined the optimal formulation of the solution [199]. Meanwhile, they used experimental methods to study the impacts of crude oil type, catalyst-oil ratio, temperature and salinity on emulsification rate and the biodegradation of dispersants, used BOD5/COD value method to measure the biodegradation of such dispersant and found that it is far much higher than 45% (see Table 7.1), meeting the requirements of industry standards. The highly efficient and environmentally friendly marine oil spill dispersants developed by [183] has more complicated synthesis process, but it has good emulsification performance and no pollution to the environment and is in line with the development trend of oil spill dispersants, laying a foundation for research and development of new dispersants.

In conclusion, the dispersant is one of the main means to deal with marine oil spills. It is generally used in combination with other measures in the

TABLE 7.1: Data of biodegradation of oil spill dispersants.

w (benzoyl peroxide) %	oil absorption ratio				
Formula	3	4	5	12	14
BOD5/COD	66	49	54	61	81

practice of dealing with oil spills, in order to better reduce the harm of spilled oil to the marine organism and marine environment.

7.2.2.2 Oil coagulant

Oil coagulant is also called oil gelatinizer or oil curing agent and is a solid powder chemical preparation capable of gelling spilled oil into agglomerates. Its mechanism of action is that it can react with oil to form oil gel. Gel is a special form of dispersion system and is composed of solid phase and liquid phase. The solid phase is formed into net structure and the liquid structure is wrapped in it. Thus the solid net structure restricts the activity of liquid particle, forming gel. In the formation of oil gel, the oil coagulant is to form a net structure and contain the large quantity of oil and a small amount of water in it to form a relatively stable oil gel. The oil content and water content of oil gel depends on the lipophilic and hydrophilic ability of oil coagulant [179]. The stronger the lipophilicity, the larger the oil content, the higher the viscosity and the greater the hardness. The stronger the hydrophilicity, the larger the relative water content, the lower the viscosity and the lower the hardness.

Oil coagulant is usually low-toxic or non-toxic, and is recovered together with oil gel. Hence, it is a chemical treatment agent effective for preventing water pollution. The advantages of using it to deal with oil spills are: (1) it can avoid the secondary pollution caused by the use of dispersants; (2) it is low-toxic or non-toxic; (3) the gelled oil is recoverable and is not affected by wind and waves; and (4) it can effectively prevent oil spill diffusion and improve the use efficiency of recovery devices such as oil booms.

In addition, the coal coagulant also has its own shortcomings, which are manifested in: (1) the price is higher; (2) oil coagulation performance is unstable (can be affected by many factors); (3) the viscosity of oil gel formed decreases as the temperature of the water body increases; and (4) the compound coagulant will be affected by the acidity and alkalinity of the water body [179]. Therefore, oil coagulant is generally used as an auxiliary means in the practice of dealing with oil spills.

Regarding the chemical composition and coagulation performance of coagulant, Sun Yunming and Chen Guohua classified it into five categories and discussed them [200]:

1. Hydrogen-bonded coagulant

Such oil coagulant mainly encapsulates spilled oil by forming hydrogen bonds between molecules, and is usually polyhydroxy or polyamino macromolecular compound, such as proteins, polysaccharides and their derivatives, and some polymers containing hydroxyl or amino groups, such as polyalcohol or polyurethane. This oil coagulant can be divided into the following categories:

(1) Oil coagulant containing oleyl alcohol group

This oil coagulant takes hydrocarbyl as lipophilic group and hydroxyl radical as the coagulating group.

(2) Oil coagulant containing oil-amino group

The oil coagulant contains amino group and lipophilic group. The amino group is often not a free $-NH_2$, but exists in the macromolecular chain link in the form of amide group or the like, such as on the peptide bond of protein. This coagulant can be divided into protein-type coagulant and amino acids-type coagulant.

(3) Polyurethane oil coagulant

The hydrolyzed urethane (containing free isocyanate) and lipophilic trichloroethane (aqueous) in the oil coagulant can rapidly solidify the spilled oil after being added into the oil and make it easy to remove the oil.

2. Chemically bonded coagulant

(1) Coordination bond coagulant

Armeen Carmamide is a typical coordination bond coagulant. During the use, it is required to spread methanol solution of long chain and let it react with the carbon dioxide in it to produce carbamate and gelatinize the spilled oil. The reaction formula is as follows: $R - NH_2 + CO_2 \rightarrow R - NH_2^+ + CO_2^-$. In addition, 30% ethanol solution of dehydroabietylamine or mixed solution with ethanol and benzyl alcohol content as 15%, respectively, may be sprayed onto the spilled oil. The usage amount is 10% of the spilled oil. And then carbon dioxide can be used to make it form oil blocks.

(2) Ionic bond coagulant

This kind of oil coagulant is made by copolymerization of copolymerisable monomer containing 10% to 25% of lipophilic groups and acrylic acid containing 10% to 60% carboxyl or other polymerisable water-soluble monomers and addition of polyvalent metal salt. This kind of oil coagulant cannot only gelatinize the spilled oil on water surface, but can also be used to plug the leakage at the beginning of the oil spill. The strong chemical bond between molecules makes this kind of oil have good oil coagulation performance. For instance, poly-oxo-aluminate modified by higher fatty acids can be used as oil coagulant and has good oil coagulation performance.

(3) Polymer cross-linked coagulant

This kind of oil coagulant mixes the polymer and cross-linking agent to coagulate the spilled oil. It is usually composed of two ingredients: liquid polymers (or solutions of polymers) containing functional groups, and cross-linking agent containing complementary functional groups. Suitable polymers are linear mono-olefins, substituted styrene, and functionalized natural polymers.

The best polymers are maleic polybutene, polyisoprene, EPDM rubber or natural rubber. Suitable cross-linking agents are compounds containing di- or polyfunctional groups, including many carboxylic acid amines.

3. Oil absorption polymer coagulant

Such oil coagulant is a copolymer containing long chain lipophilic group and coagulates the oil by oil absorption, such as a functionalized styrene-butylene block copolymer. This type of oil coagulant has great lipophilicity and the formed gel sheet does not break even in large wind and waves.

4. Long chain wax or ester coagulant

This type of oil coagulant has no oil absorption and oil coagulation properties at normal temperature. However, during the use, the oil coagulant in molten or dispersed state is mixed with spilled oil. As the temperature decreases or the solvent evaporates, the oil coagulant gradually solidifies, thereby encapsulating the spilled oil. Such as the oil coagulant prepared by adding the polyoxyethylene monoesters (30% to 70%) of unsaturated fatty acids of C10-22 into the polyoxyethylene monoester of saturated fatty acid of C8-22 and adding ethylene oxide-propylene block polymer of 0.5% to 3%.

5. Inorganic salts coagulant

Anhydrous CaSO4 powder with a particle size of 1 to 100 micron sprinkled on the spilled oil can absorb the lubricating oil, mineral oil and crude oil and other liquid hydrocarbons, and instantly form a strong block mass. After being impregnated with organic matter and dried, many porous inorganic powders can have oil coagulation performance, such as calcium oxide, diatomaceous earth, silica gel, zeolite, and the like. Oil coagulant prepared by treating calcium oxide with surfactant and styrene-butadiene latex and silicone can coagulate oil three to four times its own mass [200].

During the use, it is necessary to select the oil coagulant according to the types of spilled oil and whether the performance is suitable for the spilled oil to be treated. There are three steps to treat marine oil spills with oil coagulant: (1) Spray oil coagulant onto the spilled oil; (2) manually mix or take advantage of the wind and waves to fully mix the oil coagulant and the spilled oil so as to form gel; and (3) manually or mechanically recover the oil blocks.

Foreign countries have already conducted research on oil coagulant as early as in the 1960s and 1970s, and successively synthesized amino acids, sorbitol, protein, carboxylate and other types of oil coagulants. China began the research on oil coagulants from the end of the 1980s, and successively synthesized amino acids, sorbitol, starch, and protein type of oil coagulants. But they are far from being applied [201].

Yunming et al. modified and synthesized a series of marine oil spill coagulant by carboxymethyl etherification and esterification [182]. Experiments show that when the degree of substitution of carboxymethyl is sufficient, the performance of the oil coagulant increases with the increase of the esterification degree. When the degree of esterification is sufficient, the performance of carboxymethyl starch ester coagulant is increased with the increase of the degree of substitution of carboxymethyl. The coagulation performance is also related to the type of polyvalent metal ion in oil coagulant, the oil type and the number of carbon atoms in the aliphatic ether. The coagulation performance varies little with salinity. Namely, the oil coagulant is still effective for oil spill in freshwater surface. The coagulation performance varies greatly from oil types and the oil coagulant has stronger effect on the polar oil than the non-polar oil. The oil coagulant has better coagulation effect on the crude oil and vegetable on water surface than the fuel oil.

In the future, the main development direction of oil coagulant will be the new type of oil coagulant with fast effect, low pollution, low dosage, low toxicity, being easy to recover and less affected by the surrounding environment.

7.2.2.3 Oil-collecting agent

Oil-collecting agent is also known as oil herding agent. It is an important product for assisting the spilled oil recovery. After the liquid oil-collecting agent is sprayed on the surface of water surrounding the spilled oil with a mechanical device, the oil-collecting agent will rapidly diffuse to form a lay of monomolecular film, inhibiting the expansion of the thin oil layer on the water surface, making it shrink into a thick oil layer and reducing the area covered by the oil slick. Using oil boom, oil coagulant and oil absorption material to recover the oil at this time can significantly improve the recovery efficiency and reduce costs.

The reason that the oil-collecting agent can prevent the diffusion of spilled oil in the horizontal direction is that the surfactant it contains can greatly reduce the surface energy of water, thereby changing the tension balance on the water-oil-air-phase interface and driving the spilled oil to the thick layer [186]. This characteristic of oil-collecting agent is similar to that of oil booms. So, the oil-collecting agent is also referred to as chemical oil boom.

The United States invented the oil-collecting agent in 1971, and it was considered effective and harmless by experiments and has been approved for use. Canada, the United Kingdom, the Netherlands and other countries have also approved the use, and have a high evaluation of its practicality. Although there are several types of oil-collecting agents sold in Japan, there are still no legal provisions to approve the use of oil-collecting agents. There are also many related patents in foreign countries, such as Shell Oil Herder produced by Shell Pipeline of the United Kingdom, BP Oil Marshall produced by BP Amoco of the UK, and Corexit OC-5 coagulant produced by Exxon Chemical Americas of the United States.

In recent years, due to the aggravation of China's oil pollution, China has also begun to pay attention to the research on oil-collecting agent for dealing with oil spills. Hongshen Wang et al. synthesized five kinds of N, N-dialkylamide surfactant, and measured the state of the film formed [202]. The studies show that: (1) When the N, N-dialkylamides are formed into a film alone, the oil-collecting film is a multi-layer film, and when a suitable solvent is used to mix with it and prepare a solution, the oil-collecting film formed is then transferred into transition film or liquid coacervate film; (2) among the five N, N-dialkylamides, the longer the acyl carbon chain and the shorter the nitrogen alkyl carbon chain, the stronger the oil collecting ability; (3) four formulations formulated with N, N-dialkylamides have good oil collection capacity, and the capacity to collect crude oil is greater than the capacity to collect diesel oil. At present, the domestic QS series oil-collecting agent is prepared with N, N-dialkylamides surfactant.

Oil-collecting agent has certain characteristics and advantages in dealing with the spilled oil on sea surface and can treat the spilled oil quickly, safely and conveniently. It can be used to remove oil slicks in the areas that are not easily accessible and can be used to prevent oil slicks from expanding to the coast as long as the surface water does not flow towards the land. It can also improve the efficiency of all oil recovery systems. As with dispersants, the oil-collecting agent can be sprinkled by ships, manually or aerially.

The oil-collecting agent also has its own shortcomings which are manifested in: (1) The use of an oil-collecting agent to collect the highly volatile gasoline can easily cause a fire; (2) it is ineffective to use an oil-collecting agent when the oil layer is relatively thicker; (3) the oil-collecting agent can prevent the oil slick from diffusion but cannot fix it at a position, and the oil surface film system will move with water flow; (4) the oil-collecting agent cannot drive the oil slick off the coast or other hard-to-reach areas against the current; and (5) weathered crude oil and floating matter from industrial wastewater can interfere with the effectiveness of an oil-collecting agent. In summary, the oil-collecting agent should be used as soon as possible after an oil spill accident.

7.2.2.4   Conclusion

The chemical preparations used in the treatment of marine oil spills usually change the existing form of spilled oil in the marine environment by thickening or solidifying the spilled oil, thus reducing its degree of contamination in the seawater. However, the use of chemical preparations will also bring about some unfavorable impacts which are manifested in that they may pollute the marine environment. Therefore, the use of chemical preparations must follow the relevant regulations to avoid secondary pollution to the environment.

7.3   Biological Treatment Method

Biodegradation is used to treat the partially unrecoverable spilled oil or the spilled oil of which the existing form is changed by chemical preparations so as to finally eliminate it from the marine environment. Biological treatment is a controlled or spontaneous process that catalyzes the degradation of environmental pollutants through the use of oils as nutrient for their metabolism to reduce or eliminate oil spill pollution.

The biological treatment study of oil spills began with an early marine microbiological survey. The survey found that there are microbial populations that can grow on oil droplets and degrade petroleum hydrocarbons in sea areas that are heavily polluted by oil. These microbial populations can convert offshore oil pollutants into simple non-toxic compounds and components of bacteria [186].

Petroleum-degrading strain is such kind of microorganism and includes petroleum-degrading bacteria and fungi. Studies found that the petroleum-degrading strain usually grows at the oil-water interface instead of oil and its number is only related to the degree of oil pollution in seawater. Therefore, petroleum-degrading strain can also be used as a biological indication of oil spill pollution. At present, at least 90 kinds of bacteria and fungi are known to be able to degrade part of the petroleum component.

In China, there are many studies on petroleum-degrading strains, such as Junxiang Shi et al. isolated certain petroleum-degrading bacteria in the seawater of Zhejiang coastal areas, and proved that the degrading bacteria had obvious degradation effects on n-alkanes and the degradation rate of mixed strains was significantly higher than that of single strains [203]; the studies of Mingyu Ding et al. on 73 strains of bacteria and 10 strains of fungi isolated from the coastal waters of Qingdao showed that most of the bacteria have obvious ability to degrade oil [204]; Jinglai Zhang et al. found two highly adaptable strains (SY1 and SY2) [205]. These two strains are capable of degrading n-tetradecane, cyclohexane, benzene and xylene in oligotrophic environment, can be applied on a wide range of substrates and therefore have good development prospect in the biological treatment of marine oil spill pollution.

The petroleum-degrading strain obtains energy from the process of decomposing hydrocarbon organic matter of petroleum to maintain life activities and synthesize cellular material. The degradation process is affected by many factors, including the chemical composition of spilled oil, the concentration of nutrients, water temperature, pH value, and salinity.

The spilled oil has quite complicated chemical composition and there are many kinds of hydrocarbon substances. Any crude oil contains 200 to 300 kinds of compounds, wherein hydrocarbons account for 50% to 90%, and hydrocarbon contents of different types of oil are different. Some hydrocarbons can be degraded into CO_2 and H_2O by degrading strain and some substance can only be converted by degrading strain into another intermediate substance which is difficult to completely oxidize, and even some hydrocarbon substances are difficult to be degraded by degrading strain. Therefore, the chemical composition of different types of spilled oil determines whether it can be degraded by degrading strain. Among the degradable petroleum hydrocarbons, the chemical structure of different hydrocarbons determines the degradation rate [206].

The concentration of nutrients has a great influence on the degradation rate of petroleum hydrocarbons by degrading strain and is directly related to the degradation ability of degrading strain, where N and P are the main factors in nutritional factors affecting the degradation by petroleum-degrading strain. When using biological treatment methods, microorganisms and nutrients are generally input at the same time. In the 1990s, the UK successfully eliminated the oil slick on the sea by adding bacterial mixed cultures and N, P nutrients without the use of oil boom and chemical preparations for the first time.

The water temperature will affect the population composition of petroleum-degrading strain and also affect the physical state and chemical composition of the oil. The biodegradation rate generally decreases with the decreasing temperature. Studies of Cooney showed that the best biodegradation temperature in a freshwater environment is about 20-30 degrees [207]. The studies of Histov and Fachikov showed that the optimal degradation temperature of aerobic bacteria in the marine environment is about 15-20 degrees [208].

Changes in pH value will affect cell charge, permeability and stability of the plasma membrane, enzymatic activity and stability of nutrients in the environment. Therefore, pH changes in the environment have a great impact on the life activities of microorganisms. There is an optimum pH for growth and reproduction of each petroleum-degrading strain. When the pH of the living environment is unsuitable, its growth will be seriously affected.

The change of salinity can cause changes of biocenosis and thus may affect the biodegradation effect. In the marine environment, salinity is relatively stable and will generally not affect the activity of petroleum-degrading strain. However, in the estuary area where salinity changes drastically, its influence cannot be ignored.

In addition to the research on petroleum-degrading strain, people have begun to try to accelerate oil degradation with surfactant produced by microorganisms in recent years. Surfactant is an amphiphilic compound in which hydrophilic group and hydrophobic group structure are inside the same molecule. It can emulsify the oil, disperse it into the water body and make it form into very small oil particles, increasing the chance of its contact with O_2 and microorganisms, thereby promoting the oil biodegradation. Biosurfactant (BS) is a metabolite produced by microorganisms during its metabolism under certain conditions. It can enhance the emulsification and dissolution effect of non-polar substrates and hence promote the growth of microorganisms in non-polar substrates.

In general, except for the expensive price of biosurfactants, the biological treatment of oil spills has the following advantages: (1) economic: the cost is generally 1/5 to 1/2 that of physical or chemical methods; (2) efficient: the contaminant residues can be reduced to a very low level after biological treatment; (3) no secondary pollution: the final products of biological treatment are carbon dioxide, water and fatty acids which are harmless to humans; (4) on-site disposal can be achieved, avoiding secondary pollution in technical process and saving the processing costs; and (5) there are no damages to the soil environment and marine environment. In particular, it shows unparalleled superiority in dealing with a thin oil layer that cannot be removed by mechanical devices and the use of chemical preparations are limited. The biological treatment method also has its shortcomings which are manifested in: once a large-scale oil spill occurs or the oil layer is thick, the nutrient and oxygen supply is insufficient and the degradation will be very slow. At this time, adding lipophilic fertilizer to supplement the N and P nutrients which are lacked in the seawater can achieve good results.

References

[1] Dongyan Bei Shaojun. Lessons to be learned from the oil spill incident in the Gulf of Mexico. *China Maritime Safety*, (6):4–6, 2010.

[2] Guo Qingzhu. Analysis on the factors of oil spills from ships. *Shipbuilding Science and Technology*, (2):16–18, 2006.

[3] Meng Lu. *Research on Emergency Response Location and Emergency Materials Allocation for Marine Oil Spills*. PhD thesis, Shanghai: Master's Thesis of Shanghai Maritime University, 2007.

[4] Li Tianbiao, Cui Yuan, and Zheng Guodong. Research on management and prevention & control of the risk of oil spills from offshore oil installations. *Environmental Protection of Oil & Gas Fields*, 20(1):29–32, 2010.

[5] Pan Daxin and Huo Youli. Risk analysis on oil and gas leakage accident in offshore oil gas field engineering. *Marine Environmental Science*, 28(4):426–429, 2009.

[6] Wu Zhaocun. Numerical simulation of oil spill spread and transport in the tidal waterways. In *Proceedings of the 9th National Conference on Hydrodynamics and the 22nd National Hydrodynamics Symposium*, 2009.

[7] Maritime Safety Administration of the People's Republic of China. *Training Course for Oil Spill Emergency Response*. Beijing: China Communications Press, 2004.

[8] Wang Yaohua. *Study on the Fate and Forms of Marine Oil Spills*. PhD thesis, Dalian: Master's Thesis of Dalian Maritime University, 2010.

[9] Wang Zhixia and Liu Minyan. Study on damage of marine ecosystem by oil spills. *Journal of Waterway and Harbor*, 29(5):367–371, 2008.

[10] Liu Shengyong. *Study on the Emergency Response Organization System for Ship Oil Spill Accidents and Decision-making Treatment*. PhD thesis, Shanghai: Master's Thesis of Shanghai Maritime University, 2005.

[11] Chen Shiming. Building the emergency response system for marine oil spills and protecting marine environmental safety. *Ocean and Fishery*, (4):11–12, 2008.

[12] E. Hailiang. *Analytical Research on China's Shipboard Pollution Prevention and Control System*. PhD thesis, Dalian: Master's Thesis of Dalian Maritime University, 2008.

[13] Bai Chunjiang. Research on remote sensing monitoring technology and system for oil spills in the waters of the Bohai Sea. Master's thesis, Dalian Maritime University, Dalian, 2007.

[14] Liu Bingxin. Research on remote sensing monitoring of large-scale marine oil spills. Master's thesis, Dalian Maritime University, Dalian, 2010.

[15] Ding Qian, Zhang Yongning and Li Qijun. Study on remote sensing monitoring of marine oil spill pollution. *Journal of Dalian Maritime University*, 25(3):1–5, 1999.

[16] Zhao Dongzhi and Cong Pifu. The research on spectral features of ground objects in visual light wave-band of sea-surface oil spills. *Remote Sensing Technology and Application*, 15(3):160–164, 2000.

[17] S. Taylor. 0.45 to 1.1 μm spectra of prude crude oil and of beach materials in Prince William sound. Technical report, CRREL Special Report, 1992.

[18] Zhang Cunzhi Zhao Dongzhi and Xu Hengzhen. *Study on Emergency Response Technology for Marine Oil Spill Disasters*, pages 108–172. China Ocean Press, Beijing, 2006.

[19] Li Sihai. Marine oil spill remote sensing detection technology and its application progress. *Remote Sensing Information*, 02:53–57, 2004.

[20] M.S. Mussetto, L. Yujiri, D.P. Dixon, B.I. Hauss, and C.D. Eberhard. Passive millimeter wave radiometric sensing of oil spills. In *Proceedings of the second thematic conference on remote sensing for marine and coastal environments: needs, solutions and applications. ERIM, Ann Arbor*, volume 1, page 35, 1994.

[21] Ma Li. Research on satellite remote sensing monitoring of oil spills in the waters of the Bohai Sea. Master's thesis, Dalian Maritime University, Dalian, 2006.

[22] An Jubai and Zhang Yongning. Analysis of marine oil spill monitoring by remote sensing in developed countries. *Environmental Protection in Transportation*, 23(3):27–29, 2002.

[23] Zhao Ruxiang Cong Xudong and Wang Shumei. Application of satellite remote sensing in oil spill monitoring. *Environmental Protection in Transportation*, S1:23–25, 2003.

[24] Liu Xingquan and Su Weiguang. Research on marine oil spill extraction method based on SAR images. *Heilongjiang Science and Technology Information*, (22):56–57, 2008.

[25] Shi Lijian. *Study on Marine Oil Spill Detection Method Based on SAR and MODIS Data*. PhD thesis, Ocean University of China, QingDao, 2008.

[26] Å. Skøelv and T. Wahl. Oil spill detection using satellite based sar, phase I competition report. Technical report, Norwegian Defense Research Establishment, 1993.

[27] M.J. Manore, P.W. Vachon, C. Bjerkelund, H.R. Edel, and B. Ramsay. Operational use of radarsat SAR in the coastal zone: The canadian experience. In *27th international Symposium on Remote Sensing of the Environment, Tromso, Norway, June 8C12*, pages 115–118, 1998.

[28] L.Y. Change, K. Chen, C. Chen, and A. Chen. A multiplayer multiresolution approach to detection of oil slicks using ERS SAR image. In *17th Asian Conference of Remote Sensing, Sri Lanka*, 1996.

[29] Antony K. Liu, Chich Y. Peng, and SY-S. Chang. Wavelet analysis of satellite images for coastal watch. *IEEE Journal of Oceanic Engineering*, 22(1):9–17, 1997.

[30] Anne HS Solberg, Sverre Thune Dokken, and Rune Solberg. Automatic detection of oil spills in envisat, radarsat and ERS SAR images. In *Geoscience and Remote Sensing Symposium, 2003. IGARSS'03. Proceedings. 2003 IEEE International*, volume 4, pages 2747–2749. IEEE, 2003.

[31] Thomas F.N. Kanaa, E. Tonye, G. Mercier, V. de P. Onana, J. Mvogo Ngono, P.L. Frison, J.P. Rudant, and R. Garello. Detection of oil slick signatures in SAR images by fusion of hysteresis thresholding responses. In *Geoscience and Remote Sensing Symposium, 2003. IGARSS'03. Proceedings. 2003 IEEE International*, volume 4, pages 2750–2752. IEEE, 2003.

[32] Gregoire Mercier, Stephane Derrode, Wojciech Pieczynski, J-M Le Caillec, and Rene Garello. Multiscale oil slick segmentation with Markov chain model. In *Geoscience and Remote Sensing Symposium, 2003. IGARSS'03. Proceedings 2003 IEEE International*, volume 6, pages 3501–3503. IEEE, 2003.

[33] Xue Haojie and Chong Jinsong. Zero-Antisymmetrical Dyadic Wavelet-Based Oil Spill Detection Method in SAR Images. *Journal of Electronics and Information Technology*, 27(4):574–576, 2005.

[34] Xiong Wencheng, Wu Chuanqing, Wei Bin, Shen Wenming, and Sun Zhongping. Application of SAR images in monitoring oil spills in Korea. *Remote Sensing Technology and Application*, 23(4):410–413, 2008.

[35] Werner Alpers and Heinrich Hühnerfuss. Radar signatures of oil films floating on the sea surface and the Marangoni effect. *Journal of Geophysical Research Oceans*, 93(C4):3642–3648, 1988.

[36] Ma Guangwen, Zhao Chaofang and Shi Lijian. Preliminary research on spaceborne SAR monitoring of marine oil spill pollution. *Transactions of Oceanology and Limnology*, 2:53–60, 2008.

[37] Lv Xin. *Study on the Technique of Oil Fingerprint Identification of Moderately and Severely Weathered Oil Spills in Oceans*. PhD thesis, Master's Dissertation of Dalian Maritime University, 2004.

[38] Wang Min Wang Chuanyuan and Duan Yi. Current status and progress of study on the identification of marine oil spill sources. *Ocean Development and Management*, 25(3):84–87, 2008.

[39] Cao Lixin, Yu Chenyu, Lin Wei, and Pan Sujing. US coast guard's oil spill identification system [j]. *Environmental Protection in Transporation*, (2):11, 1999.

[40] Ma Yongan. Countermeasures for prevention and control of marine oil pollution in canada. *Marine Environmental Science*, (2):103–105, 1989.

[41] WenQiang. *Fingerprint Identification and Analysis of Crude Oil Mixture*. PhD thesis, Ocean University of China, 2007.

[42] Pang Shiping. *Monitoring of Marine Oil Pollutants by Infrared Spectroscopy*. PhD thesis, Master's Dissertation of Fuhou University, 2006.

[43] Jeffrey J. Kelly, Clyde H. Barlow, Thomas M. Jinguji, and James B. Callis. Prediction of gasoline octane numbers from near-infrared spectral features in the range 660-1215 nm. *Analytical Chemistry*, 61(4):313–320, 1989.

[44] Gy. Bohcs, Z. Ovd, and A. Salg. Prediction of gasoline properties with near infrared spectroscopy. *Journal of Near Infrared Spectroscopy*, 6(1):341, 1998.

[45] Michael J. Lysaght, Jeffrey J. Kelly, and James B. Callis. Rapid spectroscopic determination of per cent aromatics, per cent saturates and freezing point of jp-4 aviation fuel. *Fuel*, 72(5):623–631, 1993.

[46] John B. Cooper, Kent L. Wise, Michael B. Sumner, Roy R. Bledsoe, and William T. Welch. Determination of weight percent oxygen in commercial gasoline: A comparison between FT-raman, ft-ir, and dispersive near-IR spectroscopies. *Applied Spectroscopy*, 50(7):917–921, 1996.

[47] A. J. Rest, R. Warren, and S. C. Murray. Near-infrared study of the light liquid alkanes. *Applied Spectroscopy*, 50(4):517–520, 1996.

[48] Min-Sik Ku Hoeil Chung, Hyuk-Jin Choi. Rapid identification of petroleum products by near-infrared spectroscopy. *Bulletin of the Korean Chemical Society*, 20(9):1021–1025, 1999.

[49] Huiying Liu, Rui Wei, and Chanhua Xiong. Determination of gasoline properties by fourier transform infrared spectroscopy. *Chinese Journal of Analytieal Chemistry*, 29(6):731–734, 2001.

[50] X. U. Guang, Ze Liu, Yu Yang, Shi Shen, and L. U. Wan. Determination of diesel fuel composition by near infrared spectroscopy and its application. *Acta Petrolei Sinica*, 18(4):65–71, 2002.

[51] He Ying Wang Li and Wang Yanping. Identification of the types of marine oil spills with near-infrared spectroscopy combining with principal component cluster analysis. *Marine Environmental Science*, 23(2):58–60, 2004.

[52] L. Wang, L. Zhuo, Y. He, Y. Zhao, W. Li, X. R. Wang, and F Lee. [oil spill identification by near-infrared spectroscopy]. *Guang pu xue yu guang pu fen xi = Guang pu*, 24(12):1537, 2004.

[53] Wen Qiang. *Fingerprint Identification and Analysis of Crude Oil Mixture*. PhD thesis, Ocean University of China, 47–48, 2007.

[54] Zhu Lili. *Research on Oil Spill Identification Based on Synchronous Fluorescence Spectroscopy and GC-MS Fingerprint Characteristics*. PhD thesis, Ocean University of China, 2009.

[55] D. E. Nicodem, M. C. Z. Fernandes, C. L. B. Guedes, and R. J. Correa. Photochemical processes and the environmental impact of petroleum spills. *Biogeochemistry*, 39(2):121–138, 1997.

[56] Zhao Yunying and Ma Yong'an. Research progress of fluorescence spectroscopy for the identification of marine oil spill sources. *Marine Environmental Science*, (2):29–35, 1997.

[57] J. B. F. Lloyd. Synchronized excitation of fluorescence emission spectra. *Nature Physical Science*, 231(20):64–65, 1971.

[58] Zhang Dandan. Review of fluorescence spectrometry for identification of marine oil spill. *Environmental Protection Science*, 26(5):34–36, 2000.

[59] Zhou Chuanguang Li Hong, Lu Jibin. Identification of marine oil spills with fluorescence spectroscopy and capillary gc-fid. *Marine Science Bulletin*, (6):66–70, 1998.

[60] Chen Shumei and Zhao Yunying. Study on various factors affecting the identification of marine oil spills with fluorescence method. *Journal of Dalian Fisheries University*, 15(1):35–40, 2000.

[61] H. E. Xiao-Yuan, Jin Hui Shi, and Hai Hong Xin. Fluorescence spectroscopy analysis on oils from the platforms in South China Sea. *Marine Environmental Science*, 2004.

[62] Ma Yongan and Zhao Yunying. Identification of oil spill sources by GC and FS multiple linear exponential pattern recognition. *Marine Environmental Science*, (2):76–82, 1990.

[63] Xu Heng-Zhen, Chuan Guang Zhou, M. A. Yong-An, Long Sheng Shang, L. I. Hong, Zi Wei Yao, Guo Guang Zhang, Yu Hong Sun, and W. U. Zhi-Qing. Study on the indicators of spilled oil with GC-FID. *Environmental Protection in Transportation*, 22(6):5–11, 2001.

[64] Du Huaiqin and Sun Hua. Study on the application of gas chromatography in the identification of marine oil spills. *Environmental Protection of Oil Gas Fields*, 11(2):38–40, 2001.

[65] Wei Qi Chen and Luo Ping Zhang. Identification of spilled oil on sea with GC analysis of n-alkane fingerprint: a case study. *Journal of Oceanography in Taiwan Strait*, 41(3):346–348, 2002.

[66] M. Bao, Q. Wen, and W. Cui. Chromatogram fingerprint extraction of n-alkane of six product oils and identification of them. *Journal of Xi'Shiyou University (Natural Science Edition)*, 22(1):87, 2007.

[67] Linus M.V. Malmquist, Rasmus R. Olsen, Asger B. Hansen, Ole Andersen, and Jan H. Christensen. Assessment of oil weathering by gas chromatography–mass spectrometry, time warping and principal component analysis. *Journal of Chromatography A*, 1164(1-2):262–270, 2007.

[68] Li Yunlong Liu Zelong and Gao Hong. Analysis of hydrocarbon composition of diesel fraction. *The Fourth Academic Annual Conference of Petroleum Refining of China Petroleum Society*, 32(3):44–48, 2001.

[69] L. Mazeas, H. Budzinski, and N. Raymond. Absence of stable carbon isotope fractionation of saturated and polycyclic aromatic hydrocarbons during aerobic bacterial biodegradation. *Organic Geochemistry*, 33(11):1259–1272, 2002.

[70] Gao Zhechang Zhao Ruiqing, Su Danqing. GC/MS identification of on-water oil spills. *Journal of Instrumental Analysis*, 21(5):47–49, 2002.

[71] Zhang Yuanbiao Wang Haiyan, Wang Lin. Study on the application of identification of marine oil spill sources in emergency. 26(2):226–230, 2007.

[72] Ni Zhanglin. *Research on Marine Oil Spill Weathering and Identification*. PhD thesis, Ocean University of China, 2008.

[73] Cheng Haiou. *Research on Marine Oil Spill Weathering and Identification Technology*. PhD thesis, Ocean University of China, 2009.

[74] Keith A. Kvenvolden, Paul R. Carlson, Charles N. Threlkeld, and Augusta Warden. Possible connection between two Alaskan catastrophes occurring 25 yr apart (1964 and 1989). *Geology*, 21(9):813–816, 1993.

[75] Laurence Mansuy, R. Paul Philp, and Jon Allen. Source identification of oil spills based on the isotopic composition of individual components in weathered oil samples. *Environmental Science & Technology*, 31(12):3417–3425, 1997.

[76] Hoeil Chuang, Hyuk-Jin Choi, and Min-Sik Ku. Rapid identification of petroleum products by near-infrared spectroscopy. *Bulletin of the Korean Chemical Society*, 20(9):1021–1025, 1999.

[77] Wang Chuanyuan, Che Guimei, Sheng Yanqing, Li Yantai, and Qin Zhijiang. Application of carbon isotope in oil spill identification [j]. *Environmental Pollution & Control*, 7:005, 2009.

[78] Xu Yandong. Research on marine oil spill weathering process and its prediction model. *Qingdao: Master's Dissertation of Ocean University of China*, pages 13–24, 2006.

[79] Yan Zhiyu. Research and simulation of the weathering process of marine oil spills. *Dalian: Doctoral Dissertation of Dalian Maritime University*, pages 14–26, 2001.

[80] P.C. Blokker. Spreading and evaporation of petroleum products on water. In *Proceedings of the 4th International Harbour Conference, Antwerp, Belgium (1964)*, pages 911–919, 1964.

[81] J.A. Fay. Oil on the sea. *New York: Plenum Press*, pages 53–63, 1969.

[82] Wu Zhaochun. Numerical simulation of oil spill spread and transport in the tidal waterways. *Shanghai: Doctoral Dissertation of Shanghai University*, pages 16–63, 2009.

[83] Xiang Wang, John R. Campbell, and John D. Ditmars. *Computer modeling of oil drift and spreading in Delaware Bay*. Department of Civil Engineering and College of Marine Studies, University of Delaware, 1976.

[84] Mackay D., Paterson S., and Trudel K. *A mathematical model of oil spill behavior*. Environment Canada, Environmental Protection Service, Environmental Impact Control Directorate, Environmental Emergency Branch, Research and Development Division, 1980.

[85] W.J. Lehr, H.M. Cekirge, R.J. Fraga, and M.S. Belen. Empirical studies of the spreading of oil spills. *Oil and Petrochemical Pollution*, 2(1):7–11, 1984.

[86] David Shiao-Kung Liu, and J. Leendertse Jan. A 3-D oil spill model with and without ice cover, Santa Monica, Calif.: RAND Corporation, 6620, 1981.

[87] Akira Okubo. Some speculations on oceanic diffusion diagrams. Technical report, Johns Hopkins Univ., Baltimore, Md. Chesapeake Bay Inst., 1972.

[88] Lou Xia and Liu Shuguang. Overview of the theory and research of oil spill model. *Environmental Science and Management*, 33(10):33–61, 2008.

[89] Huang Lixian, Zhang Guanxi, and Wan Zhaozhong. The spreading of petroleum in the ocean. *Environmental Science Series*, 3(1):7–12, 1982.

[90] Yang Hong, Hong Bo, and Chen Sha. Research progress of marine oil spill model and its application. *Transactions of Oceanology and Limnology*, 2:157–163, 2007.

[91] Zhao Wenqian and Wu Zhouhu. Determination of the scope of spreading of the oil film caused by the instantaneous oil spill at sea. *Journal of University of Science and Technology of Chengdu*, 5(11):63–72, 1988.

[92] Zhang Yadong, Wang Xiang, and Yue Junfei. Review of oil spill spreading model. *China Water Transport*, 8(4):10, 2008.

[93] Lou Angang, Wang Xuechang, Yu Yifa, Xi Pangen, and Yu Guangyao. Study on the application of Monte Carlo method in the forecasting of marine oil spill spreading. *Marine Science*, 24(5):7–10, 2000.

[94] Wang Yaohua. Study on the fate and forms of marine oil spills. *Dalian: Master's Dissertation of Dalian Maritime University*, pages 14–28, 2010.

[95] Dongzhi Zhao, Cunzhi Zhang, and Hengzhen Xu. Study on emergency response technology for marine oil spill disasters. *Beijing: Maritime Press*, pages 128–129, 2006.

[96] Feng Shizuo, Li Fengqi, and Li Shaojing. Introduction to marine science. *Beijing: Higher Education Press*, pages 144–227, 1999.

[97] Ye Anle and Li Fengqi. Physical oceanography. *Qingdao: Ocean University of Qingdao*, pages 252–267, 1992.

[98] G.T. Csanady. Mean circulation in shallow seas. *Journal of Geophysical Research*, 81(30):5389–5399, 1976.

[99] Jacques C.J. Nihoul and Francois C. Ronday. The influence of the tidal stress on the residual circulation. *Tellus*, 27(5):484–490, 1975.

[100] Norman Stuart Heaps. Linearized vertically-integrated equations for residual circulation in coastal seas. *Deutsche Hydrografische Zeitschrift*, 31(5):147–169, 1978.

[101] Feng Shizuo. A three-dimensional weakly nonlinear dynamics on tide-induced Lagrangian residual current and mass-transport. *Chinese Journal of Oceanology & Limnology*, 4(2):139–158, 1986.

[102] Dong Wenjun. A three-dimensional step-by-step model for tidal current and suspended sediment spreading in the coastal area of the estuary. *Journal of Tianjin University*, 32(3):346–349, 1999.

[103] Xu Jindian and Jiang Yuwu. Numerical model of three-dimensional hydrodynamics and pollutant dispersion in the harbor. *Taiwan Strait*, 22(1):85–91, 2003.

[104] Bao Xianwen, Chen Bo, Shi Maochong, and Qiu Shaofang. Three-dimensional numerical simulation of tidal current in Qinzhou Bay. *Guangxi Sciences*, 11(4):375–378, 2004.

[105] Stolzenbach, Keith D., et al. *A review and evaluation of basic techniques for predicting the behavior of surface oil slicks*. Series Report of NOAA (Massachusetts Institute of Technology. Sea Grant Program); no. MITSG 77-8, 1977.

[106] G.N. Williams and R.W. Hann. Simulation models for oil spill transport and diffusion. In *Summer Computer Simulation Conference*, pages 748–752, 1975.

[107] L.E. Webb, R. Taranto, E. Hashimoto, and VIRGINIA Norfolk. Operational oil spill drift forecasting. In *Proceedings of 7th Navy Symposium on Military Oceanography, Annapolis, Maryland*, pages 114–119, 1970.

[108] Wu Zhouhu and Zhao Wenqian. Combined model of spreading, dispersion and migration of oil spills on the sea surface. *Marine Environmental Science*, 11(3):35–42, 1992.

[109] Liu Weifeng and Sun Yinglan. Discussion and improvement of numerical simulation method for the movement of marine oil spills. *Journal of East China Normal University (Natural Sciences)*, (3):91–97, 2009.

[110] Hugo B. Fischer, John E. List, C. Robert Koh, Jorg Imberger, and Norman H. Brooks. *Mixing in inland and coastal waters*. Elsevier, 2013.

[111] Yang Qingxiao. Research on the evaporation process of petroleum at sea. *Acta Oceanologica Sinica*, 12(2):187–193, 1990.

[112] Li Qiong. Development of marine oil spill weathering forecasting system. *Dalian Maritime University: Doctoral Dissertation of Dalian Maritime University*, pages 21–30, 2002.

[113] Long Shaoqiao. Study on numerical simulation of the behavior and fate of marine oil spills and their impact on the environment. *Qingdao: Master's Dissertation of Ocean University of China*, pages 13–16, 2006.

[114] A.V. Tkalin. Evaporation of petroleum hydrocarbons from films on a smooth sea surface. *Oceanology of the Academy of Sciences of the USSR*, 26:473–474, 1986.

[115] Donald Mackay and Ronald S. Matsugu. Evaporation rates of liquid hydrocarbon spills on land and water. *The Canadian Journal of Chemical Engineering*, 51(4):434–439, 1973.

[116] Peter J. Drivas. Calculation of evaporative emissions from multicomponent liquid spills. *Environmental Science & Technology*, 16(10):726–728, 1982.

[117] Warren Stiver and Donald Mackay. Evaporation rate of spills of hydrocarbons and petroleum mixtures. *Environmental Science & Technology*, 18(11):834–840, 1984.

[118] Fingas, Merv F. Studies on the evaporation of crude oil and petroleum products: I. The relationship between evaporation rate and time. *Journal of hazardous materials*, 56(3):227–236, 1997.

[119] Fingas, Merv. The evaporation of oil spills: Prediction of equations using distillation data. *Spill Science & Technology Bulletin*, 3(4):191–192, 1996.

[120] Yan Zhiyu, Xiao Jingkun, and Yin Peihai. Study on the evaporation process of oil spills. *Journal of Dalian Maritime University*, 26(4):20–23, 2000.

[121] Fan Zhijie and Song Chunyin. Weathering process of marine oil spills and its impact on the environment. *Environmental Protection of Oil & Gas Fields*, 6(1):54–57, 1996.

[122] Yang Qingxiao, Xu Junying, and Li Wensen. Study on the dissolution process of oil spills at sea. *Marine Environmental Science*, 11(3):24–28, 1992.

[123] M. Bobra. Water solubility behaviour of petroleum mixtures. *Proceedings of the 12th Arctic and Marine Oil Spill Program Technical Seminar, Ottawa: Environment Canada*, pages 91–104, 1989.

[124] Mark Reed, Erich Gundlach, and Timothy Kana. A coastal zone oil spill model: development and sensitivity studies. *Oil and Chemical Pollution*, 5(6):411–449, 1989.

[125] Gerardus Athenasius, Leonardus Delvigne, and C.E. Sweeney. Natural dispersion of oil. *Oil and Chemical Pollution*, 4(4):281–310, 1988.

[126] Mackay D., Buist I., and Mascarenhas R. Oil spill processes and models. *Environment Canada Report*, 8, 1980.

[127] Yang Qingxiao, Zhao Yunying, and Han Jianbo. Emulsification of marine oil spill under the action of breaking waves. *Marine Environmental Science*, 16(2):3–8, 1997.

[128] Zhang Cunzhi, Dou Zhenxing, Han Kang, and Wu Guan. Three-dimensional oil spill dynamic forecasting model. *Marine Environmental Science*, 16(1):22–29, 1997.

[129] Shi Yiqiang, Chen Chongcheng, Wang Qinmin, and Chen Ling. Study on the integration of geographic information system with marine oil spill model. *Environmental Protection of Oil & Gas Fields*, 15(4):17–20, 2005.

[130] Sun Jun, Yu Jiqing, and Huang Liwen. Oilmap-based oil spill management information system for port of Zhoushan of China. *Computer Simulation*, 19(4):76–78, 2002.

[131] Li Jun, Chen Rongchang, and Yan Huimin. Research on oil spill risk in Qinzhou Bay based on oil spill trajectory and fate model. *Energy Conservation & Environmental Protection in Transportation*, (1):19–24, 2010.

[132] Liu Yancheng, Yin Peihai, Lin Jianguo, and Zhang Jing. Study on GIS-based forecasting of the diffusion and drift of oil spills at sea. *Journal of Dalian Maritime University*, 28(3):41–44, 2002.

[133] Xiong Deqi, Du Chuan, Zhao Dexiang, and Yang Weiqun. Oil spill emergency forecasting Information system for the waters of Dalian and LTS application. *Environmental Protection in Transportation*, 23(3):63–66, 2002.

[134] Zhang Bo and Wu Guan. Oil spill emergency forecasting system for the Bohai Sea. *Marine Environmental Science*, 17(3):73–75, 1998.

[135] Zhang Bo, Wu Guan, Zhang Yanfeng, Zhou Ming, and Zhang Jianming. Marine oil spill forecasting system in Chinese windows environment. *Marine Environmental Science*, 16(1):37–41, 1997.

[136] Zhu Jianrong. Ocean numerical calculation method and numerical model. *Beijing: China Ocean Press*, 23–32, 2003.

[137] Changsheng Chen, Robert C. Beardsley, and Geoffrey Cowles. An unstructured grid, finite-volume coastal ocean model: FVCOM user manual. *SMAST/UMASSD*, pages 6–8, 2006.

[138] Joseph Smagorinsky. General circulation experiments with the primitive equations: I. the basic experiment. *Monthly weather review*, 91(3):99–164, 1963.

[139] George L. Mellor and Tetsuji Yamada. Development of a turbulence closure model for geophysical fluid problems. *Reviews of Geophysics*, 20(4):851–875, 1982.

[140] B. Galperin, L.H. Kantha, S. Hassid, and A. Rosati. A quasi-equilibrium turbulent energy model for geophysical flows. *Journal of the Atmospheric Sciences*, 45(1):55–62, 1988.

[141] Sun Wenxin Feng Shizuo. Physical ocean numerical calculation. *Zhengzhou: Henan Science and Technology Press*, pages 68–85, 1992.

[142] Editorial Committee of Ocean Atlas. Atlas of Bohai Sea, Yellow Sea and East China Sea. *Beijing: Maritime Press*, pages 23–32, 1992.

[143] Xu Hengzhen Zhao Dongzhi, Zhang Cunzhi. Study on emergency response technology for marine oil spill disasters. *Beijing: Maritime Press*, pages 184–185, 2006.

[144] Paul H. LeBlond and Lawrence A. Mysak. *Waves in the Ocean*, volume 20. Elsevier, 1981.

[145] Klaus Hasselmann, T.P. Barnett, E. Bouws, H. Carlson, D.E. Cartwright, K. Enke, J.A. Ewing, H. Gienapp, D.E. Hasselmann, P. Kruseman, et al. Measurements of wind-wave growth and swell decay during the joint north sea wave project (jonswap). *Ergänzungsheft 8-12*, 1973.

[146] Gerald Beresford Whitham. *Linear and nonlinear waves*, volume 42. John Wiley & Sons, 2011.

[147] David J.H. Phillips. The use of biological indicator organisms to monitor trace metal pollution in marine and estuarine environments: a review. *Environmental Pollution (1970)*, 13(4):281–317, 1977.

[148] Owen M. Phillips. On the generation of waves by turbulent wind. *Journal of fluid mechanics*, 2(5):417–445, 1957.

[149] John W. Miles. On the generation of surface waves by shear flows. *Journal of Fluid Mechanics*, 3(2):185–204, 1957.

[150] Hasselmann S. Komen G.J., Hasselmann K. On the existence of a fully developed windsea spectrum. *J. Phys. Oceanogr.*, 14:1271–1281, 1984.

[151] Peter AEM Janssen. Wave-induced stress and the drag of air flow over sea waves. *Journal of Physical Oceanography*, 19(6):745–754, 1989.

[152] Peter AEM Janssen. Quasi-linear theory of wind-wave generation applied to wave forecasting. *Journal of Physical Oceanography*, 21(11):1631–1642, 1991.

[153] Peter AEM Janssen. Consequences of the effect of surface gravity waves on the mean air flow. In *Breaking Waves*, pages 193–198. Springer, 1992.

[154] J. Ian Collins. Prediction of shallow-water spectra. *Journal of Geophysical Research*, 77(15):2693–2707, 1972.

[155] J.D. Madsen, L.W. Eichlerq, and W. Boylen. Vegetative spread of Eurasian thermilfoil in Lake George, New York. *J. Aquat. Plant Manage*, (26):47–50, 1988.

[156] Klaus Hasselmann. Spectral dissipation of finite-depth gravity waves due to turbulent bottom friction. *J. Marine Res.*, 26:1–12, 1968.

[157] Ivar G. Jonsson. A new approach to oscillatory rough turbulent boundary layers. *Ocean Engineering*, 7(1):109–152, 1980.

[158] Jurjen A. Battjes and JPFM Janssen. Energy loss and set-up due to breaking of random waves. In *Proceedings of the 16th International Conference on Coastal Engineering, American Society of Civil Engineers*, pages 569–587, 1978.

[159] Klaus Hasselmann. On the non-linear energy transfer in a gravity-wave spectrum part 1. General theory. *Journal of Fluid Mechanics*, 12(4):481–500, 1962.

[160] K. Hasselmann. On the non-linear energy transfer in a gravity wave spectrum part 2. Conservation theorems; wave-particle analogy; irrevesibility. *Journal of Fluid Mechanics*, 15(2):273–281, 1963.

[161] K. Hasselmann. On the non-linear energy transfer in a gravity-wave spectrum part 3. Evaluation of the energy flux and swell-sea interaction for a Neumann spectrum. *Journal of Fluid Mechanics*, 15(3):385–398, 1963.

[162] Susanne Hasselmann, Klaus Hasselmann, J.H. Allender, and T.P. Barnett. Computations and parameterizations of the nonlinear energy transfer in a gravity-wave specturm part ii: Parameterizations of the nonlinear energy transfer for application in wave models. *Journal of Physical Oceanography*, 15(11):1378–1391, 1985.

[163] Yasser Eldeberky and Jurjen A. Battjes. Spectral modeling of wave breaking: Application to Boussinesq equations. *Journal of Geophysical Research: Oceans*, 101(C1):1253–1264, 1996.

[164] Gus S. Stelling and Jan J. Leendertse. Approximation of convective processes by cyclic AOI methods. In *Estuarine and coastal modeling*, pages 771–782. ASCE, 1992.

[165] Herbert L. Stone. Iterative solution of implicit approximations of multi-dimensional partial differential equations. *SIAM Journal on Numerical Analysis*, 5(3):530–558, 1968.

[166] René Laprise. The Euler equations of motion with hydrostatic pressure as an independent variable. *Monthly Weather Review*, 120(1):197–207, 1992.

[167] Rene Laprise. The resolution of global spectral models. *Bulletin of the American Meteorological Society*, 73(9):1453–1455, 1992.

[168] Michele Iacono, Laura Villa, Daniela Fortini, Roberta Bordoni, Francesco Imperi, Raoul J.P. Bonnal, Thomas Sicheritz-Ponten, Gianluca De Bellis, Paolo Visca, Antonio Cassone, et al. Whole-genome pyrosequencing of an epidemic multidrug-resistant Acinetobacter baumannii strain belonging to the European clone ii group. *Antimicrobial Agents and Chemotherapy*, 52(7):2616–2625, 2008.

[169] Giada Iacono Marziano, Fabrice Gaillard, and Michel Pichavant. Limestone assimilation by basaltic magmas: an experimental re-assessment and application to Italian volcanoes. *Contributions to Mineralogy and Petrology*, 155(6):719–738, 2008.

[170] James A. Fay. Physical processes in the spread of oil on a water surface. In *International Oil Spill Conference*, volume 1971, pages 463–467. American Petroleum Institute, 1971.

[171] Mark Reed, Øistein Johansen, Per Johan Brandvik, Per Daling, Alun Lewis, Robert Fiocco, Don Mackay, and Richard Prentki. Oil spill modeling towards the close of the 20th century: overview of the state of the art. *Spill Science & Technology Bulletin*, 5(1):3–16, 1999.

[172] D. Mackay, K. Trudel, S. Paterson. *A Mathematical Model of Oil Spill Behaviour*. Publisher: Research and Development Division, Environmental Emergency Branch, 1980.

[173] Kolluru V.S. Influence of number of spillets on spill model predictions. *Applied Science Associates Internal Report, Narragansett, RI*, 1992.

[174] Warren Stiver and Donald Mackay. Evaporation rate of spills of hydrocarbons and petroleum mixtures. *Environmental Science & Technology*, 18(11):834–840, 1984.

[175] Gerardus Athenasius, Leonardus Delvigne, and C.E. Sweeney. Natural dispersion of oil. *Oil and Chemical Pollution*, 4(4):281–310, 1988.

[176] Gerad A.L. Delvigne and Lumber JM Hulsen. Simplified laboratory measurement of oil dispersion coefficient– application in computations of natural oil dispersion. Proceedings of the 17th Arctic and Marine Oil Spill Program (AMOP) Technical Seminar. Vancouver, British Columbia, 1:173–187, 1994.

[177] Mackay D. and Zagorski W. Water-in-oil emulsions. *Environment Canada Manuscript Report EE-34*, 1982.

[178] Mark Reed, Erich Gundlach, and Timothy Kana. A coastal zone oil spill model: development and sensitivity studies. *Oil and Chemical Pollution*, 5(6):411–449, 1989.

[179] Maritime Safety Administration of the People's Republic of China. Training course for oil spill emergency response, pages 25–72, 2004.

[180] Zhan Dexin. A performance test of an oil boom. *Wuhan Shipbuilding*, (1):16–18, 2001.

[181] Badesha S. Structural analysis of oil boom. *Environmental Protection in Transportation*, (3):41–49, 1995.

[182] Yunming Sun, Huiluan Liu, Guohua Chen, and Jinming Song. Preparation and properties of starch oil gelling agent on seawater surface. *Marine sciences/Haiyang Kexue. Qingdao*, 25(8):37–41, 2001a.

[183] Cheng-Yi Zhang, Lei Wang, and Mei-Wang Yao. Experimental research on solid float type oil containment boom [j]. *Research and Exploration in Laboratory*, 4, 2005.

[184] Gui-Feng Yu, Wan-Qing Wu, and Xing Feng. Numerical experiment on applicable conditions of typical oil boom based on the software fluent. *Journal of Dalian Maritime University*, 36(2):117–120, 2010.

[185] Yin-Dong Zhang, Wen-Hua Li, Jie-Min Hou, and Yu-Qing Sun. Improvement design of the charging valve of boom. *Ji Xie She Ji Yu Yan Jiu(Machine Design and Research)*, 22(4):116–119, 2006.

[186] Zhang Cunzhi, Zhao Dongzhi, and Xu Hengzhen. Study on emergency response technology for marine oil spill disasters. pages 335–348, 2006.

[187] Yang Fan, Yang Changzhu, and Zhou Lixin. Principles and performance of oil skimmer [j]. *Industrial Safety and Dust Control*, 5, 2004.

[188] Lixin Zhou, Wenhong Pu, and Fan Yang. Study on sea oil spill recovery technology. *Environmental Protection of Oil & Gas Fields*, 15(1):46–49, 2005.

[189] Li Fasheng, Gu Qingbao, Wu Bing, et. al. Development and investigation of biodegradable oil sorbents. *Environmental Protection in Petrochemical Industry*, 25(2):23–25, 2002.

[190] Xing, Heqin, et al. *Preparation and Properties of Environmentally Friendly Water Treatment Composite.* Fourth International Conference on Intelligent Computation Technology and Automation. Vol. 2. IEEE, 2011.

[191] Shu Wubing, Li Yunyuan, and Zan Linna. The recent research on high oil-absorbing resin and its applications. *Fine and Specialty Chemicals*, 15(20):1–32, 2007.

[192] Lin Xi, Guo Sanwei, Jiang Tiekun, et. al. Synthesis of nano oil-absorbing resin by emulsion polymerization and its properties. *Journal of Wuhan Institute of Technology*, 32(11):77–80, 2010.

[193] Cao Wenjie, Wang Shulei, Xue Yan, et. al. Synthesis of high oil-absorbing resin for the treatment of jet fuel. *ACTA Petrolei Sinica (Petroleum Processing Section)*, (S1):157–159, 2010.

[194] Zhao Ruxiang. In-situ burning technique in oil spill response. *Environmental Protection in Transportation*, 23(3):39–42, 2002.

[195] Li Jincheng Xia Wenxiang, Lin Haitao and Zheng Xilai. Application of dispersants in pollution control of oil spill. *Techniques and Equipment for Environmental Pollution Control*, 5(7):39–42, 2004.

[196] Li Bin. Characteristics and application of oil spill dispersants. *China Water Transport*, (11):48–49, 2005.

[197] Li Pinfang and Chen Luling. Discussion chemical dispersants. *Environmental Protection in Transportation*, 23(3):30–32, 2002.

[198] Pei Zhisong, Wang Zongting, and Wang Hanqing. Study on oil spill dispersants of concentrated and low freezing point type. *Marine Environmental Science*, 23(1):44–46, 2004.

[199] Shen Benxian, Ling Hao, and Chen Xinzhong. Preparation and properties of high efficient drag reducing agent for crude oil. *Speciality Petrochemicals*, 2, 2005.

[200] Sun Yunming and Chen Guohua. The chemical compositions and structures and their coagulations of spilled oil gelling agents at sea. *Marine Science*, (8):37–41, 1999.

[201] Zhang Ning, Zhao Xin, Bao Zhiyu, et al. Preparation and properties of ionic coagulant of spilled oil on water surface. *Environmental Science and Technology*, 29(5):84–86, 2006.

[202] Sun Mingkun, Wang Hongshen, Chen Guohua, et. al. Study on film state of n, n-dialkylaclamides and their application in oil collecting agents on water surface. *Periodical of Ocean University of China*, 32(5):795–803, 2002.

[203] Hu Xiang, Shi Junxian, Chen Zhongyuan, et. al. Ecological effect of the blue-green algae on sustainable reproduction of choleraic vibrio in the estuaries of zhejiang i. *Donghaihaiyang*, 18(2):52–57, 2000.

[204] Y DingM, J Huang, and YQ Li. The degradation of crude oil bymarine-microorganism s. *Acta Scientiae Circum Stantiae*, 21(1):84–88, 2001.

[205] Jinglai Zhang, Zhengyao Li, and Li Wang. Biodegradation of crude oil in ocean. *Journal of University of Science and Technology Beijing*, 25(5):410–413, 2003.

[206] Zhu Junhuang Zhao Jianqiang. Discussion on biodegradation technology of oil spill. *Environmental Protection in Transportation*, 17(3):4–6, 1996.

[207] R. Ted Cooney. Bering sea zooplankton and micronekton communities with emphasis on annual production. *The Eastern Bering Sea Shelf: Oceanography and Resources*, 2:947–974, 1981.

[208] Jordan Hristov and Ludmil Fachikov. An overview of separation by magnetically stabilized beds: State-of-the-art and potential applications. *China Particuology*, 5(1):11–18, 2007.

[209] Dai Yuncong and Xu Xueren. Marine Environmental Science. *Beijing: China Ocean Press*, 23–26, 1984.

[210] Zhao Yunying and Wang Jingfang. Marine Science Bulletin. *Beijing: China Ocean Press*, 22–26, 1993.

[211] Gong Jingxia. Identification of Marine Oil Spill Sources with n-alkane GC Fingerprint. *Fujian Environment*. 19(6):53–54, 2002.

[212] Pál, Róbert, Miklós Juhász, and Árpád Stumpf. Detailed analysis of hydrocarbon groups in diesel range petroleum fractions with on-line coupled supercritical fluid chromatography–gas chromatography–mass spectrometry. *Journal of Chromatography A*, 819(1-2):249–257, 1998.

[213] Su Huanhua. *Application and Research of Modern Organic Mass Spectrometry*. Beijing: Press of People's Public Security University of China, 358–364, 1999.

[214] Xu Hengzhen, Zhou Chuanguang, Ma Yong'an and Cheng ye. Research on Special Biomarkers as Indicators of Oil Spills. *Environmental Protection in Transportation*, 22(6):5–11, 2001.

[215] Han Yunli. Study on Oil Fingerprint Identification of Marin Oil Spills. *Dalian Maritime University*, 60–61, 2008.

[216] O. Johansen. Dispersion of oil from drifting slicks. Spill Technology Newsletter, 134:149, 1982.

[217] L. Anderson. Simultaneous spectrophotometric determination of nitrite and nitrate by flow injection analysis. *Analytica Chimica Acta*, 110(1):123–128, 1979.

[218] M. Reed, M.L. Spaulding, P.C. Cornillon. *An oil spill fishery interaction model: development and applications*. US Department of Energy, 1980.

[219] Freegarde, M., C.G. Hatchard, and C.A. Parker. Oil spilt at sea: its identification, determination and ultimate fate. Laboratory Practice 20(1):35–40, 1971.

Index